D1230236

Mathematics and Culture I

Springer

Berlin
Heidelberg
New York
Hong Kong
London
Milan
Paris
Tokyo

Michele Emmer
Editor

Mathematics and Culture I

Translated by
Emanuela Moreale

Editor
Michele Emmer
Dipartimento di Matematica "G. Castelnuovo"
Università degli Studi "La Sapienza", Roma, Italy
e-mail: emmer@mat.uniroma1.it

Translator
Emanuela Moreale
Research Fellow
Knowledge Media Institute (KMI)
Open University
Milton Keynes, MK7 6AA, UK
e-mail: E.Moreale@open.ac.uk

Cataloging-in-Publication Data applied for

A catalog record for this book is available from the Library of Congress.

Bibliographic information published by Die Deutsche Bibliothek
Die Deutsche Bibliothek lists this publication in the Deutsche Nationalbibliografie;
detailed bibliographic data is available in the Internet at <http://dnb.ddb.de>.

Translation from the Italian language edition: *Matematica e cultura 2000* by Michele Emmer
Copyright © Springer-Verlag Italia, Milano 2000

Mathematics Subject Classification (2000): 00Axx, 00B10, 01-XX

ISBN 3-540-01770-4 Springer-Verlag Berlin Heidelberg New York

Springer-Verlag is a part of Springer Science+Business Media
springeronline.com

Printed in Germany

Engraving on cover and part beginnings by Matteo Emmer, from the book: M. Emmer "La Venezia perfetta", Centro Internazionale della Grafica, Venezia, 2003; by kind permission.
Typeset: perform electronic publishing GmbH, Heidelberg
Cover design: Erich Kirchner, Heidelberg

Printed on acid-free paper 46/3111/LK - 5 4 3 2 1 SPIN 11413127

Preface

It was on March 27, 1999 that the bombings started in Yugoslavia. It was the opening day of the third congress on "Mathematics and Culture" at the Santa Margherita auditorium of the Ca' Foscari University in Venice. We were all in anguish thinking of what now seemed inevitable. Just as it had happened during the Gulf War, we were soon going to hear about intelligent bombs, high precision targeting systems, clean war and ... mathematics. And we, a group of mathematicians and people with an interest in the relationship between mathematics and culture, felt powerless and useless.

The afternoon session scheduled a talk by film director Peter Greenaway, who was expected to arrive from the Netherlands, where he lives. The Marco Polo airport in Venice had been closed to traffic: the planes were taking off from the US base in Aviano on their bombing mission and they were using civil routes. We all thought that Greenaway was never going to join us. In the early afternoon, we received a phone call from the airport of Amsterdam: it was him letting us know that the departure of his flight was being delayed again and again. I have no idea how he managed to find the phone number. The departure was expected to be in about one hour and he was going to phone again. The congress proceeded and we got another phone call from Greenaway: his departure had been delayed again. No one knew what was happening and when and whether the civil air routes were going to be reopened. Between phone calls and departure delays, it got to 8pm.

I told Greenaway not to bother boarding the plane, even in the unlikely case that the plane was to have taken off, since he would have had to leave early next morning anyway, due to important engagements. Greenaway – whom I had never met in person – told me that he wanted to come and that he did not want the war to prevent his participation in the congress. Finally, the plane was allowed to depart: the air space had been reopened and the bombing had ceased for the day. I told Greenaway to wait and asked the congress participants if they felt like waiting for his arrival, which was to be no earlier than 11pm. They all agreed to do so. Greenaway arrived at the airport, and Matteo and Francesco escorted him to Santa Margherita by taxi boat. He arrived at 11:30pm, shortly after the end of the film "Drowning by numbers". Greenaway talked for about half an hour, fascinating the whole audience.

It is rhetoric to say that culture was not stopped by war? Future editions of "Mathematics and Culture" will feature a section dedicated to "mathematics and war". We, however, hope that no one will have to join us late because of bombings.

Michele Emmer

Table of Contents

Mathematical Centers

Mathematics and Literature

Mathematics and Technology

Homage to Venice

Mathematics and Music

Mathematics and Medicine

Mathematics

Research and Teaching in Mathematics: Conflict or Synthesis?

CLAUDIO PROCESI

Translated from the Italian by Emanuela Moreale

When I chose the title of this chapter, I had no idea what I was going to talk about. I only clearly knew that the way the link between teaching and research is perceived is an important element in the current debate on new teaching methods in mathematics (and other disciplines) as well as in the discussion on the organisation of new degree courses and the introduction of university diplomas.[1]

In fact, this theme – already rather vast in itself – is related to even more general themes, such as the definition and acquisition of knowledge and the very difficult one of the spreading of mathematics. However, here, I will limit my discussion to the topic I have chosen for this article.

University rules require lecturers to carry out the double role of constructors and transmitters of new knowledge. Now, anyone who has carried out these activities will have realised the existence of many contradictions between/across the two roles.

Let us limit our discussion to mathematics for a moment: it is sufficient to have a look around any – even the least well furnished – library, to realise how much new mathematics is produced every year, every month and every single day. Those among us who have had the rare idea of subscribing to an automated electronic preprints service, soon had to give it up, since it is simply impossible to read all the works and even just all the abstracts and titles.

It is therefore clear that the problem of transforming research into utilisable information affects all levels, from elementary school teaching to university research. In mathematics, the matter is particularly complex, as mathematics – unlike other disciplines – has a "linear" history: mathematical knowledge has always added itself to the knowledge inherited from the past, improving on it, simplifying it, but hardly ever removing past knowledge because wrong or backward. In some way, mathematics claims a substantial unity and specialists in mathematics must always face inputs of ideas coming from other disciplines (physics, economics and so on).

I have identified three levels that should be analysed in our current discussion: pre-university teaching, university teaching and university research.

1 *A page discussing university diplomas can be found at the URL: http://www.mat.uniroma1.it/ordinari/procesi/home.html - contributions welcome.*

As regards introduction to research, the greatest difficulty consists in having sufficient resources to entice the best students to choose this activity, intelligent methods to select them and an organisational and normative adjustment of our doctoral programs (even though there are difficulties relating to understanding research in progress and the formation of closed or impenetrable schools). Globally, this is a non-trivial problem, but I do not believe this holds as much general interest as the other two levels. I will therefore proceed to deal with these two levels.

Before going into detail, I would like to highlight a general point: in mathematics, the transition from research product to teaching usually takes place at a much slower pace than in other disciplines and this causes a series of rather difficult problems. Let's look at some examples.

The first example – well known to those who work in this discipline – is Euclidean Geometry, a product of work done two thousand five hundred years ago but that nevertheless remains at the core of mathematics. Its teaching in secondary schools (which is based essentially on Euclid's original *Elements)* has provided the best example of a deductive logic system for many generations. It is a theory that has a strong value of knowledge to the students, thanks particularly to its connection with the description of the physical world, even if this knowledge is weakened by the practice of teaching only plane geometry and almost no spatial geometry.

At the beginning of the twentieth century, this theory underwent a thorough reformulation, not so much of its theorem proofs, as of the precise formulation of its fundamental axioms (Euclidean postulates). Yet, this important work (by Hilbert and others) had virtually no influence on elementary school teaching and, anyhow, had only limited influence on its teaching in universities.

A strange contradiction can be seen here: on the one hand, the fundamental value of Euclidean geometry in secondary teaching consists in its illustration of the fundamental ideas of a deductive logic system (postulates, definitions, lemmas, theorems, proofs, corollaries); on the other hand, the critical review that brings it into line with our standards of formal rigour is, or seems to be, too complex to be taught!

There is another interesting example: a few decades ago (in the Sixties), there was a serious attempt to revolutionise the teaching of mathematics by introducing set theory *(new math* in the United States), but I believe this to have been mostly a failure. In my opinion, the main reason for this failure is to be found in the very nature of this theory: to contemporary mathematics, it is the natural answer to the problem of having – and therefore letting interact – several axiomatic systems, such as Euclidean geometry, non-Euclidean geometry and arithmetic.

Set theory is a primitive theory of which all the others are subtheories. This gives rise to two problems: the first is caused by the fact that set theory – as an axiomatic system – is extremely difficult (and, in fact, very few professional mathematicians have examined this approach in detail). The second consists in the deep link between the axioms of geometry and the physical intuition that is lost in set theory.

Moreover, to do justice to these ideas, it is necessary to take a rather long step resulting from twentieth century studies: that of introducing symbolic logic and a symbolic and "machine verifiable" logic definition of mathematical theory and proof. This teaching is not currently performed in secondary schools and has only recently been introduced in university.

This is certainly a very instructive example. On the one hand, there is the negative fact that, clearly, set theory has little utility in demonstrating the deductive method (unless one takes out of the hat a bunch of axioms, whenever they are needed), even if it can be useful in getting students accustomed to abstraction and to introduce arithmetic. On the other hand, there is the positive fact that symbolic logic, lost as a basis for the formalisation of mathematics, magically reappears in a different reincarnation, i.e. in programming languages.

I am persuaded that teaching a simple programming language should and could be the true contribution of this theoretical revolution in secondary school teaching. I believe that this could be a very useful cultural experience: a machine does not forgive errors; students, measuring themselves against the steel and possibly stolid logic of the program, could understand that logic discipline which neither grammar nor syntax can impose. The best bit is that writing useful programs also imposes the same logic.

As to university teaching, the contradictions are present here too, and in not smaller measure, although they do take place over different temporal scales: if in secondary schools the problem is how to review the mathematics of two thousand five hundred years ago, at university the main problem is that of incorporating into the teaching the most important developments in mathematics over the last century. Among the ideas to be included are, in particular, probability theory, computer science and numerical analysis: in general, these are mainly ideas acquired fifty or more years ago.

The solution must take into account two basic needs: on the one hand, it is necessary not to arbitrarily remove "old" teachings to make room for "new" ones and, on the other hand, it is important not to increase learning times by overloading the curriculum.

Mathematics has always met the first requirement the same way, that is by abstracting and generalising, so as to unify theories, thus supplying general ideas and methods, that can then be turned into precise algorithms to solve problems or to create models. For example, already in the nineteenth century, it is possible to find a whole book offering a treatise on conics; but now this theory is a short chapter in a book on linear algebra or an exercise in a book on algebraic geometry. From this point of view, abstraction is a knowledge compression mechanism which, when used, should be decompressed by a scientist into its utilisable form.

This process (of abstraction) has an important premise and consequence. The premise consists in finding the right method to prepare the mathematics students for abstraction. The consequence is that, if the necessary preparation is not completed properly, the abstract notions lose all their unifying and formative value only to become useless mnemonic exercises with a background of logic sequentiality. Anyone who has examined a mediocre student must have experimented the intellectual darkness produced when the student is asked to illustrate a general theorem with a simple example, possibly one in which one of the hypotheses is not even necessary (a classic example is the Hahn-Banach theorem for finite-dimensional spaces).

The second need becomes extremely relevant when we consider mathematics as a tool in other disciplines. The difficulties are multiplied: the student of engineering, physics, chemistry and – even more – biology, medicine and the like, see mathematics as a tool which must be easily learnt and utilised: this totally clashes with mathematicians' idea of abstraction and unification. Moreover, the demand for mathematics teaching is potentially on the rise: in a world that is more and more dependent on sophisticated technologies, a greater scientific literacy is necessary (– incidentally, for this problem take as standard reference the meeting of Benigni and Troisi with Leonardo in the movie "Non ci resta che piangere"). I am not only thinking of our usual "users" (physicists, chemists and even economists and statisticians), but of the use of graphics, statistics, simple quantitative analysis in medicine, of the scientific background necessary to magistrates, the police or the army. Finally, I am also reminded of the obscurantist winds that blow more and more in our societies (horoscopes, fortune-tellers, lottery numbers and general diffidence towards scientists).

This conflict is slowly leading to a crack whose final result could be the creation of a species of sub-mathematics practiced by non-mathematicians, a kind of rough lingua franca to be spoken in an ungrammatical but functional way. In my opinion, this would represent a big intellectual defeat and it is up to mathematicians to find a way to teach mathematics even just as a tool and to pass on the image of a live and constantly changing science.

The phenomenon, which I am talking about, is happening worldwide, but in Italy this (separatist) trend is reinforced by the country's corporate structure, both in general and in the academic world in particular.

The signs are extremely clear: division in faculties, the chair system, discipline groups – all strongly push towards a teaching that is more and more functional to sector-based objectives and therefore to the reduction or total abolition of everything that is not seen as immediately related to the profession. As regards mathematics, this is combined with our tradition to regard as secondrate all "service" teaching, i.e. the teaching not within a degree course in mathematics. The chair system and the division in faculties do the rest, by creating a deadly barrier to didactic renewal. Finally, discipline groups – aiming at defending algebra, analysis, physics, mathematics, geometry and so on – complete the picture by producing considerable friction to the need for renewal.

A curious aspect is that, in the meantime, there has been a great increase in the offer of university teaching – even from countries which used to be almost without – with the creation of teaching positions connected with research activities (following typical western standards). This has produced a large increase in the number of researchers and quantity of research produced.

Perhaps the most remarkable consequence of this change consists in the greater and greater interaction between universities in different countries and in the development of a common culture (and the internet plays no secondary role in this process). But I am trying to concentrate on the Italian experience.

I believe that we are really starting to experience a crisis in the pre-university and university teaching model. We could define it "nineteenth century" teaching model, without implying any negative connotations such as old and antiquated. A good book on mathematical analysis, physics or rational mechanics from the end of the nineteenth century might just as well be adopted today, without requiring many changes in the curriculum. Analysis is the product of about three of the most fascinating centuries in scientific history; it also constitutes the theoretical and practical foundations of the mathematical physics that is used in physics and engineering.

The nineteenth century passed on a science with many characteristics of completeness and definitive structure. But today – after a hundred years of development have brought into play so many new ideas and so much content – is it still possible to continue to teach using simple superimposition? To be clearer and more concrete, could we think of an institutional mathematics taught in the first two years (or in the future diploma) – a "nineteenth century" mathematics – and then of a more advanced one?

Nineteenth-century mathematics mostly envisages a deterministic model of science: after all, it describes the idea that an infinitely capable mathematician can write and solve all the equations describing nature. Stochastic ideas, models, and all the problems relating to efficient computation (which obviously depend on the fact that we have powerful, but not almighty machines) are the product of the twentieth century.

The question is then: how can we introduce something new if we do not want to simply add it to the existing material? Allow me a "philosophical" digression here.

My last sentence, as banal as it may seem, contains one of the most fundamental problems in real life: in its inscrutable mechanisms, nature proceeds through the alternation of life and death as well as evolution. Human cultures abhor death and sometimes let some of their institutions rot before changing them (the dynamicity and good health of a society are measured mostly by its ability to renew itself without traumas). Unfortunately, probably out of very deep reasons, our society cannot let any institution die and delegates this task to nature (and nature does not fail society in this respect). In society, there is a background spirit, which we could define "reactionary" – both politically, culturally and intelligencewise – whose constitutional meaning in our culture escapes me.

I am thinking, for instance, of the discussions I sometimes read about the teaching of Greek and Latin and the foundation of western culture. These are certainly important considerations, yet a certain rebel inclination always prompts me to be a bit of a rebel. Why do we never talk about the role of barbarians in founding our culture? Do we ever consider that, if those marvellous cultures had not died, we would not exist now? Latin was a great language and it disappeared; after all, now we are fond of Italian and of the culture that has been produced in this language, even if it is probably destined to disappear.

Do not misunderstand me, I am not proposing the disappearance of any part of mathematics! But you will have heard those discussions on the teaching of history, geography and English that are always full of half-truths and half-lies and soaked with a reactionary spirit! I will give you three examples:

- History. I wonder how the history of the last 5,000 years will be presented in the year 20000! I imagine it will be along these lines: "after the discovery of writing, a turbulent period of wars and technological development led to the globalisation of commerce and culture. We recall the names of Euclid, Confucius, Christ, Einstein, Woody Allen..." (Unfortunately, I have no idea about the next 5,000).
- English. It is often said that Latin or German are "more formative" than English because of their grammar, as if this (undoubtedly important) aspect of a language were the only important aspect. On the other hand, the plasticity of the English language often allows the seemingly effortless production of extremely useful words to describe reality in its evolution (format, icons, reboot). Is this plasticity not a formative element and an element of deep reflection on the constitutive aspects of a language?
- Geography. It has always been my passion, but learning the names of all the Alps chains ("ma con gran pena le recan giù"[2]) and of all the tributaries of the Po has turned out to be totally useless. On the other hands, no one has ever explained to me how come two places at the same latitude can have such different climates.

This is a digression, but not totally: it also highlights a punitive and immobile view of teaching, stressing how one should not have fun while learning.

Let us now go back to the question: how should we introduce the new, avoiding simply adding it to what is already existing? What follows is an attempt to outline an answer, starting with the teaching of analysis. I am for it, starting with "calculus" that, if taught properly, constitutes a greater intellectual challenge and offers greater cultural rewards than analysis. When presenting it, all proofs connected to the topology of real numbers should be ignored and developed in an intuitive form (anyway, we are already aware of Cauchy, Dedekind, Weierstrass and others and know that everything is settled and ends well). If the attempt to shortcircuit all the technical part and the epsilon-deltas is successful, the intuitive content and the real power of infinitesimal calculus should emerge by themselves. This is partly because, to be fussy, even the most extravagant things (such as the formal sums of divergent series) have in fact deep and amazing meanings, which take us back to physics! One can relatively easily overcome the (in my opinion harmful) cultural break between single-variable and multi-variable analysis (with an obvious maximum of three) and between real and complex analysis. At the same time, one could give some intuitive and simple elements of so-called superior topics (dating back to the eighteenth century) such as the calculus of variation.

2 *This is a mnemonic that was previously used in schools to help remember the names of all the Alps chains: Ma = Marittime, con = Cozie, Gran = Graie...*

There are several questions of elements on which one could discuss (for instance, the notion of angle and trigonometric functions), but I am persuaded that complex numbers are to be presented as soon as possible (read the speech by Feynmann to the general public on QED) and that the separation between real and complex analysis has some pernicious aspects.

The situation is different for geometry and algebra and it is worth understanding the reason for this.

Geometry is a discipline the teaching of which has undergone changes. The reason is obviously twofold: on the one hand, it is clear that some aspects of traditional teaching (projective geometry and design) did not have the same fundamental value for mathematics as infinitesimal analysis; on the other hand, it is necessary to take into account the debate on topology. In my opinion, this debate has led to an unsatisfactory solution in Italy, where it has separated "geometric topology" from "analytic topology" in a rather artificial way. The reason is the same as in all academic phenomena: corporative spirit fuelled by obsolete legislation.

One could say the same things about differential geometry: there is a theorem of Stokes in geometry and one in analysis (not to say one in physics), which really defies the principle of the unity of thought. Thus, one of the fundamental constructions of multi-dimensional analysis – the theory of external differential forms – is not able to penetrate into the teaching of either mathematics or physics with all the resulting damage, I believe, since external calculus is a marvellous simplification.

The reason for the usual absence of discussion on non-Euclidean geometry is less clear: not only is this discovery a turning point in the history of mathematics, but in fact these geometries provide important models for both relativity and, in general, for variety theory.

On the contrary, some progress has been made in the teaching of linear algebra. In my time, there existed (at least) a linear algebra in geometry, one in analysis and one in mechanics. Then came the worst of all: a linear algebra in algebra. All these different sciences distinguished themselves for the use of arrows or dashes and more or less obscure ways of explaining elementary concepts in verbose accounts or with horrible notations. The "algebraic" linear algebra then arrived at such sublime abstractions that even the task of solving two equations with two variables was quite an undertaking. Now, a mix of algorithms (the elimination of Gauss) with a pinch of vector calculus and quadratic forms should provide all fundamental elements.

When I was in my third year at university, algebra was introduced as a fundamental subject in the first year of a mathematics degree: I believe it is worth reflecting on this event and its consequences.

Algebra replaced chemistry. I do not care to say whether this was a good or bad choice, but I will limit myself to observing that the introduction of algebra was only possible because this did not affect the corporation of mathematicians. Had the introduction of algebra been an unavoidable scientific necessity, but had there been no chemistry to sacrifice, algebra would have not been introduced or it would have simply been added.

One of the main reasons that led to the introduction of algebra in the university curriculum is linked to the great defeat of science in our country. Italian mathematics had basically created algebraic geometry but then had totally lost touch with its transformation and refounding. One of the main causes of this detachment was the non-participation in the development of modern algebra in the Twenties and Thirties (and then of algebraic topology, homological algebra, sheaf theory and so on). Italian geometers of the Sixties and Seventies have had to learn algebraic geometry *exnovo* from the French and the Americans.

Even today, the fate of algebra in Italy remains rather odd. For instance, at the moment the algebrisation of high energy physics is in fashion in the scientific world, but we are again almost totally excluded from this because in Italy algebra is almost only commutative algebra – the one of algebraic geometry – and group theory.

There is a fascinating debate on the limitations of the algebrisation of mathematics, its great successes and also the damages it can produce, but in Italy I notice only the most complete silence.

Even in the first couple of years of the mathematics curriculum at university level, the teaching of algebra causes at least two problems. In fact, algebra used to carry out two distinct tasks: on the one hand providing a discussion of the "principles" as intuitive set theory, induction principle, construction of numeric systems etc. and, on the other hand, introducing some algebraic structures, both basic and more complex (groups, rings and the like).

But if the "principles" are not dealt with at the most appropriate time and with the appropriate thoughtfulness, they do not provide a good foundation at all (we all know how difficult the programme to found mathematics was and what surprises it caused). On the other hand, if algebraic structures end up not being used, they are quite useless; moreover, some of the most interesting algebraic structures (such as Lie algebras or simple groups) are far from elementary.

There have been some cases in which Galois theory was introduced in the first year with partial success and I would like to give my opinion on this. Galois theory is a marvellous theory (I believe that its discovery is more or less contemporary with the creation of Beethoven's fifth symphony), but in order to appreciate it, one needs to be interested in mathematics as a cultural artefact. It is necessary to dedicate it time and energy, even more so than needed to start to appreciate the fifth symphony. The symphony can even be used to sell tiles (instead of the Fate, it is the vendor who knocks on the door!) but Galois theory cannot: it is clear that studying Galois theory must be a choice and that there can be no obligation.

Let me give a last example. One of the great revolutions of the scientific thought of this century is connected to the discovery of quantum mechanics: not only has it deeply transformed our way of seeing reality, but it has also transformed mathematics. How have Italian mathematicians lived it revolution? I remember the dispute between mathematical physicists of type A and those of type B: each group wanted its own slice (now there is even a third group, type C, that utilises algebraic geometry).

Quantum mechanics is to be tackled with even greater respect than Galois theory: it should be one of the important cultural offers of a possible curriculum. Of course, one could ask: how can one do quantum mechanics without knowing functional analysis? I believe this can be done and that we should partly rethink foundation courses: apparently, the creators of quantum mechanics were not totally familiar with matrices. Functional analysis has been strongly influenced by quantum mechanics.

It is therefore necessary to redirect teaching towards contents and problems rather than towards tools; after all, the best scientific research has always put problems and contents first.

In short, this is my conclusion: mathematics is too complex and vast to be framed in a definitive way, even if only its institutional aspects are taken into account. I remember that the first time I entered the library of the Institute of Mathematics, my first thought was: but how can I possibly read all these books? Of course, the answer was quite obvious, but I was very young and I was still convinced that it was possible to have a comprehensive knowledge of mathematics. Therefore, the true challenge lies in trying to create strong dynamics in teaching and a large availability of individual choices. Only this way is there a hope that the preservation of science will indeed happen out of its own internal dialectic and not because imposed by decrees.

The Discovery and Application of Linear and Nonlinear Programming

Harold W. Kuhn

Students who learn a result, or a technique, or a subject in mathematics are often given the impression that it has always existed, perhaps handed down by God at the same time that he gave Moses the Ten Commandments. This should not be the case for linear and nonlinear programming. For these subjects, many of the main results, which are now considered classical, were discovered in the last half century or so. In many cases, the discoverers are still among us and can still shed light on the subjects by an examination of their historical origins. For example, George Dantzig, the father of linear programming is still active and participated in the celebration of the 50th anniversary of the Simplex Method at a meeting in Lausanne, Switzerland in 1997. He has written extensively [2] on the origins of linear programming. Although I regret that my colleague, Al Tucker, who died in 1995, cannot be with us today, I am pleased to have this opportunity to share my memories of the founding of the subject of nonlinear programming.

This very short lecture will have two major objectives. Firstly, there will be a brief narrative account of the founding of the subjects of linear and nonlinear programming about 50 years ago, with some indications of the motivations of the principal participants. Secondly, I shall propose an answer to the question: What were the influences economic, sociological, and mathematical that led to the extraordinarily rapid and widespread application of linear and nonlinear programming?

Although these subjects have reached the curriculum in mathematics in many high-schools, it seems necessary that we establish a common understanding of the nature of the subjects: linear and nonlinear programming. To describe them in the most accessible manner, it is easiest to start with canonical forms of linear programs.

As a *canonical maximization problem,* consider the choice of nonnegative activity levels x that maximize a linear objective function cx subject to the linear constraints $Ax \leq b$. These constraints express the conditions that the activities do not use more than the available quantities of scarce resources b.

As a canonical minimization problem, consider the choice of nonnegative activity levels v that minimize the linear cost function vb subject to the linear constraints $vA \geq c$ that express the conditions that the activities produce outputs that are not less than the required quantities c.

This brief description leads the Pure Mathematician to pontificate: "What a trivial subject that is! The set of variables that satisfy the constraints clearly forms a convex polytope in a finite dimensional Euclidean space and an optimal answer can be found at one of its extreme points. The extreme points are finite in number; just enumerate them, using elementary linear algebra and pick the best."

How wrong the Pure Mathematician is! In order to solve even moderately large problems, not to mention realistic applications with potentially millions of variables and tens of thousands of constraints, you need an efficient algorithm. This is what George Dantzig provided in 1947: the Simplex Method. Briefly and without formal detail, the Simplex Method finds an extreme point in a concise algebraic description, then moves from extreme point to adjacent extreme point in an efficient manner, always improving the objective function until no further improvement is possible. In practice, only a very small fraction of the astronomically large number of extreme points is examined in this process.

Historically, the first significant application to be solved by the Simplex Method was a canonical minimization problem: *The Diet Problem.* In terms of the notation introduced above, the activities v are quantities of foods to be purchased at unit prices b. A diet v produces quantities of nutritive constituents vA and these must not be less than the daily requirements given by c. When I first met George Dantzig in the summer of 1948, the walls of his office were papered with the results of a computation using approximately 120 man days of work on hand operated desk calculators to produce a least cost diet chosen from 77 foods in order to satisfy 9 daily requirements. There is a somewhat apocryphal story that the calculation produced a diet that was composed entirely of canned spinach and condensed milk, but that is not true! Oddly enough. although the diet problem is not much used in human settings today, almost all animal feed products are blended optimally using linear programming.

The term "nonlinear program" was introduced in a paper [9] that Tucker and I wrote in the spring of 1950. What is a nonlinear program? We intended it to be a straightforward generalization of a linear program and perhaps the most accessible formulation is a canonical maximization problem, which expresses a classical statement of the problem of a firm that is attempting to maximize profits while using scarce resources. Mathematically, find the choice of nonnegative activity levels x that maximize a general objective function $f(x)$ subject to the general constraints that the amounts of scarce resources utilized by these activities do not exceed the endowment. In formal terms: Maximize $f(x_1, \ldots, x_n)$ subject to

$$
\begin{gathered}
a_1\,(x_1, \ldots, x_n) \leqq b_1 \\
\ldots\ \ldots\ \ldots \\
a_m\,(x_1, \ldots, x_n) \leqq b_m \\
x_1 \geqq 0, \ldots, x_n \geqq 0.
\end{gathered}
$$

Our motives for formulating nonlinear programming in this manner were very clear in our minds. With David Gale, Tucker and I [6] had provided a rigorous proof of the duality of linear programming, which had been broached by von Neumann in a short, unpublished, typescript. The dual variables can be interpreted as a kind of Lagrange multipliers and the duality can be expressed as a saddle value property. When the linear objective function and the linear constraints are generalized to be nonlinear, somewhat miraculously, a necessary condition for a local optimal solution is the existence of generalized Lagrange multipliers that satisfy conditions that are dual to the original constraints. These are often called the "Kuhn-Tucker conditions" and form the basis for most algorithms to solve nonlinear programs.

I have described on several previous occasions the fact that similar mathematical results had been obtained by William Karush [8] and Fritz John [7]. We learned of John's work when our paper was in galley proofs; internal evidence of this is provided by the fact that, when we inserted the reference to his work, we did not renumber the bibliography correctly. I did not learn of Karush's work until 1974 when Takayama's book [11] on mathematical economics appeared. The following passage tells the story very well:

> "Linear programming aroused interest in constraints in the form of inequalities and in the theory of linear inequalities and convex sets. The Kuhn-Tucker study appeared in the middle of this interest with a full recognition of such developments. However, the theory of nonlinear programming when the constraints are all in the form of equalities has been known for a long time – in fact, since Euler and Lagrange. The inequality constraints were treated in a fairly satisfactory manner already in 1939 by Karush. Karush's work is apparently under the influence of a similar work in the calculus of variations by Valentine. Unfortunately, Karush's work has been largely ignored."

Although known to a number of people, especially mathematicians with a connection to the Chicago school of the calculus of variations, it is certainly true that Karush's work had been ignored up to 25 years ago. I conducted a thorough search of the literature prior to 1974 and found only four citations [10], [3], [4], [5] to add to Takayama's book. Of course, one reason is that Karush's work had not been published.

Karush's work was done as a master's thesis at the University of Chicago under L. M. Graves, who also proposed the problem. It was written in the final years of the very influential school of the classical calculus of variations that had flourished in Chicago. One may presume that the problem was set as a finite-dimensional version of research then proceeding on the calculus of variations with inequality side conditions. As a struggling graduate student who was trying to meet the requirements for the Ph.D., the thought of publication never occurred to Karush and he was not encouraged to publish by Graves. Also, and most important for our story, the work was done as pure mathematics without any economic context; no one anticipated the future interest in these problems and their potential practical application.

To introduce the related work by F. John, we shall again quote Takayama:

> "Next to Karush, but still prior to Kuhn and Tucker, Fritz John considered the nonlinear programming with inequality constraints. He assumed no qualifica-

tion except that all functions are continuously differentiable. Here the Lagrangian expression looks like $v_0 f(x) - va(x)$ and v_0 can be zero in the first order condition."

Questions of priority aside, what led Fritz John to consider this problem? His motivation came from an attempt to prove a geometric theorem that asserts that the boundary of a compact convex set in n -dimensional Euclidean space lies between two homothetic ellipsoids of ratio not greater than n. In dimensions 2 and 3, the result had been proved in the 1930's; using tools from the theory linear inequalities, John was able to prove the general result as an application of the theorem cited by Takayama above. The resulting paper was rejected by the *Duke Mathematical Journal* and so very nearly joined the ranks of unpublished classics in our subject. However, this rejection only gave more time to explore the implications of the technique used to derive necessary conditions for the minimum of a quantity (here the volume of the circumscribing ellipsoid) subject to inequalities as side conditions.

Fritz John was aided in solving this problem by a heuristic principle often stressed by Richard Courant that in a variational problem where an inequality is a constraint, a solution always behaves as if the inequality were absent, or satisfies strict equality. It is poetic justice that it was the occasion of Courant's sixtieth birthday in 1948 that gave John the opportunity to complete and publish his paper.

For both of the subjects of linear and nonlinear programming, applications came almost immediately. Practical industrial applications had to wait for the development of computers that could handle problems of realistic size. These applications began in 1951 but a most remarkable growth occurred in the years 1955–60. This growth was so rapid that Dantzig could write in the early 60's: "Since 1957 the number of applications has grown so rapidly that it is impossible to give an adequate treatment." At that point in time, it was estimated that more than half the time spent on electronic computers was spent solving linear and nonlinear programs. By the time of the Princeton Symposium on Mathematical Programming in 1967, A. R. Colville [1] presented a comparative study of nonlinear programming codes, in which 34 different codes were evaluated. Most of these originated in large industrial organizations, where they had been written to solve particular applications.

At the same time, the subject of mathematical programming was entering the realm of popular culture. This can be illustrated by an excerpt from an Academy Award winning film that was made in 1976. In this film, a actor portraying the head of a television network says:

"What do you think the Russians talk about in their Councils of State? Karl Marx? They get out their linear programming charts, statistical decision theories, minimax solutions and compute the price cost probabilities of their transactions and investments, just like we do." (Ned Beatty in NETWORK, screenplay by Paddy Chayefsky, 1976)

The speed with which this passage entered into popular culture is astounding. It is just as if Jonathan Swift, writing *Gulliver's Travels* in the early 18th century had Lilliputians using the Calculus which was invented in the late 17th century. This is a very rare occurrence for a subject in mathematics, applied though it may be, to appear in a popular context.

Having completed our brief account of the discovery and application of linear and nonlinear programming, we can now propose an answer to the question: What were the factors that led their extraordinarily rapid and widespread application? The answer is the conjunction of three factors. All three of them are necessary; no two of them would have been sufficient.

Firstly, the models of linear and nonlinear programming were flexible enough to embrace a wide class of real-life problems that had not been attacked by any satisfactory techniques. Sociologically, after the success of operations research models in the Second World War, major industries were ready to try to apply such theoretical models.

Secondly, Dantzig discovered an algorithm for solving linear programs, the Simplex Method. This was a set of rules or procedures that could be implemented on an electronic computer to provide answers to increasingly large instances. A plethora of such algorithms for nonlinear programs implemented the search for solutions to the necessary conditions established by Karush, Kuhn and Tucker.

Thirdly, and most importantly, concurrently with the first two factors, the early 1950's saw the development, and rapid expansion, of the electronic computer. In 1953 there was only one computer that was programmed to solve linear programs with 25 variables and 25 constraints. Today, even desktop computers can solve practical problems of a much larger size.

The occasion of this lecture gives me the opportunity to pay homage to the late Albert W. Tucker, who was my friend and colleague for 47 years. A creative mathematician and a superb expositor, he was notable for providing financial support and a congenial atmosphere for a large number of mathematicians working in what is now called "finite mathematics", a term that he coined. It was my good fortune to be in the right place at the right time to work with him.

Bibliography

[1] R. Colville (1970). A Comparative Study of Nonlinear Programming Codes. H.W. Kuhn (ed). *Proceedings of the Princeton Symposium on Mathematical Programming,* Princeton University Press, Princeton

[2] G.B. Dantzig (1991) Linear Programming, in: J.K. Lenstra, A.H.G. Rinnoy Kan, A. Schrijver (eds) *History of Mathematical Programming,* North-Holland, Amsterdam

[3] M.A. El-Hodiri (1967) *The Karush Characterization of Constrained Extrema of Functions of a Finite Number of Variabiles,* Research Memorandum A.3, UAR Ministry of Treasury

[4] M.A. El-Hodiri (1971) *Constrained Extrema: Introduction to the Differential Case with Economic Applications*, Springer-Verlag, Berlin Heidelberg New York. [Republished from: M.A. El-Hodiri (1966) *Constrained Extrema of Functions of a Finite Number of Variables: Review and Generalizations,* Krannert Institute Paper 141, Purdue University, Purdue]

[5] V. Fiaco, G.P. McCormick (1968) *Nonlinear Programming: Sequential Unconstrained Optimization Techniques*, Wiley, New York
[6] D. Gale, H.W. Kuhn, A.W. Tucker (1951) Linear programming and the theory of games, in: T.C. Koopmans (ed) *Activity Analysis of Production and Allocation*, Wiley, New York, pp 317–329
[7] F. John (1948) Extremum problems with inequalities as subsidiary conditions, *Studies and Essays, Presented to R. Courant on his 60th Birthday, January 8, 1948,* Interscience, New York, pp 187–204
[8] W. Karush (1939) *Minima of Functions of Several Variables with Inequalities as Side Conditions*, M.Sc. Thesis, Department of Mathematics, University of Chicago, Chicago. [A summary of this work has appeared as an appendix to: H.W. Kuhn (1976) Nonlinear Programming: a Historical view, in: R.W. Cottle, C.E. Lemke (eds) *Nonlinear Programming*, SIAM-AMS Proceedings 9, pp 1–26]
[9] H.W. Kuhn, A.W. Tucker (1951) Nonlinear programming, in: J. Neyman (ed) *Proceedings of the Second Berkeley Symposium on Mathematical Statistics and Probability*, University of California Press, Berkeley, pp 481–492
[10] L.L. Pennisi (1953) An Indirect Sufficiency Proof for the Problem of Lagrange with Differential Inequalities as Added Side Conditions, *Trans Amer Math Soc 74*, pp 177–198
[11] A. Takayama (1974) *Mathematical Economics*, Drydale Press, Hinsdale

Mathematics and history

Italian Mathematics, Fascism and Racial Policy

Giorgio Israel

Translated from the Italian by Emanuela Moreale

Modernisation or Backwardness?

The historical analysis of the transformations of science and its institutions during the Fascist period is permeated by a question which characterises the whole of the fascist historiography; that is, whether the influence of regime politics determined only involution and decadence or whether it also favoured processes of modernisation. In our opinion, both of these aspects were present. Yet, it should also be pointed out that views have sometimes been extreme: in fact, the regime has been clearly placed at either of these extremes, thus indulging in uncritical and unfounded reevaluations. Such attitudes are often determined by valuations that have little to do with objective historical analysis and are rather determined by political considerations. On the one hand, they could be due to fear that more balanced valuations could lead to a reevaluation, or even absolution, of the regime. On the other hand, they could be aimed at the hurried establishment of the historiographic basis for the need of a national reconciliation that, by distributing blame and fault equally, leaves everyone happy and reassured.

In the context of contemporary Italian culture, it is the latter temptation that appears to be the most dangerous and ambiguous. Far from closing some of those currently still open wounds, it risks reopening them in the most unpleasant of ways and acting as an obstacle to the development of historical analysis by making it hostage to contingent opportunities (and opportunisms). It also risks giving rise to a homegrown, and unrefined, version of "historiographic revisionism". On the one hand, nowadays there is a wide availability of the knowledge and the historiographic materials needed to establish a balanced evaluation of the role of fascism in the modernization of our country and in the degenerative processes of structures, institutions and culture inherited from the liberal period. On the other hand, it is evident that the sheer existence of material knowledge is not sufficient if there is room for misunderstandings in the interpretation of basic issues. For instance, it should by now be obvious and accepted that consensus with the regime, particularly among cultured people, was then impressive and nearly universal: any attempt to contradict this given fact, that is that Italy, and particularly Italian culture, was "Fascist" in an almost totalitarian sense at least until towards the end of the Thirties, would represent a useless obstacle to historical analysis. The fact that most intellectuals were engaged in the regime's work does not by itself imply a positive valuation. Such a consideration should be considered as obvious and superfluous. Still, the idea that a universal consensus itself represents a justification lingers on. This is an idea which is clearly wrong and which is explained by it being a remnant of totalitarian and populist views: as long as it stands, it will be difficult to discuss in a balanced way the history of totalitarian regimes endowed with a wide consensus base, because this is exactly where the cause of confusion between historical and political judgment lies.

We must therefore start off from the given fact that the scientific community, as a part of the Italian cultured community, sided almost unanimously with the regime and collaborated fully and heartedly to the realization of its projects. This community never had the characteristics of a separate and autonomous island and even the few anti-Fascist fringes, which may well have gained admiration for the convinced coherence with which they opposed the regime, still did not succeed in modifying the course of events that transformed Italian science together with the whole of society.

In considering the position of Italian science in the nineteen-twenties – the period during which Fascism came to power – it is necessary to first of all stress one of its characteristic aspects: Italian science was based on an excellent tradition (that of Galileo), which was, however, too far away to provide a real basis for modern national scientific research. In fact, there existed no "Italian" scientific community as such at the time in which the political union of the country took place. On the contrary, the supremacy of the big and scientifically advanced nations, such as France and Germany, was based on the existence and coherence of their national scientific communities. It was a great merit to scientists living in the years just after the unification of Italy, particularly mathematicians such as Enrico Betti, Luigi Brioschi, Luigi Cremona, to have understood that the setting-up of a national scientific community was the most urgent necessity to ensure that the new nation could find its place in advanced research. The results obtained in this direction were not only extraordinary but were also achieved very quickly. Because it was mainly mathematicians that worked on this, it was in this field that Italy quickly managed to find its place among the scientifically advanced nations. Towards the end of the nineteenth century, Italian mathematics was in third position, behind Germany and France, while Italian schools of geometry and analysis enjoyed a recognized and undisputed reputation. This development acted as a driving force in other sectors of research, such as physics: the foundation of the Società Italiana di Fisica (the Italian Society of Physics) was due to a mathematician, Vito Volterra. Also chemistry and biology (with personalities such as Stanislao Cannizzaro and Giovanni Battista Grassi) were starting to reach high qualitative levels.

Of particular interest is the attention which was given to applied sciences and engineering. Even in this field mathematicians acted as a driving force: it is sufficient to think of the foundation of the *Reale Scuola degli Ingegneri* (Royal School of Engineers) by Luigi Cremona in Rome or to the refounding of the *Società Italiana per il Progresso delle Scienze* (Italian Society for the Progress of Sciences) by Vito Volterra in 1906. This latter aimed at establishing a common ground for interaction between pure and applied sciences as well as science and culture, by putting in touch all the various disciplines and cultural segments.

However, these operations of cultural and scientific policies had to rely on the study and acceptance of foreign models, particularly the German and French models, given their position of hegemony in world science. It is neither possible nor appropriate here to analyse the characteristics of such an imitational process and its consequences on the structure of Italian science, particularly since this has already been done in contributions listed in the recommended readings at the end of this article. Yet, one crucial aspect of the relationship between pure and applied science should be explained here. Even though the influence of the German model was of great importance, particularly in the geometric sciences, most Italian mathematicians and physicists-mathematicians ended up choosing the French model as their preferred cultural model of reference, while not appreciating the novelty represented by that particular branch of German science known as the "Göttingen School". Founded by mathematician Felix Klein, who for a long time was the main reference point for Italians, the Göttingen School, under the influence of David Hilbert, turned its efforts towards those which will were to become the leading topics of twentieth century science: set theory, functional analysis, quantum mechanics and mathematical logics. It did so by taking on as its methodical principle the axiomatic method that was to revolutionise the science of that century, from the theory of probabilities to theoretical physics. Just like French science, Italian science also remained alienated from, if not diffident towards, the developments of axiomatic science.

In this context, it is necessary to highlight an aspect of primary importance as regards the relationship between science and applications. The Göttingen model proposed a completely new and original vision of this relationship. In fact, it saw basic science as abstract and "pure", which seemed to detach it from applications and, on the contrary, presented itself as the most appropriate to the new requirements of engineering. The nineteenth-century view coincided with the traditional view of the relationship between science and engineering as a direct and immediate interaction in which there was no sense of the new role that more abstract (even when seemingly less immediately applicable) research could play. It is exactly the figure of the traditional engineer, seen as the only direct mediator in the relationship between science and engineering, which seemed antiquated and inadequate. Technology, that is engineering based on science, needed a much deeper and complete contribution from scientific research and from all scientific research, even the more abstract and seemingly remote from "concreteness". The Göttingen School was the most advanced interpreter of this view and was a model of it also for science in the United States, which in fact took it in completely, thanks also to the emigration of the most important representatives of this school to the United States, after Nazism rose to power in Germany.

Moreover, the Göttingen School represented a model of scientific research which was totally "internationalist" and which overcame the old barriers of national schools, or at least seemed to suggest that it would be useful to overcome them. A scientific research founded on themes and methods aiming at unification and appropriate to the requirements of technology could not remain closed within national cultures, but had to propose itself as a great and boundaryless world undertaking. The big paradox in German science during the first decades of the twentieth century was determined on the one hand by the forming of an internationalist view of research and by the fact that some research environments (particularly that of Göttingen) became the gathering point of researchers of many nationalities. On the other hand, this was happening just at a time when German nationalism was resurrecting and the germs of anti-Semitic racism were being stimulated in the most virulent of ways.

It was political reasons, and in particular the siding of Italy with France during World War I, that determined a fracture between Italian scientists and their German colleagues. Italian science, which had so successfully striven to become a national school, and which was impregnated with patriotic feelings in the wake of the "Risorgimento"[1] , failed to understand the need for the new trend towards an international model of research. Italian science therefore remained dependent on the French model and, in particular, retained the view of a scientist as a "man of general culture", who is particularly attentive to the relationship between science and humanistic culture and is interested in the applications of science, as in the nineteenth century view of engineering.

Vito Volterra intended the *Società Italiana per il Progresso delle Scienze (SIPS)* to be a place where university lecturers and secondary school teachers, engineers, technicians, doctors and scholars in the humanities could meet to compare their points of view and their achievements. This was an interdisciplinary intent aiming at ranging both horizontally and vertically so as to put in touch higher and lower levels of scientific culture as well as theoretical and applicative approaches.

However, Volterra did not believe that the diffusion of technology was a simple and linear route to progress. With development, "new needs are created in human society, needs which each nation must meet if it does not want its intellectual life to die or languish and the very roots of its own prosperity to wither". It was therefore necessary for scientists to do their share by breaking old institutions, creating associations and places for meetings as well as for more "liberal" and open discussions. Such places were therefore in a position to foster the cooperation of tendencies capable of fighting the evils of excessive specialisation and also to enable a rapprochement between scientists and ordinary cultured people, facilitating a mass diffusion of scientific culture.

> Old academies are too closed up, teaching institutions have predetermined intents, each scientific society is too small to work on these aims. These goals can only be achieved within a large association gathering specialists in all disciplines [...]. All that the public cannot learn from books and from speeches will be understood when it attends and participates in discussions by scientists, since it is spontaneous and lively discussions which show under the most natural and truest of lights the germinating expressiveness of those thoughts which are usually divulged through learned artificiality.[2]

[1] *Risorgimento: movement that began in the early nineteenth century and led to the proclamation of the Kingdom of Italy (1861) and eventually its unification (1871).*

[2] *See Volterra V, Il momento scientifico presente e la nuova Società Italiana delle Scienze (The Present Scientific Time and the New Società Italiana per il Progresso delle Scienze). For precise references, relating to this or any following quotes, please see rec-ommended readings at the end of this article.*

Today, such a humanistic view of science and technology cannot but arouse sympathy in those of us who fear the excesses of the single line of thinking dominating contemporary society: techno-scientific culture. However, from a historical viewpoint, it must be said that this viewpoint was not to prevail. Vito Volterra was the main representative of a liberal-democratic scientific culture whose inspiration was derived from the ideals of the Risorgimento and the Enlightenment and had the French model as its polar star. Of this model, it accepted both the view of the relationship between science and culture and between science and engineering, which had its roots in the nineteenth century tradition of polytechnic sciences. The events of World War I consolidated its adherence to this model, while identifying only negative aspects in the German model. So much so that this lead to the reevaluation of "Latin" culture as the supporting beam of European culture and as the bulwark against the "barbarism" and obscurantism coming from north and central Europe. This attitude had negative consequences and, in science, precluded any attention to novelties proposed by German science, in particular concerning the relationship between science and technology. Therefore, the liberal-democratic model personified by Volterra, also known as "Mr Italian Science", to indicate how abroad he was considered to be the major representative of national science, was on the decline by the 1920s. One of the most significant episodes of this crisis is Volterra's useless opposition to the teaching reform promoted by Giovanni Gentile through a Commission he had set up through the Accademia dei Lincei[3]. The observations and proposals of this Commission were not devoid of sharpness and great interest, but ended up going back to the old school model introduced by the Casati reform and therefore, in spite of all its good intentions, they appeared more like a defence of the existing system than an alternative proposal. The Gentile reform was certainly little appreciative of scientific culture and, on the contrary, it subordinated scientific culture to humanistic culture, but drew its strength from identifying the need for subordinating the whole school structure to a single *formative axis*. It cannot be denied that the Gentile reform was one of the most original, innovative and solid results of the regime's cultural policies, as can be seen by its own endurance against the passing of time and by the difficulty in outlining an alternative proposal with the same amount of coherence.

Regime policies were not so effective in restructuring its cultural institutions, particularly scientific ones. Pietro Nastasi was right in commenting, in his wideranging and detailed close examination[4], that the scientific politics of the regime was episodic, chaotic and lacked a coherent direction. After all, it was a policy dictated by improvisation. This was due not only to the fact that the dominant neoidealist ideology suggested a devaluation of scientific culture and therefore neglected, when it did not quite hinder, any attempt to promote a wide diffusion and critique of scientific culture. According to this ideology, science could

3 *See Israel G, Vito Volterra e la Riforma Scolastica Gentile*
4 *See Nastasi P, Il Contesto Istituzionale.*

not aspire to the role of "cultural axis" of national life, and particularly of school life, because this had to be carried out solely by humanistic culture, seen as totally separate from scientific culture. It is, however, true that the corporate Fascist view tended to promote a separation of pure scientific activity from applied science, which were seen as interacting but institutionally distinct. This can also be considered to be an innovation with respect to the integrated vision "à la Volterra", which goes back to an antiquated nineteenth century model. Of course, this view was partly determined by political reasons. The regime saw university as the most appropriate place to rebuild the corporate unity of the scientific community, as well as a politically controllable institution. Here scientists, "segregated from mundane noise", to use the words of minister Giuseppe Bottai, would have been able to dedicate themselves to pure research. Instead, other institutions, such as the Consiglio Nazionale delle Ricerche (the National Research Council), had to supervise applied research or manage the contributions of pure research to applications.

It would, however, be inappropriate to go as far as to identify in this articulated view of the functions of pure and applied science a coherent and coherently followed plan. The modernisation factor implicit in this restructuring of scientific institutions was not totally fulfilled and, on the contrary, it was compromised on the one hand by the confused and amateur management of the regime and, on the other hand, by its obsession with detailed political control. In this context, exemplary is the fate of SIPS, which progressively lost its role as a place of interdisciplinary cultural interaction, wanted by Volterra, and ended up becoming a collection of distinct cultural "sections" of scientific research. More importantly, it turned more and more into a catwalk for all of the regime's triumphs, including even the dullest and most vulgar expressions of respect. Independently of any more or less defined scientific policy, the regime was definitely obsessed with control and certainly could not like a scientific community, whose activity was open to interaction with other sectors of cultural and productive life.

Talking at the refounding of SIPS in 1906, Volterra observed in a lyrical tone:

> The wires, which we see extending like a net above our abodes and shooting far away, are the most eloquent evidence of our economic prosperity. In the solitary Roman countryside, they run parallel to the superb aqueducts. To Lord Kelvin's genius, who admired them in a bright twilight, they spoke a language which was as solemn as the majestic remains of the power of ancient Rome[5].

Nothing was further from the Fascist view than the idea of a network of electric wire taking the symbolic place occupied by the ruins of ancient Rome. It would be stretching attributing to fascism a solely technical view of science,

[5] *Volterra V, Il Momento Scientifico Presente e la Nuova Società Italiana per il Progresso delle Scienze.*

which was rather congenial to the thesis of neo-idealism "à la Croce". However, it cannot be doubted that the distinct separation of pure and applied science was not so much the expression of a "modernisation" project as the expression of a twofold need: excluding science from a cultural role capable of contending with humanistic sciences for supremacy and, above all, controlling the various sections of the cultural and scientific world through their corporate division into sections.

This twofold aspect of the regime's actions, in the case of mathematics, will be taken up again later. On the one hand, it managed to introduced significant novelties as concerns the restructuring of scientific institutions; on the other hand, it cancelled out the innovative value of these reforms, partly with the incoherent and occasional character of its action, but mostly by subordinating them to its plan for totalitarianism and autarchy. The contradiction at the root of this negative outcome can be summarised through a rhetorical question: how could an autarchic and nationalist view of science be reconciled with the irreversible trend towards an international and barrierfree structure for research and technology? Was such a restriction not destined to cancel out the innovative effect of any reform?

The question in the title of this section can now be answered: the regime did not stay still in its policy of reform of science and institutions, but the deep goals animating its actions ended up by dragging the whole of Italian science into a backward position of which its racial policy represented the final event and the dramatic seal of a failure.

The Case of Mathematics

The state of Italian mathematics at the beginning of the Fascist period was effectively described by Pietro Nastasi who summarised its two main aspects[6]. On the one hand, Italian mathematics during World War I "had become an institutional discipline, with a definite position within the scientific and educational systems" and had conquered a well-respected position within the international mathematical community: a "third place" after German and French mathematics. On the other hand, according to French mathematician André Weil, it seemed affected by a process of "sclerosis analogous to that which was menacing France, but which had even prompter and more destructive effects than in France".

This negative situation was much more evident on a contents level than on an institutions level. The glorious school of Italian algebraic geometry appeared as if stuck in the trails of its own tradition and incapable of innovation and attention to the new algebraic and topological methods and approaches that had been

[6] *Nastasi P, Il Contesto Istituzionale.*

circulating in Germany. The school of mathematical analysis was still showing some very excellent results, but seemed incapable of accepting the achievements of functional and abstract analysis also coming from the German school. The situation was better in the field of mathematical physics, represented mainly by Tullio Levi-Civita, which was not only keeping up with the latest advancements in the field but also managed to be one of its protagonists. It should be noticed that, for a series of reasons we have analysed elsewhere, Italian science, mainly thanks to mathematicians such as Levi-Civita, Enriques and Castelnuovo, appeared more open to the novelty represented by the theory of relativity than French science was. After all, it was Levi-Civita who prepared the mathematical language of the theory of general relativity, that is the absolute differential calculus. In spite of the single exception of Levi-Civita (who, however, continued to see himself as a "conservative"), in general, Italian mathematicians did not appear to be fully aware that the structure of physics was radically changing and that this change was reflected in the creation of the new branch of *theoretical physics:* this latter did not see itself as part of the classical *corpus* of mathematical physics and, in fact, originally developed from quantum mechanics. On the contrary, it was in this area that Italian mathematicians showed a silent opposition that was all the more pernicious since the mathematical community had (and was to maintain until World War II) a position of preeminence within the Italian scientific community. An indication of this was the opposition to the setting-up of a Faculty of Mathematical, Physical and Natural Sciences (destined to Enrico Fermi) at the University of Rome: such opposition was an example of unjustifiable refusal of the new trends of research in physics.

Thus, in the Twenties, Italian mathematics went through a stagnant phase of decline, while preserving a respectable position in the international scene and while being characterised by the presence of prominent personalities.

As regards institutions, the position of Italian mathematics appeared strong. There were numerous journals dedicated to pure and applied mathematical research and, besides the prestigious *Circolo Matematico di Palermo* (Mathematical Circle of Palermo), there were numerous institutions whose mathematicians had prominent, if not even hegemonic, positions: the above mentioned SIPS, the *Consiglio Nazionale delle Ricerche* (set up thanks to Volterra himself in the first instance), the *Accademia Nazionale dei Lincei and the Scuola Normale Superiore* in Pisa. It is exactly this solid institutional presence which explains the delay with which a professional association for mathematicians, the *Unione Matematica Italiana* (Italian Mathematical Union, UMI) was promoted, following the model of what had been happening in most European countries. It also explains the coldness with which some circles (such as the Roman circle) collaborated with the new institution founded in 1922. Later, Italian mathematics was enriched with the addition of two more prominent institutions: the *Istituto Nazionale per le Applicazioni del Calcolo* (National Institute for the Application of Calculus, *INAC),* founded by Mauro Picone in Naples in 1927 and then transferred to Rome in 1932 and the *Istituto Nazionale di Alta Matematica* (National Institute of Advanced Mathematics, *INDAM)* founded by Francesco Severi in Rome in 1939.

This rich institutional structure in Italian mathematics was the expression of its academic strength, which in turn was a consequence of its highly prestigious tradition as well as of inertia, rather than of a forward-thrusting force. The setting-up of institutions such as *INAC* and *INDAM* and the strengthening of *UMI's* role could be seen as evidence that the regime was undertaking a modernisation of the leading sector of Italian science, but this would be a superficial and incorrect deduction. In fact, the regime was not aware which the more promising sectors in scientific research were and decided to favour the less expensive undertakings which had a good "payback" in terms of propaganda and which did not hinder its policy of "Fascistisation" of institutions. This is particularly obvious if we consider the attitude the regime had taken towards projects by Orso Maria Corbino and Enrico Fermi, which aimed to set up an "Istituto Nazionale di Fisica" endowed with large autonomy from the University and considerable financial means. The project was failed by the regime, both because it was expensive and because it could have weakened the centralising function of the University. This shows how it had not been understood that the miraculous supremacy that Italy had gained in the field of nuclear physics with the discoveries of the "school of via Panisperna" could not last long without massive support and without the acceptance of the principle that research of this sort requires large autonomy from the traditional academic setting. This is the view that had been taken in the United States and it was the main reason that research in nuclear physics moved to that country. It was "reasons of ideals and balance sheet" which sank the Corbino-Fermi project, while similar reasons did not hinder the strengthening of Italian mathematics through the foundation of *INDAM* and *INAC*. This was not only because these were structures which cost less, but also because they did not seem to bring into question the hegemony of the academic world and the possibility of easy political control on them. In this respect, the mathematical community met the requirements of the regime, even though it would be incorrect to say that it was one of the most "Fascist" sectors of Italian science.

As mentioned earlier, if one looks only superficially at the institutional initiatives of the regime in the field of mathematics, particularly the foundation of *UMI, INDAM* and *INAC,* one could easily mistake these for a sign of a modernisation process. Instead, one should take into account that usually initiatives in the scientific field encountered the favour of the regime only when they met three criteria: inexpensiveness, adherence to the Fascist institutional view and therefore strict political control and payback in terms of image, that is capacity to extol the regime and its policies and, in general, the values of the nation and of its people. The second need implied the strengthening of the role of University in the control of research (the University being, in turn, a structure on which it is easy to exert political control) and a real distinction between pure and applied research that, however, did not give the latter an excessive autonomy and did not build independent clusters. The third need was being felt more and more as the regime developed an autarchic policy which was firmly opposed to any view of international research, while, on the contrary, it tended to extol the role and supremacy of the "Italic genius" in science.

The initiatives of the Italian mathematical community effectively answered these three needs. The political control appeared total, after the marginalisation of anti-Fascist islands, which were to be found mainly around the figure of Volterra, and after the final defeat of the liberal-democratic view of research and of its relationship with the cultural and productive world. UMI was to become more and more a platform of Fascist mathematics and, at its congresses, many prominent political figures of Fascism were present, particularly Giuseppe Bottai. *INDAM* and *INAC* represented the two branches of pure and applied mathematical research, but the latter did not make any financial or structural-organisational demands such as those made by the Istituto di Fisica conceived by Corbino and Fermi. The two founders – Severi and Picone – fought to prove that they were the most faithful interpreters of the regime's policies. Picone declared himself an example of "Fascist mathematician" and called the research that was taking place in his institute "Fascist mathematics" aiming at strengthening the nation's productive and military structures. Severi, during the inaugural speech of *INDAM*[7], which remains an impressive show of political servitude, was very polemic against any view which tended to subject pure research to direct application needs, stressing however that he could see how the regime fully supported his vision in this respect. In fact, the regime was trying to please all of his supporters and competitive champions, by claiming with Solomon-like equality that science was both to take place in an environment segregated from "worldly noise" and also, as regards applications, to answer the practical needs created by the development of the Fascist nation.

The most important aspect concerning interpretation is given by the disastrous interaction between the regime's autarchic and nationalist view and the tendency of Italian mathematics to fall back on itself and lose interest in the research development outside our country. One could grant that regime-approved initiatives might have had, or possibly even had, favourable effects on Italian mathematical research. *INDAM* represented a model of scientific institution which, for some aspects, was somewhat remindful of the Göttingen mathematics institute and which could be seen as gathering point for young researchers to attain research experience and prepare them for advanced research. The excellence of many lecturers (such as Severi himself) was a further point favouring this possibility. But the push towards an autarchic view favoured the consolidation of the tendency to ignore the latest research achievements, particularly in geometrical and algebraic mathematics, due to the wrong belief that Italian mathematics, thanks to its glorious tradition, had no need for anything but itself. The seal on this unpropitious tendency was set by the 1938 racial laws whose outcome can be easily imagined by remembering that most preeminent geometers, but also most analysts and physicists-mathematicians, were Jewish.

7 *Severi F, L'Inaugurazione del R. Istituto di Alta Matematica (Roma).*

A similar argument can be applied to *INAC*. Although the lack of development of numerical calculus in Italy, at least to levels similar to those in other nations, could not be attributed to Fascism, but rather to postwar research policies (and had several causes which cannot possibly be analysed here), it is certain that the limited and lifeless educational structure which had been inherited from the previous period was greatly responsible for this failure.

In short, the involution and declining trends in Italian mathematics, which were already evident in the Twenties, worsened due to the regime's autarchic and nationalist political choices which accelerated the decadence process. Postwar recovery was slow and difficult: it was only at the beginnings of the 1960s that algebra really became a full-blown topic in the curriculum for the degree in mathematics and the weight of tradition in Italian algebraic geometry seriously hindered the development of analysis. Even in the 1950s, Francesco Severi, from his high position as president of *INDAM*, warned that in mathematics "geometry is thought and analysis is calligraphy".

It has been said that the regime's racial policy, the authentic apex of its nationalist autarchic view, represented the seal on such destructive trends and gave a serious blow to the mathematical community, which, however, did little to defend itself against it. To illustrate these aspects, it is necessary to deal with the origins, the nature and characteristics of this racial policy in which all the segments of the university and research world, albeit in different forms, were involved.

Racial Policy and Scientific Community

The literature on racial policies promoted by the nazi regime is very wide and deals with both its theoretical and sociological issues. The former relates to the contributions that various scientific disciplines (anthropology, biology, eugenics and demography) made to the definitions of the "foundations" of its racial policies. The latter concerns the role that the German scientific community had in the realisation of that policy and the consequences to the scientific community itself, mostly due to the emigration of Jewish and anti-Nazi scientists which impoverished German science to the point of causing it to lose its forefront position to the United States[8].

This particular direction of research was not followed as much in the case of Italy. The reasons for this lack of attention can be traced to the fact that the German case is so relevant and produced such dramatic effects to keep the Italian case in the background. Moreover, the anti-Semitic racial policy had a central *constitutive* role in the Nazi racism, that is it had a basic function in Hitler's political view. The same certainly cannot be said in the case of Italy: the

[8] *For a detailed bibliography, see Israel G and Nastasi P, Scienza e Razza nell'Italia Fascista.*

Mussolini regime only adopted a racist legislation in 1938 and, in the preceding years, it repeatedly distanced itself from the measures and racial policies of its German ally, opposing the idea that it was possible to define races from a point of view which was purely or exclusively biological. This difference has led some historians to state that Fascism's anti-Semitic racial policy was marginal, reducing it to an act of deference towards its German ally, a concession towards the requests received from the latter.

A more accurate historical analysis, however, shows that this was not the case. Italian Fascism elaborated its own racial policy, with its autonomous and original characteristics, reflecting the legal measures adopted in 1938. The disagreement with Nazism as regards the racial policies it adopted starting from 1934 did not diminish, but rather increased in the period between 1938 and 1943. The element of divergence could be captured by the phrase "discriminate, not persecute": Fascists did not propose hitting Jews directly, but "rather" marginalising them from national life. This was a clearly hypocritical formula, because their cruel, pedantic and accurately-followed measures were, in fact, harshly persecuting. Renzo De Felice himself, after proposing an oversimplification of Fascist racism[9], sensibly corrected this view in his biography of Mussolini[10].

In our opinion, both "simple" interpretations of Fascist racism are wrong: the one claiming it was marginal and the result of a contingent political choice and the one stating that the anti-Semitic racial policy had a constitutive role in the Fascist ideology. In Italy, next to what is traditionally considered the determining factor (that is, choices dictated by foreign policy), all the factors (with no exception) that contributed to determining the choice of developing a racial policy in Italy should also be considered. Fascism's racial policy can be understood only as the arrival point of a series of routes or tracks that, although distinct and independent, produced an explosive cumulative effect. Next, these various factors will be described briefly.

It is usually agreed that Italy is a European country where anti-Semitism has never been intense. This is true, but some clarifications are required. Anti-Semitic demonstrations in Italy were largely connected to religion and therefore to Catholic traditional anti-Semitism. This statement is fully supported by an examination of the situation in the first decades of the twentieth century: the most violent anti-Jewish demonstrations came from religious circles and particularly from the Compagnia di Gesù (Italian Gesuits): their publication, *La Civiltà Cattolica,* provides the best evidence of this. This phenomenon explains the peculiar characteristics of anti-Semitic behaviours in Italy: Catholic anti-Semitism did not aim at persecuting the Jews on the basis of race, neither did it seek their physical elimination, but rather its goal was dissolving their religious identity. Their main objective was therefore the conversion of each and every Jew, the dispersion of old Israel into the new Catholic Israel. Conversion would

[9] *De Felice R, Storia degli Ebrei Italiani sotto il Fascismo.*

[10] *De Felice R, Mussolini Il Rivoluzionario, Il Fascista, Il Duce, L'Alleato, La Guerra Civile.*

redeem each Jew of any sin and integrate him or her fully into the Christian faith: persecution is therefore only one of the tools through which to exercise pressure to reach this objective. It is now clear how Catholic anti-Semitism, apart from some notable exceptions coming from Jesuit circles, could not agree with German, biologically-based racism which, claiming that races were inevitably different, did not offer any choice other than the pure and simple destruction of the "Jewish race". Hence Italian racism, under the influence of the Catholic Church, simply had to have a mild character; a racism which could only propose itself through a definition of race not in the biological sense, but as a cultural and spiritual identity.

It is interesting to note how this view influenced also some attitudes towards Jews in the liberal-democratic sector of the Italian lay world. Jean-Paul Sartre, in his famous essay on anti-Semitism, had thrown light on a peculiar aspect of liberal thought on the Jewish question: in spite of all its unquestionable merits in favouring the process of Jewish emancipation in the fight to end discriminations, the liberal thought showed an obvious incapacity to understand the right of Jews to keep their own cultural, spiritual and religious specificity. According to liberal thought, Jews had to become emancipated by giving up its own "Jewishness" and by dissolving all their own "diversities" into the notion of abstract person ("citizen") on which the democratic-liberal society is founded. This shows a strange parallelism to the Catholic point of view: in both cases, as a condition for discriminations against them to stop, Jews were required to wipe out their Jewish characteristics and, either convert to Catholicism or feel like a citizen devoid of any other specification. In the Italian context, these two requests found points of contact and even convergence. In this respect, Benedetto Croce's position seems symbolic: his opposition to anti-Semitic persecutions was undoubtedly strong, but his position on the Jewish question was not as clear. Croce claimed several times that Jewish identity was a cultural heritage to be lost as soon as possible, in case its persistence was used as a pretext to the perpetuating of anti-Semitism itself. Now, if we take into consideration that, to Croce, the identity of citizens in a western country such as Italy included their being Christian ("Why Can We Not Fail To Call Ourselves Christians?" is the title of his famous essay), it is easy to understand the ambiguous superimposition of the two refusals. These odd liberals, who looked at Europe as a democratic-representative reincarnation of the Sacred Roman Empire, asked Jews to give up their own identity not for the "citizen" ideal, but for the "Christian citizen" ideal. They did so by resorting to the very argument used for centuries to justify religious anti-Semitism, that is the obsolescence of the Jewish religious message compared to the Christian one.

The example of Croce is all the more significant in that he was one of the few who did not derive from such a position an attitude of indifference towards, or worse, of justification of persecutions. On the contrary, most intellectuals did make this deduction: their sometimes icy indifference towards the Jewish question and Jewish identity, with the emergence of the racial campaign, turned into silent acquiescence and often in more or less open approval.

One of the peculiar characteristics in Fascist Italy was therefore: a wide lack of participation of the population in racial policy but, on the contrary, the participation and often shameful compromise of a large majority of the intellectual world in that policy.

It is not difficult to grasp the differences between the policies and racial legislation of Italian Fascism with respect to those developed in Germany. In Italy there was a pronounced diffidence towards the theme of biological racism and a rather ambiguous notion of race was preferred. The most popular term was "stirpe" ("descent"). In comparison with the Italic "descent", the Jewish "descent" was characterised not so much by its biological diversity, as by a set of spiritual, cultural and religious elements as well as customs that appear totally alienated and different from those dominating the country. A perfect rendition of this idea can be found in a 1939 news report by the Guf (Gruppo Universitario Fascista, Fascist University Group) in Catanzaro:

> Racism is not only a biological problem, but also and mainly an ethical problem touching the conscience of all of us and sharpening the sense of nation, the sense of this human community united by a common language, religion and history, but which has its main element of cohesion in that identity of thought and action which are independent from the psychological conformation of a unitary race.

According to pathologist Nicola Pende, one of the main theoreticians of Italian racism, the concept of race must be "dynamic-synthetic-evolutionary" and this concept, when applied to the case of Italy, provides us with a notion of "Roman-Italic unity". This latter was the result of the merging of Aryan-Italic people in a superior unit "unmistakable with any other European Aryans... with its own anthropological and psychological characteristics". Of course, this unity ended up determining also its unitary and distinct physical and biological characteristics. The Jews were obviously alienated from this unity and therefore "breeding the superior biological state of the Italian nation" implied a "strong prohibition" against marrying Italians with "people who, like the Jews, Ethiopians and Arabs, are so (particularly spiritually) far from people of Romano-Italic descent"[11].

It is this "theoretical" vision that provided the grounds for exclusion and discrimination measures rather than measures aiming at their physical elimination. Fascist racial legislation clearly reflected this formulation and in this it differentiated itself from the Germans.

In order to understand how those policies and measures could have been arrived at in such a climate of acquiescence, if not approval, particularly among intellectuals, several factors should be taken into account. From what has been described so far, it is clear how one of these factors was the peculiarity of Italian

[11] *For these quotations see Israel G, Nastasi P, Scienza e Razza nell'Italia Fascista.*

nationalism going back to the myth of Rome, mother of peoples, capable of unifying different peoples in a common ethical and spiritual vision and not the myth of blood and race which was the main myth used by German nationalism. But one of the more interesting aspects, which have been less extensively explored so far, is the role of the policy of population, demographics, eugenics and orthogenetics in the diffusion of a racial view which, although within the spiritualistic characteristics mentioned above, ended up taking deeper and deeper roots and polluting minds, until it eventually proposed itself as a real policy.

It is appropriate to reflect on how population issues were perceived in Western countries in the first few decades of the twentieth century. One of the most worrying problems was the general demographic decline in European countries. On top of this quantitative problem, there was a qualitative problem: the assumed decline in the anthropological quality of Europeans with respect to other peoples. Some thought its causes were a sort of biological degeneration, or a decreasing reproductive power and a debate took place about the possibility of some "spiritual factors" being involved (such as the loose morals of an increasingly more well-to-do population) or an inevitable genetic decadence due to inexorable "scientific" laws governing the ascent and decline of peoples. The problem of handicapped people, of the "faulty" and degenerate elements and of the negative influence they could have on the decline of population became an obsessive theme in all European societies and in the United States and involved also the scientific world. It was in this period which eugenics rose to the foreground and caused almost agonizing interest. Scientists were divided between supporters of socalled "positive" eugenics, which aimed at identifying the hygienic-sanitary practices capable of improving the general state of the population; and "negative" eugenics, which aimed at gaining the same result through coercive and violent practices, from the prohibition of marrying "faulty" subjects to the sterilisation of such subjects. Even in democratic countries, such as Scandinavian countries, and even before totalitarian regimes, negative eugenics was used to justify violent practices such as forced sterilisation of psychologically disturbed women or the kidnapping of their children.

The population problem was therefore considered as an essentially physical and biological problem and the population as a material that could be handled with unprejudiced techniques justified in terms of a "superior" end and national ethics. In other terms, the improvement of the quantity and quality of the population was motivated by the objective of preserving the continuity and the superiority of the European civilisation or, to be more precise, of each of its constituting nations. Hundreds of pages could be filled with documentations for the arguments, sometimes expressed in harsh and violent ways, with which the need to preserve the overall health of the population was supported even by prominent scientists. As can be easily deduced from what has been said so far, negative eugenics was less successful in Italy, partly due to the opposition of the Catholic Church and also due to the general spiritualist stance of "Italic" racism. It is very easy to find quotes in which scientists supported the role of the state in its main objective improving the race. For instance, in 1939 Ferruccio Banissoni (one of Pende's followers) wrote:

> Individualism, seen as the sense of the prevailing of arbitrariness and not the free will, which implies thoughtout and accepted self-limitations in the relationship with the collective, is on its way out. Also on its way out is the "bourgeoisie", seen as a satisfied and fearful quietness. Medicine, if it accepts and continues to accept with pity the egocentric individualism of the ill and the suffering, tends towards the masses of people, workers and soldiers, towards wider and brighter horizons, towards the big task of preparing a healthy, efficient and powerful generation.[12]

In the first few decades of the twentieth century, anthropology coincided with "physical anthropology", that is the study of the physical appearance of an individual to draw conclusions as to the individual's social and psychological behaviour. Although even the most serious followers of this approach tried to follow scientific and objective methods in their research, physical anthropology contained within itself a dangerously racial inclination. The study of physical characteristics almost invariably ended up aiming for the identification of racial "diversities" and therefore of "superiority" and "inferiority" characteristics. The unanimous adherence of anthropologists to such approach illustrates the dark and unhealthy climate in which this kind of scientific research took place. The inclined plane, from the study of races to racism, on which it was so easy to slip, is illustrated by numerous examples. Suffice to quote the illustrious and scientifically valid figure of Sergio Sergi. It would be difficult to claim that his participation in the *Consiglio Superiore della Demografia e della Razza* founded by the regime in 1938 and which gathered a distinguished group of university lecturers, was a reproachable political act disjoint from his scientific views. This is because Sergi was also a supporter of what he himself called "colonial anthropology", which was to act as support and theoretical justification to the imperial and colonial aims of Fascism. To Sergi, racism was not an external imposition or a concession made out of fear, but rather something in keeping with his scientific view. It is also painful to observe that prominent scientists, such as Corrado Gini (previously president of the *Istituto Centrale di Statistica* and distinguished Italian statistical-demographic scientist) participated in person in missions such as the 1935 mission involving the Dauada indigenous populations of the Libian lakes of Fezzàn: plaster moulds were taken of their faces and their basal metabolism was measured for the declared purpose of not only "finding out the laws which regulate human bio-psychological phenomena, but also basing on them measures of contingent and practical order". All this for the obvious and declared purpose of identifying the bio-psychological "diversities" of the indigenous population and proposing measures to adopt for the best "management" of this human material acquired in property by the Fascist Empire.

[12] *Again, see: Israel G and Nastasi P, Scienza e Razza nell'Italia Fascista*

Just as significant is the perseverance with which an impressive number of scientists (from Nicola Pende to Sabato Visco and Sergio Sergi to Corrado Gini, under the auspices of the president of the *Consiglio Nazionale delle Ricerche* Guglielmo Marconi) threw themselves onto the "human breeding ground" (the unfortunate phrase used at the time) represented by the farming population coming from the north-east of Italy who had colonised the reclaimed drained? lands in the Agro Pontino. They seemed to believe that this "human breeding ground" (or "laboratory of human biology", another common phrase at the time) was the ideal context in which to apply a form of rational eugenics and realise those improvements that, applied on a large scale, would have enabled the "weaving of physical, moral and intellectual habits for the new great homeland", that is the creation of a new "superior biological state of the nation" (in Pende's words).

This fanatic climate, the favour and wide circulation of these views starting from the mid-Twenties, gave a strong contribution to the forming of a racist way of thinking that manifested itself through political and legislative measures in the Thirties. The first occasion was the transformation of the Kingdom of Italy into a colonial Empire. The occupation of vast African lands and their colonisation through a remarkable migratory movement of Italian colonisers to the socalled "quarta Sponda" (fourth shore) established a direct contact with other "races". The need to avoid promiscuity that could have belittled the image of masters and dominators that occupants had to project onto themselves, pushed the regime to introduce racial legislative measures. These were the first Italian racial laws, justified by the theories that most of the scientific world (anthropologists, demographers, statisticians, biologists and geneticists) had worked to set up and that were now being used directly as theoretical support. The second racial laws were those which hit the other "race" Italians were in contact with and which they had to distinguish themselves from to avoid dangerous and debilitating contaminations: the Jews.

In this crucial step from theory to practice, up to their concrete, heavy and dramatic consequences, one can observe a determinant factor: the colonial and imperial aims of a totalitarian regime such as Mussolini's. It is not a coincidence that, between 1935 and 1938, Fascism decided to steer towards a policy which, in a sense, brought it back to its original inspiration: the polemic against the embellished and decadent bourgeoisie and the effort to make a definitive break with any form of compromise with liberal-bourgeois democratic-looking behaviour. In 1937, Mussolini was exasperated by the difficulties in instilling in its people a national conscience and the sentiment of being a strong, dominating and, if necessary, cruel people. He therefore decided to "throw three strong punches in the stomach of the bourgeoisie". These three punches were: the abolition of the "lei" form in favour of the stronger and dignified "voi"; the introduction of the Roman goose step in military parades and, finally, racism and the anti-Jewish racist legislation.

Such political and ideological choices were a decisive catalyst in the taking up of racial policy as a founding and characteristic element of the new revolutionary phase of Fascism, but Fascism already had a theoretical grounding and a wide support, especially in the scientific and intellectual world, mostly thanks to its long work on these themes.

The racism of Italian Fascism was therefore an autonomous movement endowed with peculiar characteristics and not a simple copy of foreign trends or concessions to its German ally. It was, instead, an extreme expression of Fascist autarchic nationalism. Reliable evidence of this was provided by Giorgio Almirante in 1938 through his contribution to the theoretical publication of Fascist racism, *La Difesa della Razza:*

> Racism is the most far reaching and courageous recognition of itself that Italy has ever attempted. Those who fear, even today, that it is a foreign imitation (and the young are also represented in this group) do not realise that they are thinking irrationally. It is really absurd to suspect that a movement meant to give Italians a race conscience, that is something like a nationalism developed to its five hundred percent, can possibly lead to enslavement to foreign ideologies[13].

Starting from the end of the Twenties, Mussolini had considered the demographic and racial problem as the nation's main problem. At first it was a question of using coercive means to block the phenomenon of emigration which was impoverishing "proletarian" Italy of its only resource and riches. Then it was a question of facing the problem of its quantitative growth: the problem of problems, as Mussolini himself had defined it in a famous 1927 article entitled "Strength in numbers". He observed that the demographic question was "the hardest touchstone that the conscience of Fascist generations will be measured against" and that was to show "if the soul of Fascist Italy is or is not irreparably hedonistic and affected by stuffy bourgeois attitudes". A large number of scientists promptly answered the appeal: the decade starting from that moment in time and going up to the 1938 racial laws represents a blossoming of treatises and practical realisations (such as the Agro Pontino reclamation) on the theme of demography and race. This theme was not considered purely quantitatively (as numerical expansion of the population), but also qualitatively, that is as improvement of the population in terms of biology, genetics and psychology: it was mainly anthropologists, geneticists, biologists and doctors working on this front. It is only this commitment and this background which can explain how, at a time when the regime embarked upon its racial campaign and legislation, it also found so many eminent scientists and academics ready at hand. For instance, the Consiglio Superiore della Demografia e Razza was no gathering of unimportant people, but rather gathered eminent figures in Italian science. Moreover, the racial campaign was launched through the *Manifesto degli scienziati razzisti.* Finally, a propaganda and political action publication such as the

[13] *See: Israel G and Nastasi P, Scienza e Razza nell'Italia Fascista*

Ufficio Razza of the Ministero della Cultura Popolare (Ministry of Popular Culture) was entrusted to the lead of Sabato Visco, dean of the Faculty of Science of the University of Rome and eminent member of the *Consiglio Nazionale delle Ricerche* and many other institutions.

Undoubtedly, in this active participation of such a huge chunk of Italian science in the racial policy, one can identify some characteristics that were typical of a racial view of a certain Italian political and cultural context described earlier under the heading of "spiritualistic" view. Within Fascism, there were influent trends going back to the dogmas of German biological racism and that, therefore, put the Jewish question in the same context as Hitler's theoreticians. But this was not the dominant trend. A historical reconstruction which does not limit itself to the surface, would throw light on the conflict between German-like biological trends and the majority view, represented by Pende and Visco, supporting a "spiritualist" view of racism that could be easily acceptable to the Catholic world and the Vatican and which translated itself into a racial policy of total exclusion of the Jewish component from national life (possibly in the hope of causing a massive emigration), without resorting to direct suppression measures.

A good historical analysis must therefore take into account this evident and significant phenomenon: the strict correlation existing between Fascist political choices in the area of population, demography, eugenics and race and the practical and theoretical views of the majority of the scientific world on these themes. It certainly cannot be said that the racial questions obsessing most of the scientific world could by themselves determine a choice in the area of racial politics and the elaboration of a racial legislation. In order for these choices to be made, the determination of a totalitarian and imperialistic regime such as the Fascist regime was necessary. But this determination alone would not have been sufficient if a large enough consensus had not existed, particularly in the nation's ruling class and in the section which was directly involved in demographic-anthropological-racial questions, that is the scientific and university world. That background was the essential basis of this consensus on which the totalitarian policy was able to base itself and from which it drew its themes and its theoretical elaborations. It is not a coincidence that Italian racism had some peculiar "spiritualistic" characteristics that could not be found in other countries, such as Germany. The different routes which racism takes in the two Iron Pact countries' political arenas correspond very precisely to the two countries' different dominant views on race as well as to the different views of the scientific world in this area.

When talking about the theoretical characteristics of Fascist racism, those scientific circles that were more directly involved were anthropological, biological and social sciences. Biologists, anthropologists, physicians, geneticists, statisticians, demographers, jurists were called upon to build Italian Fascism's theoretical framework and they replied zealously and in unison to this appeal. But what can be said of the rest of the scientific community? In what measure, was it involved in this action? If physicists, mathematicians, chemists and scientists in other branches had kept away from these themes, or even opposed them, we could consider the themes described in this paragraph as an incoherent digression from the situation among the rest of the mathematical community. At first sight, one could get the impression that this was indeed the case, because it is difficult to find explicit contributions to the racial policy by scientists not belonging to the abovelisted categories. But here, too, one should avoid drawing hasty and superficial conclusions. As it happens, even though there was no explicit involvement in the specific theme of race, the scientific community as a whole allowed itself to be dragged into the racial policy by its most active sections, i.e. biology, anthropology and demography. The objective that Italian science accepted without any complaint, and even zealously and faithfully strove for, was the following: the *Italian population, culture and, in particular, Italian science had to be purified of its Jewish elements.* After all, the regime had set out a crucial doctrinal and practical point: to rebuild the Italian people, it was necessary to start from the foundation, by focusing on the education of the young, the brandnew Italians. Therefore, the realisation of a racially pure school was the main tool to realise this objective. It certainly was no coincidence that the first legislative measures for the "defence of the race" (promulgated in 1938) had as their objective the school and university system and the elimination of their Jewish elements.

This objective was accepted and zealously followed by the academic and university community and in particular by the scientific community. On 5th October 1938, *Vita Universitaria,* the magazine of the University of Rome, published ahead of time the full "authorised" list of lecturers who, by force of the new laws, were to abandon teaching on 16th October.

The university community was therefore all set to fill the gaps which were to be created by the leaving of about a hundred lecturers and many more assistants and researchers of Jewish descent. From that point onwards, it was like a choir that, in unison, continued to state that the sending away of Jewish colleagues would not create problems to the university, but, on the contrary, would start a new glorious phase in the academic and scientific life of the university, now that it was finally free of racial contamination.

The university world adhered to the racial policy of the regime in a dual way: theoretically and practically. On a theoretical level, all the specific and superior characteristics of Italian science were highlighted and the need to free it from the presence of contaminating contributions from totally different races was stressed: hence the development of a campaign aiming to claim the values of the Italian race and science, its undisputed superiority and the necessity of its autarchic development. On a practical level, the regime's measures were scrupulously applied, thus punctiliously excluding all Jewish colleagues from participating in university and scientific life. This is the way in which the university world actively collaborated in the policy of race. The next section will briefly examine how Italian mathematicians were implicated, and allowed themselves to be involved, in that policy.

Italian Mathematics and the Racial Laws

The events that characterised the behaviour of the Italian mathematical community towards the racial laws have already been described and analysed elsewhere[14].

Without doubt, the most significant event from an institutional viewpoint was represented by the decision taken by the Commissione Scientifica dell'Unione Matematica Italiana shortly after the promulgation of the racial measures. At a meeting on 10th December 1938, the Commission published a statement amounting to one of the most shameful and compromising act of the scientific world with regard to the anti-Semitic racial campaign:

> The Scientific Commission of U.M.I. is meeting on 10th December in a hall of the Mathematical Institute of the University of Rome. Present: Berzolari, Bompiani, Bortolotti Ettore, Chisini, Comessatti, Fantappié, Picone, Sansone, Scorza, Severi. Apologies: Prof. Tonelli. Meeting chair: Prof. Berzolari. Secretary: Prof. Bortolotti. After a friendly, exhaustive discussion, the following decision has been taken: a group of representatives of U.M.I. will visit the Ministry of National Education to communicate the Commission's vote to the effect that none of the chairs in mathematics left vacant by the application of the measures for the integrity of the race is taken away from mathematical disciplines. The vote continues: the Italian mathematical school, which has gained vast fame throughout the scientific world, is almost totally the creation of scientists of Italian (Aryan) race: suffice to quote the names of late Lagrangia, Arzelà, Battaglini, Bellavitis, Beltrami, Bertini, Betti, Bianchi, Bordoni, Brioschi, Capelli, Caporali, Cesàro, Cremona, De Paolis, Dini, D'Ovidio, Genocchi, Morera, Peano, Ricci Curbastro, Ruffini, Saccheri, Siacci, Trudi, Veronese, Vitali. Even after the elimination of some lowly Jewish scholars, the Italian mathematical school has preserved enough scientists, in both quantity and quality, to maintain the high standard of Italian mathematical science with respect to other countries. These master scientists, with their intensive work of scientific conversion assure the Nation of the existence of enough worthy elements to cover all necessary university chairs.

It was a document that, with its impudent topic – the pupils rising to the role of "masters worthy of covering all necessary university chairs" while declaring the marginality of the *masters* demoted to the standing of simple "lowly Jewish *scholars*" – proclaimed the intention of immediately rushing to take on the positions left uncovered. This was done scrupulously and, given this behaviour, the tears shed by some people appeared to be pathetically hypocritical crocodile

[14] *See, in particular: Israel G, Politica della Razza e Antisemitismo nella Comunità Scientifica Italiana, in: Le Legislazioni Antiebraiche in Italia e in Europa; Israel G and Nastasi P, Scienza e Razza nell'Italia Fascista; Nastasi P, La Comunità Matematica Italiana di Fronte alle Leggi Razziali; Nastasi P, La Matematica Italiana dal Manifesto degli Intellettuali Fascisti alle Leggi Razziali. Subsequent quotes in this chapter are taken from these works. Note that all quotes are contained in the second book.*

tears. One such example was that of Antonio Signorini who, after occupying the chair of Tullio Levi-Civita, wrote of his Jewish master: "I am still very troubled by the recent events and I ask myself with apprehension if I didn't do wrong by you by accepting the offer of the Roman mathematical group". The expulsion of Jewish lecturers happened also through punctilious measures, such as denying access to the Istituto Matematico Romano (Roman Mathematical Institute) to people such as Federigo Enriques (who had helped so much to create it) or Levi-Civita. The latter, when putting Max Born's name forward for the Nobel Prize in his December 1938 letter, declared his difficulty in providing precise details in his proposal "because of the anti-Semitic campaign raging here" which did not allow him to have "sufficient contact with the Italian academic world" so as to be "fully informed". Moreover, with similar hypocrisy to that of Signorini, Severi sometimes visited Levi-Civita at home to give him some issues of scientific reviews that he could no longer read.

So much zeal on the part of the mathematical community earned it (through the 13th July 1939 law) the foundation of the *Istituto Nazionale di Alta Matematica (INDAM)* that was covered earlier. Its lead was given to one of the mathematicians most faithful to the regime, Francesco Severi, and the institute was inaugurated in the presence of the Duce himself[15]. The institute joined another pure mathematics institute, *INAC,* led by Mauro Picone. The two great Fascist mathematicians were therefore competing for the title of champion of Fascism in mathematics (as well as champion of mathematics in Fascism), arguing over whether the preferred and most convenient approach for Fascist society was the "pure" approach (Severi) or the "applicative" one (Picone).

The absence of later self-criticisms or repentance statements confirms the sordid nature of the vote by the Scientific Commission of *UMI.* Unless one decides to considers a form of self-criticism, rather than further hypocrisy, statements like Mauro Picone's 1946 obituary notice for Guido Ascoli (written in 1948):

> ... Unfortunately, Guido Ascoli's university life was painfully interrupted for seven years, from 1938 to 1945, because of those senseless racial measures which deprived Italy, in that long and difficult period, of the precious work of citizens of very high moral, spiritual and intellectual standing. They always loved Italy in its dangerous and bloody undertakings, tirelessly fighting in its defence, just like Ascoli himself had done in World War I. Also, the great mathematician Eugenio Elia Levi, who, during this war, heroically perished in those unfavourable days in Caporetto, hit in the forehead by an enemy with whom he was contending his beloved Soil, after all our defences had all been swept away.

[15] *Duce = Mussolini*

These words were spoken by Picone, the same Picone who had participated in and voted for the motion of the Scientific Commission of *UMI* in December 1938.

The hypocrisy of these attitudes was proportional to the committed crimes, which were not limited to that vote.

One of the most significant episodes was the October 1938 decision to officially replace the only Italian representative in the editorial office of the *Zentralblatt für Mathematik,* the main international journal reviewing mathematical publications, published in Germany: the only national representative was a Jew, Tullio Levi-Civita. The mathematicians they had decided to replace him with were Enrico Bompiani and Francesco Severi. It should be noted that this exclusion was an excess of zeal. In fact, although anti-Semitic persecution in Germany had much more violent characteristics than in Italy, the management of the *Zentralblatt* had turned a blind eye to the presence in its office of some Jewish mathematicians such as Richard Courant, famous German Jewish mathematician who had long emigrated from Göttingen to the United States. This shows how, even in Germany, an effort was being made to maintain some of the remaining relationship between the German scientific world and the international scientific community.

The top of the Italian mathematical community was even more extremist than German mathematicians: they chose to break this last delicate thread of international relationships. It was however evident how, if and when the problem of the presence of Jewish scientists in the editorial office of the *Zentralblatt* was brought into light, Nazi authorities would not have been able to ignore this situation any longer. The reaction of the editorial office director of the *Zentralblatt,* Otto Neugebauer, to the decisions of the top of the Italian mathematical community was prompt. First of all, he got confirmation of the exclusion from Levi-Civita and from the journal's editor, Ferdinand Springer. When the situation was clear and it was evident that the reason behind the replacement was racial, Neugebauer resigned and so did Courant, American mathematicians Oswald Veblen and J.D.Tamarkin, the Danish H. Bohr and the English G.H. Hardy. In a letter to the editor, Veblen brought to light the gravity of this tragedy: international scientific solidarity had been wounded and the remaining threads linking the world of international mathematical research to German and Italian research had been severed and the *Zentralblatt* could no longer consider itself "a useful scientific undertaking". The reviewing activity had to move elsewhere, to the United States, not for nationalistic reasons, or a concept such as "national mathematics" proposed by Veblen, but because only in that country could it benefit from the necessary freedom of expression. It was the birth certificate of a new international review journal, the *Mathematical Reviews.*

The excess of zeal with which Italian mathematicians collaborated to the "Aryanisation" of the *Zentralblatt* was no isolated case. Francesco Severi was the protagonist of another, just as serious episode. In fact, at the first wind of the anti-Semitic measures, he actively started working towards the "Aryanisation" of the most prestigious review in Italian mathematics, the *Annali di Matematica Pura e Applicata,* whose editorial committee was composed of Severi himself and three Jews: Guido Fubini, Tullio Levi-Civita and Beniamino Segre. As shown in a letter sent on 16th October 1938 by Beniamino Segre to Tullio Levi-Civita, Severi had asked for the backing of the President of the Accademia d'Italia in exonerating Levi-Civita from his position of editor in chief of the *Annali* and expelling the other two members of the Jewish race. The President gave the goahead to Severi who acted accordingly. As in the case of the university chairs, the *Annali* were completely "Aryanised": Severi proclaimed himself editor in chief and the new Aryan members of the editorial committee were Enrico Bompiani, Michele De Franchis, Antonio Signorini and Leonida Tonelli.

The activism of figures such as Enrico Bompiani deserves a special mention. He prepared a large universal exhibition called E42, which should have taken place in the new Roman quarter of the EUR but was cancelled at the outbreak of the war. As part of this show, an "Exhibition of Science" had been scheduled and the mathematical part was entrusted to a special subcommittee. Its first installation meeting took place on 17th November 1939 under the presidency of Ugo Bordoni and with the participation of a large number of prominent mathematicians such as Enrico Bompiani, Ettore Bortolotti, Francesco Paolo Cantelli, Giovanni Giorgi, Giulio Krall, Mauro Picone, Giovanni Sansone, Filippo Sibirani, Francesco Severi, Antonio Signorini and Leonida Tonelli. They were later joined by Attilio Frajese, Fabio Conforto and Roberto Marcolongo.

Bompiani soon distinguished himself by stressing the need to develop an autarchic and vindicating viewpoint, stating that it was necessary "not only to illustrate the principles, but also to claim Italians' research excellence, *as much as possible without distorting the history of science*".

The first product of the subcommittee's work, "Index and norms for the presentation of Mathematics in the Italic Civilization Exhibition" specifies that such index has the purpose of indicating the people who "must not be forgotten". Moreover, it must have as its main guide the principle "that the Italian contribution constitutes, in several essential moments, one of the highest symbols of the intellectual value of the Italic race and that it therefore should be put in the forefront; particularly since it is systematically ignored in foreign works on the history of Mathematics". The result of this intention is clear: there are no names belonging to Jewish mathematicians in that list, not even by mistake. This "ethnic cleansing" line was all the more evident in the long historical essay by Enrico Bompiani, which was seen as a "theoretical" contribution to the exhibition and whose title was *Italian Contributions to Mathematics*. It contained no explicit anti-Semitic statements, but the effort to produce an image of Italian mathematics purified of any Jewish contributions went beyond any limit inspired by decency. The introduction by the editorial committee clarified, beyond any shade of doubt, the intentions by stating that "the Italian contribution to Mathematics constitutes, in several essential moments, one of the highest symbols of the intellectual value of the Italic race". The fact that the text mentioned no names of Jewish mathematicians led to a farcical representation of some research developments. For example, in functional analysis, the name of Volterra (creator of the concept of "functional" and of the concepts in this discipline) was omitted. Even more sensational was the contribution of Tullio Levi-Civita to the foundation of absolute differential calculus that, according to Bompiani, was the exclusive work of Gregorio Ricci Curbastro. A peak was eventually reached in the presentation of the contribution of the Italian geometric school that, as Bompiani observed, "holds a position of absolute supremacy in the algebraic approach". This supremacy was gained also and mainly thanks to the research by Jewish mathematicians like Corrado Segre, Guido Castelnuovo and Federigo Enriques, whose names were left out. Forgetting his promise to "claim Italians' research excellence, as much as possible without distorting the history of science", Bompiani stated, without heeding the ridicule, that the "extraordinary blossoming of talent and abundance of results which account for Italy being called the "aquilifer[16] of Geometry" [sic] had prepared the ground for the further conquests reached by the present generations of Italian geometricians".

But the work of "historiographic review" did not take place only in the context of the preparation to the E42 and Bompiani was certainly not the only one capable of such acts. On the contrary, his review work was practically nothing compared to that which was done in the monumental work *A century of Italian Scientific Progress* published by SIPS in 1939. "The working force of tradition – wrote mathematician Annibale Comessatti in the first essay – works on historical destiny, when, just like in the case of the Italian geometric school, that tradition is grafted onto its distinguished race qualities, creating a type of thought which is the precious heritage of intellectual autarchy". However, the effort of carrying out such careful cleansing operation removing all Jewish contributions turned out to be so difficult that a note taking up a whole page at the beginning of the book stated:

> For a better understanding of the following Articles, the most relevant contributions by Jewish mathematicians, previously professors at Italian universities, have also been quoted. This is because their work, given the official position they occupied, could not fail to determine reciprocal exchanges between the contributions they made and those made by Aryan mathematicians. The same criterion was applied to the Articles in all other Sections.

There followed a list of the names of the Jewish mathematicians with which Aryan mathematicians could not avoid reciprocal exchanges: Emilio Almansi, Alberto Mario Bedarida, Guido Castelnuovo, Federigo Enriques, Gino Fano, Guido Fubini-Ghiron, Beppo Levi, Eugenio Elia Levi, Tullio Levi-Civita, Salvatore Pincherle, Giulio Racah, Beniamino Segre, Corrado Segre, Giulio Supino, Alessandro Terracini and Vito Volterra. In other words, what followed was a list of the leading figures in Italian mathematics.

[16] *The Aquilifer was the special standard bearer carrying the Aquila (or Eagle), symbol of a Roman Legion, found at the head of the Legion while on the march. The Aquila, as the preminent symbol for a legion, was introduced by Marius in 106BC, when he took over and reorganized the Roman military into a "standing" professional army. The "fall" or loss of a Legion's Aquila standard was considered a disgrace to the Legion's reputation.*

The "rewriting" operation on the history of mathematics aiming to make it autarchic and racially pure found a central actor in the person of historian Ettore Bortolotti. Not only did he distinguish himself for contributions obsessively aimed at claiming the superiority of Italian mathematics, but he also aimed his fierce arrows at another historian of mathematics, a (Jewish) colleague named Gino Loria, whom he accused of little loyalty to the values of national science and of undeservedly exalting the achievements of foreign mathematics.

The theme of the superiority of Italian mathematics and of the substantial irrelevance of any Jewish contributions to it appears therefore as a leitmotiv in Italian mathematics at the time of the Fascist racial campaign. In occasion of the Second National Congress of the *Unione Matematica Italiana* in 1940, president Luigi Berzolari sent the prefect in Bologna a letter asking for help for this organisation. In this letter, the themes of the famous decision arrived at by the scientific commission of the *UMI* are taken up again:

> This congress will really have a national interest, because it will be a review of Italian mathematical production in the last three years and will demonstrate that, even after the departure of our Jewish professors, the scientific production in our country has not declined and, on the contrary, it has taken on new life and vigour in the Fascist climate.

The regime was sensitive to so much activism on the racial front and delegated to Giuseppe Bottai, its main representative in the field of autarchic and racial cultural policy, the task of inaugurating the congress. He was greeted by huge applause when, remembering the first congress of *UMI* in 1937, he listed as Aryan achievements just those sectors in which the contributions of "geometricians belonging to other races" had been the strongest.

That congress affirmed the supremacy of Italy in the fields of algebraic geometry, calculation of variations, projective differential geometry as well as its forefront position in functional theory, differential equations, algebras, relativity, thermoelastic transformations, study in probability and actuarial calculus, history of mathematics and history of numbers. More than a triumph, it was a revelation: Italian mathematics, no longer the monopoly of geometricians belonging to other races, found again its own geniality and eclecticism which had made great people such as Casorati, Brioschi, Betti, Cremona, Beltrami and took up again, with the power of a purified and freed race, its ascending path.

What ascending path… In spite of the triumphant claims of the *Unione Matematica Italiana* and the support given by the regime, it was by no means easy to "fill the gaps" and show the "power of a purified race". The application of the racial policy to scientific research represented the greatest expression of an autarchic nationalist view that would have isolated Italian mathematics from international mathematics, closing it in a "ghetto" and causing it nearly irreparable damage. The isolation of Italian mathematics during the war was extreme, as witnessed by the international congress organised by Severi's *INDAM* and held in Rome in November 1942. The proceedings of this congress, which was attended only by mathematicians from Fascist or "fascistised" countries, show a total absence or only a marginal and occasional presence of Italian mathematics in central sectors of research such as number theory, topology and topological groups or commutative algebra. We noticed from the beginning that the first signs of this involution of Italian mathematics had appeared even before the advent of Fascism. However, since they were strictly linked to a condescending attitude of disinterest towards the new developments in research, the process of involution could not fail to worsen at each push aiming to stimulate the isolation of Italian mathematicians in their small vegetable garden. The autarchic policy of the regime encouraged the worse feelings of loathing and refusal towards all that came from abroad, bearing out the illusion of an indisputable and unshakeable supremacy of "Italic" mathematics. The racial policy was the most extreme expression of this trend: not only could foreigners be done without: even the most eminent and illustrious Jewish mathematicians could. It was sufficient to annul their presence and cancel their memory.

In conclusion, the racial laws constituted the final blow that worsened and made irreversible a crisis in Italian mathematics that could have been healed simply by a closer relationship with international research. The isolation and decline that had been showing for a long time, both in terms of internal and institutional dynamics, were accelerated and made worse by the (material and intellectual) autarchy and the policy for the purity of the race. A complete balance of the effects that such sad events have had on the evolution of Italian mathematics in the second part of the twentieth century has not yet been made.

Recommended Readings

Brigaglia A, Ciliberto C (1998) Geometria algebrica, in: AA.VV., *La Matematica Italiana Dopo l'Unità. Gli Anni tra le due Guerre Mondiali, Milano,* Marcos y Marcos, pp. 185–320; English version: Brigaglia A and Ciliberto C (1995) "Italian Algebraic Geometry Between the two World Wars"/A Translation by Jeanne Duflot, in *Queen's Papers in Pure and Applied Mathematics;* n 100, Kingston, Ontario : Queen's University

De Felice R (1961) *Storia Degli Ebrei Italiani Sotto il Fascismo,* Einaudi, Torino; English translation: De Felice R (2001) The Jews in Fascist Italy: A History, Enigma Books .

De Felice R (1965–97) *Mussolini il Rivoluzionario, Il Fascista, Il Duce, L'Alleato, La Guerra Civile,* Einaudi, Torino

Di Sieno S (1998) Storia e Didattica, in: AA.VV., *La Matematica Italiana dopo l'Unità. Gli Anni tra le due Guerre Mondiali,* Marcos y Marcos, Milano, pp 765–816

Finzi R (1997) L'Università Italiana e le Leggi Antiebraiche, Editori Riuniti, Roma

Guerraggio A, Nastasi P (1993) *Gentile e i Matematici Italiani. Lettere 1907–1943,* Bollati Boringhieri, Torino

Israel G (1984) Le due vie della Matematica Italiana Contemporanea, in: *La Ristrutturazione delle Scienze fra le due Guerre Mondiali* (Atti del Convegno, Firenze/Roma 28 Giugno – 3 Luglio 1980), a cura di G. Battimelli, M. De Maria, A. Rossi (2 voll.), Editrice Universitaria di Roma La Goliardica, Roma, vol. I (L'Europa), pp 253–287

Israel G (1989) Politica della Razza e Antisemitismo nella Comunità Scientifica Italiana, in: *Le Legislazioni Antiebraiche in Italia e in Europa,* Atti del Convegno nel Cinquantenario delle Leggi Razziali (Roma, 17–18 Ottobre 1988) Camera dei Deputati, Roma, pp 123–162

Israel G (1998) Vito Volterra e la riforma scolastica Gentile, *Bollettino dell'Unione Matematica Italiana, Sez. A, La Matematica nella Società e nella Cultura* (8), 1-A, pp 269–288

Israel G (2002) La Questione Ebraica Oggi. I Nostri Conti con il Razzismo, Il Mulino, Bologna

Israel G, Millán Gasca A (1995) *Il Mondo come Gioco Matematico, John von Neumann scienziato del Novecento,* La Nuova Italia Scientifica, Roma

Israel G, Nastasi P (1998) *Scienza e Razza nell'Italia Fascista,* Il Mulino, Bologna

Israel G, Nurzia L (1989) Fundamental Trends and Conflicts in Italian Mathematics between the Two World Wars, *Archives Internationales d'Histoire des Sciences,* 39, 122:111–143

Olff-Nathan J (1993) *La Science sous le Troisième Reich,* Editions du Seuil, Paris

Nastasi P (1991) La Comunità Matematica Italiana di Fronte alle Leggi Razziali, in: M. Galuzzi (ed) *Giornate di Storia della Matematica,* Atti del Convegno, Cetraro, Settembre 1988, Editel, Cosenza, pp 365–464

Nastasi P (1998) Il Contesto Istituzionale, in: AA.VV., *La Matematica Italiana dopo l'Unità. Gli Anni tra le due Guerre Mondiali,* Marcos y Marcos, Milano, pp 817–944

Nastasi P (1998) La Matematica Italiana dal Manifesto degli Intellettuali Fascisti alle Leggi Razziali, *Bollettino dell'Unione Matematica Italiana,* Sez. A, *La matematica nella Società e nella Cultura* (8), 1-A, pp 317–346

Pucci C (1986) L'Unione Matematica Italiana dal 1922 al 1944: Documenti e Riflessioni, in: *Symposia Mathematica,* vol. XXVII, Academic Press, London, pp 187–212

Sarfatti M (1994) Mussolini Contro gli Ebrei. Cronaca dell'Elaborazione delle Leggi del 1938, Zamorani, Bologna

Severi F (1940) L'Inaugurazione del R. Istituto di Alta Matematica (Roma), *Scienza e Tecnica,* vol 4, pp 272–276

Volterra V (1907) Il Momento Scientifico Presente e la Nuova Società Italiana per il Progresso delle Scienze, *Rivista di Scienza,* vol 2, pp 225–237

Mathematics and Fascism – The Case of Berlin

JOCHEN BRÜNING

Introduction

When Adolf Hitler came to power on January 30, 1933, a long and prosperous era of scientific and cultural life in Germany came to an abrupt end. The first World War and the reverberations of the Russian Revolution had certainly shaken, but not seriously damaged the social basis which supported intellectual and artistic achievements of the highest quality; the Nazis managed to destroy it in a few years. The main tools used were the laws against political and "racial" enemies, notably the *Gesetz zur Wiederherstellung des Berufsbeamtentums* (Civil Service Law) of April 7, 1933, and constant mass propaganda against all "Feinde des Reiches". This prepared the ground for measures of persecution and destruction of unprecedented thoroughness and cruelty.

Under these circumstances, even a profession of such minor political value as mathematics could not remain untouched and, indeed, many German mathematicians became victims of persecution, expulsion, and even murder; and German mathematics suffered enormously from the disappearance of the majority of its leading representatives. In this article, we want to take a closer look at the German mathematical community during the days of Fascism. We are interested in the various ways its members dealt with the Nazi regime, which included outright collaboration as well as cautious resistance, but the main theme of this brief report is emigration.

We include in our considerations academic mathematicians, school teachers, mathematically educated engineers and philosophers, and also graduates who decided to work in industry. We emphasize that this community is quite homogeneous with respect to its educational background and, to a lesser extent, with respect to professional routine. Also, it is clear that the members of this mathematical community were socially privileged, by the status of their professions. Moreover, many of them were internationally recognized for their research work

[1] *This report is based on the exhibition "Terror and Exile", shown during the ICM 1998. On behalf of the Deutsche Mathematiker-Vereinigung, this exhibition and its catalogue were prepared by Dirk Ferus, Reinhard Siegmund-Schultze, and the author. This catalogue can be obtained from Geschäftsstelle der DMV, Mohrenstr. 39, 10117 Berlin, Germany.*

and had, by and large, a reasonable chance to emigrate and to continue their lives under bearable circumstances. This does not apply to those who were not active researchers, though, making it much more difficult for them to cope with the Nazi terror. It should be understood that this remark does not intend to belittle the hardships of emigration; we simply have to put the various fates considered here into perspective.

As of now, we know of 145 German speaking mathematicians who had to leave their positions and their homes after 1933. Many of them emigrated but fourteen mathematicians were killed by the Nazis or driven to suicide. In what follows, though, we will restrict attention to mathematicians from Berlin who contributed 53 of the emigrants and two murder victims. This narrowing of perspective is legitimate to some extent since the Berlin mathematical community represented well the professional trends of its time and was quite typical in its social and political behavior. The history of mathematical research up to 1933 is basically well known – even though many details remain to be fully analyzed – but the extent of the involvement of mathematicians in the intellectual and cultural activities of their time, beyond their professional aims, came as a surprise to us. In quite a few cases we find a truly astonishing range of interests and activities, as detailed in the case of Richard von Mises below, but only rarely do we witness outright political activities.

Almost ninety percent of the persecuted mathematicians were "Jewish" by Nazi standards, which is an overwhelming majority. But there was a certain number of "Aryan" mathematicians who were rebuffed, disadvantaged, or persecuted for political resistance, which often simply took the form of solidarity with their Jewish colleagues. This occasional brave attitude did not change the course of affairs in any essential way. But it may give us some comfort in considering this dark period, as does the solidarity experienced by many emigrants in their host countries.

Due to lack of space, we can present here only a fraction of the material covered by the above mentioned exhibition "Terror and Exile" and its catalogue [1], but we hope that the most important aspects become visible.

Some Biographical Sketches

To illustrate the preceding statements, we first present brief accounts of individual fates. These may be viewed as typical, to some extent, within the group of mathematicians who interacted with the Nazi regime in a serious way. And even beyond this narrow group, these examples shed some light on the mental landscape of the Weimar Republic, hence on the conditions which brought the Nazis to power and made their system run. It must be kept in mind, however, that each individual case is special; thus we will refrain from general judgements and simply present examples of suffering, of resistance, and of collaboration.

Five Leading Non-Jewish Mathematicians

The following group of five people was most influential on the Berlin scene and probably beyond, certainly in the case of Theodor Vahlen who was responsible for the field of mathematics in the Reichserziehungsministerium, the ministry of education.

Theodor Vahlen, 1869–1945
Theodor Vahlen set out for an academic carreer as a research mathematician. He studied in Göttingen with Klein and in Berlin with Kronecker, specializing in Algebra. In 1905, he published a book entitled "Abstrakte Geometrie", which was not well received; in fact, Max Dehn wrote a devastating review [2]. Vahlen then turned to applied mathematics, notably to ballistics – on which subject he published the first monograph in German –, and became professor at the University of Greifswald in 1911; but his research remained uninfluential. The radical change of mathematical interest corresponded, however, to an expansion of Vahlen's activities into the field of politics, based on a strongly nationalistic ideology. Thus he called mathematics a "mirror of race" already in 1923, and he launched a carreer in the Nazi party NSDAP in the same year. He held the party position of "Gauleiter" (in the northern state of Pommerania), from 1924 till 1927. In this year, he was dismissed as professor, without pension rights, because of his actions directed against the state. Even though his dismissal did not prevent the damage to the Weimar Republic caused by Vahlen and his comrades, it shows nevertheless that the state was not entirely helpless against its enemies; indeed, it could be overthrown only with the strong support of state employees.

In 1933, Vahlen continued his carreer in the *Reichserziehungsministerium,* under the new minister Bernhard Rust, becoming responsible, among other things, for the hiring of all professors in mathematics. He soon became active in dismissing Jewish mathematicians, mainly on grounds of the Civil Service Law. Since this law in §3, its most powerful clause, excluded civil servants who had fought in World War I or had been hired before August 1, 1914, he had to construct other reasons in many cases. For this, a pretext was given in the very unspecific §6. This applied in particular to Max Dehn, whose review mentioned before was paid back by Vahlen in 1935, when he dismissed Dehn as a "measure to cut public spending". As further reward for his services, Vahlen became successor of Richard von Mises at Berlin University in December 1934, then being already 65 years old. He joined Bieberbach in editing the journal *Deutsche Mathematik,* and became eventually ordinary member and even acting president of the *Preussische Akademie der Wissenschaften.*

Ludwig Bieberbach, 1886–1982
Ludwig Bieberbach was without any doubt a very gifted mathematician and a brilliant geometer. His classification of the generalized cristallographic – or Bieberbach – groups will always be connected with his name, and probably also his conjecture on "schlichte Funktionen", even though it became a theorem a few years ago. Bieberbach did his studies in Göttingen where he was decisively influ-

enced by Felix Klein. With his impressive Habilitation of 1912 and other significant work, Bieberbach was quickly promoted to positions of high distinction: after positions in Zürich, Königsberg, Basel, and Frankfurt he was chosen as successor of Constantin Carathéodory at the University of Berlin in 1921. The development of his relationship with Nazi ideology and later with the Nazi party organisation is less clear, cf. [3], [4]. It started apparently with the conviction that mathematics arises from two very different sources: the abstract algebraic formalism building on the properties of the integers, and the geometric reasoning derived from our perception of space. Adapting, like Vahlen, the popular racist theories of the time, Bieberbach claimed that geometric, concrete thinking was typical of the "Aryan" race [5], and eventually tried to propagate this conviction as managing editor of the journal *Deutsche Mathematik*. Whereas Bieberbach's statements were, at least before 1933, free of antisemitic bias and regarded by himself as strictly "scientific", he became increasingly political along the way. It seems clear that Bieberbach tried to get a hold of all of German mathematics by subjecting the *Deutsche Mathematiker-Vereinigung* (DMV), the professional organization of the German university mathematicians, to the *Führer-Prinzip*, in accordance with the requirements of the Nazi government. However, his position was not at all widely accepted among German mathematicians. The struggle for complete control over the DMV, launched by Bieberbach and some followers in 1934, ultimately failed (a detailed account can be found in reference [3]). Bieberbach had to resign the position of managing editor of the *Jahresbericht der Deutschen Mathematiker-Vereinigung*, and, as a consequence, his political influence diminished considerably. The DMV, however, also had to pay a price. While the autonomy of research was basically saved and mathematicians thus had found a pretext to survive as "unpolitical specialists", the organization had to conform with the overall goals of the Nazi state. Formally, this was expressed by a bylaw saying that the chairman of DMV had to be approved by the government.

Even though Bieberbach joined the Nazi party early and wholeheartedly, he kept high standards of professional quality and, in the beginning, also of personal honesty. On April 1, 1933, he delivered a speech at the university at the occasion of what was called "Boycott day", directed against all Jewish professionals, in particular the university professors. After some general approving statements he said: "A drop of remorse falls into my joy because my dear friend and colleague Schur is not allowed to be among us today." [6]. But only a year later he fully endorsed the dismissal of Edmund Landau at Göttingen, who had to leave not because of the Civil Service Law but because of boycotts organized by Nazi students, saying among other things "that representatives of different races do not mix as teachers and students" [7]. In the following years, Bieberbach remained one of the most active defenders of Nazi ideology among mathematicians in Germany, and he engaged in the dismissal of Jewish or "politically unreliable" colleagues in quite a few cases, including ultimately even his friend Issai Schur.

Bieberbach survived the war. He apparently kept his "scientific" convictions about the connections between race and mathematical thought, but he wrote an "apology", in a letter to Heinz Hopf in 1949. He was not reinstituted as professor and actively resisted certain attempts to do so after he learned about mathematicians who had died in the concentration camps [8].

Georg Hamel, 1877–1954

Hamel was a student of David Hilbert, under whose influence he studied the foundations of mechanics; in this subject he earned a certain reputation. He became a professor at the Technische Hochschule Berlin-Charlottenburg, the other institution in Berlin employing research mathematicians; naturally, their teaching duties and often their research interests were focused on applications in engineering.

Hamel was an experienced and skillful administrator. Therefore, he became the permanent chair of the *Reichsverband deutscher mathematischer Gesellschaften und Vereine,* a professional pressure group for all aspects of mathematics (founded in 1921), but mainly with the goal of securing the status of mathematics and mathematicians in all educational areas. Thus, as a political organization, the *Reichsverband* was more directly subject to political pressure from the Nazi government than the DMV. Hamel did not hesitate in the least to implement the *Führerprinzip* in his organization at the first request. In the fall of 1933 he described the position of the *Reichsverband* as follows:

> "We want to cooperate sincerely and loyally in accordance with the total state. Like all Germans, we place ourselves unconditionally and happily in the service of the National Socialist movement, behind its Führer, our chancellor Adolf Hitler...Mathematics as a teaching of spirit, of spirit as action, belongs next to the teachings of blood and soil as an integral part of the entire educational process. The unity of body, mind, and spirit in the human parallels the unity of body hygiene, mother tongue, and teachings of blood, soil, and active creative spirit in education. Mathematics is the central core of the latter." [9]

Quite obviously, Hamel had the backing of the Nazi officials so that, among other functions, he became the compromise chairman of DMV after Bieberbach's failed attempt to install the *Führerprinzip.* But abiding completely by the new Nazi rules on the side of the *Reichsverband* he took the pressure off the other scientific organizations of mathematicians like the DMV. Beyond lending his administrative abilities to such dubious collaborative efforts, Hamel is not known to have caused any damage to others.

Erhardt Schmidt, 1876–1959

Erhard Schmidt studied with Hermann Amandus Schwarz in Berlin and David Hilbert in Göttingen. He obtained his Habilitation 1906 at the University of Bonn and held positions in Zürich, Erlangen, and Breslau before he came to Berlin in 1917. He was an elegant and profound geometer and analyst. He combined Hilbert's various ideas and results into the single concept of Hilbert space and developed much of its theory, so fundamental for many discoveries of the 20th century. A good part of his later work was devoted to the isoperimetric problem which he solved in spaces of constant curvature.

Schmidt was by no means attracted by the Nazi ideology. So he called the police against rioting Nazi student groups while he was Rektor of the University in 1929. In 1933, he worked hard against the (unlawful) dismissal of his colleague Issai Schur, which was eventually reversed. When Schur was separated from the institute by other means later on, Schmidt continued to visit him regurlarly at home; among the other Berlin professors, only Max von Laue is known to also have kept this same habit – which was not entirely without danger.

On the other hand, Schmidt certainly shared several widespread positions which are referred to as *deutsch-national.* Their main content was the desire for a thorough revision of the Versaille treaty which was generally perceived as humiliating for Germany. Menahem Max Schiffer, a student of Schur's who himself emigrated to the USA, reports the following account of a conversation between Schmidt and Schur in 1938, communicated to him by Schur:

> "When he (Schur) complained bitterly to Schmidt about the Nazi actions and Hitler, Schmidt defended the latter. He said, 'Suppose we had to fight a war to rearm Germany, unite with Austria, liberate the Saar and the German part of Czechoslowakia. Such a war would have cost us half a million young men ... Now Hitler has sacrificed half a million Jews and has achieved great things for Germany. I hope some day you will be recompensed but I am still grateful to Hitler' " [10] The bluntness and naivety of this reported statement is hard to reconcile with what we know otherwise about Schmidt's political utterances. But the basic tendency is credible; it certainly formed one of the main pillars upon which Hitler built his power.

About Schmidt's personal integrity there is little doubt. This is convincingly documented in a speech delivered at the occasion of Schmidt's 75th birthday in 1951. The speaker was the Jewish mathematician Hans Freudenthal, Schmidt's former student, who had just managed to live through the German occupation of the Netherlands; he said:

> "It is easy to practice the honesty that mathematics demands in mathematics itself. If you don't, you will be punished quickly and bitterly. It is so much more difficult to stick to this virtue, proven with numbers and figures, against humans and friends. That we outside, excluded for years from a hostile Germany, know this, and never doubted you, is evident from a large number of contributions from abroad that have reached the editors of the *Festschrift.*" [11]

Rudolf Rothe, 1873–1942

Rudolf Rothe was a student of Hermann Amandus Schwarz. He worked in differential geometry and complex function theory; he is known for completing the edition of the collected works of Karl Weierstraß after the death of Johannes Knoblauch. Rothe became professor at the Technische Hochschule Charlottenburg in 1915, and also held a teaching assignment at the University for some time. Rothe offers another example of the *deutsch-nationale* position seen with

Schmidt; in an address to students as Rektor of the Technische Hochschule he said in 1921:"From your ranks shall rise the leaders of our nation to guide it in its cultural and spiritual renewal ... Let us all remember the aim of preparing this and the coming generations for the time when our saviour shall appear." [12] Nevertheless, Rothe stood up for Jewish colleagues later – for example in the case of Ernst Jacobsthal – much more forcefully than most other Berlin mathematicians.

Richard von Mises, the Leading Representative of Applied Mathematics in Berlin

Richard von Mises was born in Lemberg (then an Austrian city) in 1883, into a noble family. After earning a degree in engineering in Vienna, he got his *Habilitation* in Brünn in 1908. Already in 1909 he became professor at the *Reichsuniversität* Straßburg, then he moved on to Aachen in 1911, and to Berlin in 1920. Von Mises worked in almost all areas of applied mathematics, which developed rapidly in the wake of the first World War. He did groundbreaking work practically everywhere, based on a strong intuition but with occasional lack of mathematical precision; a well known example is his 'limiting frequency' theory of probability which, despite its enormous intuitive appeal, never reached a theoretical maturity comparable with Kolmogorov's measure theoretic approach.

During the first World War, von Mises constructed and built a large airplane (bearing his name) and served as a pilot in the Austrian army. In 1913 he gave what seems to have been the first university course on the mechanics of powered flight.

In Berlin, he became the first director of the newly founded Institute of Applied Mathematics, one of the oldest such institutes in Germany, and he masterminded very successfully the buildup of this new field not only in Berlin but in all of Germany. Beyond his interests in mathematics, which were already very broad, he was deeply involved in philosophic and artistic questions. So he wrote a profound study on positivism and was internationally recognized as an authority on the poet Rainer Maria Rilke; his *Rilke Sammlung* is now at Harvard.

Although Jewish – by Nazi standards – von Mises was not affected by the Civil Service Law since he had fought in the war. Nevertheless, he felt that the actions of the new government would affect him sooner or later and at the very least destroy his opportunities to work. Thus he decided to emigrate already in 1933. He accepted an offer from Turkey (which had been rejected before by Richard Courant and James Franck) and took over a newly founded chair in Istanbul. He thus was in a position to negotiate with the ministry, represented by Vahlen, and he requested that his pension rights, earned in 24 years of service, would remain untouched. This, however, was not granted even though Vahlen had promised him full rights – but never in written form. Ironically, when von Mises applied for reparation after the war, the German authorities asked Bieberbach, of all persons, to give an expert opinion on the case. Bieberbach wrote in favour of von Mises, but the compensation was ultimately approved only after his death.

The situation in Turkey was unsatisfying for von Mises and other emigrants at least for two reasons: the university was in the process of a serious buildup but had not yet reached the standards they were used to, and the influence of Nazi Germany increased continuously. Thus von Mises decided to move on to the USA. In 1939 he arrived there together with Hilda Geiringer, his former assistant whom he married in 1943, and in 1944 he became professor at Harvard. ("The risk of being overtaken by the third Reich was too grave" he wrote to Theodor von Kármán in 1939 [13].)

As mentioned before, Vahlen, now already 64 years old, was eager to take over the position of director at the Institute of Applied Mathematics in Berlin. Von Mises was asked by Schmidt, as chairman of the Department of Mathematics, to give a detailed recommendation of possible successors. Somewhat surprisingly, he recommended Vahlen as the best possible choice even though "with deliberate renunciation, he addresses relatively elementary tasks"; but he did not see a more qualified scientist who was also enough of a politician to guide the young institute through the difficult years to come. It is conceivable that von Mises thought also of his pension rights while wording this recommendation but if so, to no avail. And also his institute declined steadily over the remaining years of the Nazi period.

Richard von Mises incorporated two rather different traditions: he displayed the characteristics of a typical "German Professor", as seen at Berlin University in the late 19th century, and he was also very conscious of the traditions of the Austrian empire. Thus it is by no means surprising that he, too, shared the *deutsch-nationale* position; for example, in a speech at Berlin University in 1930 he remembered the war saying:

> "We commemorate with deepest respect the immeasurable stream of dead, who were with us in the war but did not return, who tried so courageously, with unshaken discipline, and loving enthusiasm to keep the horror of war away from the Rheinland; but who were not able to spare it the humiliating occupation by the enemy after the war. We remember with sadness the lost and not yet liberated country, on which still we cannot set foot."

Nevertheless, von Mises abstained all his life from uttering political convictions or joining any kind of political movement.

Issai Schur, a Great Mathematician and a Great Man

In 1933, the Berlin University had four professors of mathematics: Erhard Schmidt, Richard von Mises, Ludwig Bieberbach, and Issai Schur, the one with the longest connections with the University among them. He was born in Bjelorussia, in the city of Mogilev, but spoke German, which he considered his mother tongue, without any accent. He was emotionally very strongly attached to Germany and to the University of Berlin, which made him refuse many cordial invitations to emigrate early in the Nazi years. Schur had studied with Frobenius and had become an internationally leading algebraist. But his work was very widespread: it comprised important contributions to group theory and number theory, invariant theory, the theory of functions, integral equations, and many other fields. Schur was maybe the last mathematician to cover all aspects of 'classical' mathematics in the sense of the 19th century; but this position made it difficult for him to accomodate revolutionary developments like the "new algebra" put forth by Emmy Noether.

Besides being a brilliant scientist, Issai Schur was also an outstanding and unusually successful teacher. His lectures were admired and unexpectedly crowded, as reported by Walter Ledermann: "Schur was a superb lecturer. His lectures were meticulously prepared...(and) were exceedingly popular. I remember attending his algebra course which was held in a lecture theater filled with about 400 students. Sometimes, when I had to be content with a seat at the back of the lecture theatre, I used opera glasses to get at least a glimpse of the speaker." [14]. Schur also created a "school" through his enormous influence on those mathematicians who did their studies with him. Between 1917 and 1936 no less than 22 students completed their dissertation under Schur's guidance; 6 other students started to work with him but could not finish because of Schur's dismissal. The choice of dissertation projects corresponded to the wide scope of Schur's approach to mathematics. Beyond his mathematical guidance, Schur was caring about his students also in other respects, helping them along wherever he could. Among Schur's students, many were forced to emigrate later, notably Alfred Brauer, Käte Fenchel (born Sperling), Kurt August Hirsch, Walter Ledermann, Bernhard Neumann, Hanna Neumann (born von Caemmerer), Rose Peltesohn, Felix Pollaczek, Richard Rado, and Menahem Max Schiffer.

As already mentioned, the Civil Service Law did not apply to Schur since he became professor before 1914. He was, nevertheless, dismissed in the spring of 1933. This caused a heavy protest among students and colleagues:

> "When Schur's lectures were cancelled there was an outcry among the students and professors, for Schur was respected and very well liked. The next day Erhard Schmidt started his lecture with a protest against this dismissal and even Bieberbach, who later made himself a shameful reputation as a Nazi, came out in Schur's defense. Schur went on quietly with his work on algebra at home." [10]

The first dismissal had to be revoked as unlawful but, as in many other cases, continuous pressure was exerted by repeated official requests to seek early retirement. In addition, the daily life of the Jews remaining in Germany became increasingly unbearable. No less than 2000 administrative rules against Jewish citizens were enacted during the 12 years of Nazi terror, comprising the exclusion from theaters, libraries, public parks, hotels, and the purchase of many items, including even paper. It is not surprising that, eventually, Schur accepted his retirement as of August 31, 1935. He did not emigrate before 1938, though, because he was so much attached to his home country, and also because he was concerned that he might block a position which could be used to help younger mathematicians. This meant, of course, that he had to live through even more humiliations.

One of the stigmatizations which hurt Schur in particular was the prohibition against using libraries. Alfred Brauer remembers:

"When Landau died in February 1938, Schur was asked to give an address at his funeral. For this, he needed some mathematical details from the literature he had forgotten. He asked me to look them up in the library. Of course, I was not allowed to use the library of the mathematical institute which I had built up over many years. I applied to the *Preußische Staatsbibliothek* and was allowed to use the reading room for a week, but not to borrow books ... So I could answer at least some of Schur's questions." [15].

After his retirement, Schur was more and more isolated; only his students and Schmidt and von Laue kept in contact with him. But he remained a member of the *Preußische Akademie der Wissenschaften* until, in 1938, Bieberbach triggered some action against him (cf. Fig.8); soon after, Schur resigned "voluntarily" from the academy. This thoroughly disgraceful action may have caused Schur eventually to emigrate to Palestine. As a final humiliation, he had to pay the *Reichsfluchtsteuer,* a tax (of 25% of the total fortune) for those who fled Nazi Germany. After living through this ordeal he still remained attached to German mathematics:

"When Schur could not sleep, he would read the *Jahrbuch über die Fortschritte der Mathematik.* When, later in Palestine, he was forced to sell his personal library and the Institute for Advanced Study in Princeton showed interest in the *Jahrbuch,* he sent a telegram saying that it was not for sale, only a few weeks before his death. Only after his death, the Institute bought his copy." [15].

Issai Schur died in Jerusalem in 1941.

Murdered by the Nazis: Robert Remak (1888–1942) and Kurt Grelling (1886–1942)

As pointed out above, the group of Nazi victims under consideration is not typical because many of them had a good chance to save at least their lives through emigration. Generally speaking, emigration was difficult to achieve but it was a lot easier for scientists with an international reputation and with a widespread network of friends and colleagues. Lacking either one, the situation of a Jewish scientist could become desperate very soon. This was the case for Robert Remak and Kurt Grelling: both were highly qualified mathematicians, both were murdered at Auschwitz.

Robert Remak was the grandson of a famous Jewish physician and physiologist of the same name, one of the first Jews to be appointed as professor at Berlin University before 1918. He studied with Frobenius and specialized in group theory and the geometry of numbers. His dissertation of 1911 made an important contribution to group theory (the "Wedderburn-Remak-Schmidt-Krull Theorem"). His standing was attested in a letter by Issai Schur to Oswald Veblen of July 10, 1936: "I consider Dr. Robert Remak to be an outstanding researcher who is distinguished by his versality, originality, strength, and brilliancy ... He may, without doubt, be called a leading scholar in the splendid and important field of geometry of numbers." In addition, Remak had extended his interests into rather applied areas like economics with a view to relevant mathematical models. Thus he is considered a precursor of activity analysis. He even anticipated the importance of computers for significant progress in this field; in his essay "Can economics become an exact science?" he stated:

> "I emphasize...that I have not made any political or economic statements, I merely stated some problems and indicated some calculational schemes...it is still open whether the outcome of the calculation will favour capitalism, socialism, or communism ... These equations are very awkward to handle mathematically. There is, however, work in progress concerning the numerical solution of linear equations with several unknowns using electric circuits." [16].

It seems that Remak was a difficult person and lacked a certain smoothness in social interactions which may have held up his academic career. In 1933, he was dismissed according to the Civil Service Law. He then went on to work in private; notably, he was leading a group trying to understand the new algebra as put forth in the book of van der Waerden. After the *Kristallnacht* pogrom in 1938, he was taken to the concentration camp Sachsenhausen, near Berlin, but set free again after eight weeks. Remak and his (non-Jewish) wife tried desperately to emigrate to the USA after this nightmare, but they did not succeed. In 1939, Remak fled to the Netherlands where his wife did not want to follow him; they were divorced soon afterwards. In Amsterdam, among others, Hans Freudenthal helped him along; this, according to Freudenthal's own testimony, was not an easy thing to do in view of Remak's complicated personality. In 1942, Robert Remak was captured by the German occupiers and sent to Auschwitz, where he died.

Kurt Grelling was one of the very few mathematicians in the group we consider who was politically active. The foundations of his socialist convictions were laid by his father, the Jewish lawyer Richard Grelling, well known as one of the cofounders, in 1892, of the *Deutsche Friedensgesellschaft.* Grelling studied mathematics, physics, and philosophy in Berlin and Göttingen, and graduated in 1910 with Ernst Zermelo in Göttingen. Already in 1908 he had coauthored with Leonard Nelson the article *Bemerkungen zu den Paradoxien von Russell und Burali-Forti,* which contained a thorough analysis of the different attempts to resolve the paradoxes of set theory, and presented the *Grellingsche Antinomie.* Between 1911 and 1914, Grelling was in charge of the philosophy column in the socialist periodical *Sozialistische Monatshefte;* in 1919, he took part as a delegate in the SPD party convention in Weimar.

Grelling had to leave the university for financial reasons. He became a school teacher, with a permanent position at the *Walter-Rathenau-Oberrealschule* in Berlin-Neukölln from 1923 on. In 1933, he, too, was dismissed using the Civil Service Law. Again, Grelling was exempted from the main racial clause (§3) as a war participant, so the very "flexible" §6 (simplification of the administration) was used instead. Paradoxically, in spite of daily humiliations, Grelling now found the time for intensive work again, living on a small fortune he had just inherited. In particular, he became very active in the *Gesellschaft für empirische Philosophie,* an interdisciplinary philosophically oriented association founded by the philosopher Hans Reichenbach.

Under the dreadful impression of the *Kristallnacht,* Grelling did not return to Germany from a visit to Belgium. After the German attack on Belgium in May 1940, Grelling was deported by the Belgians to the intern camp Gurs in the south of France. His non-Jewish wife Greta had refused to divorce him and followed him everywhere. Tragically, Grelling could not accept the offer of an Associate Professorship at the New School for Social Research in New York which reached him in 1941; he was not allowed to enter the USA. In a letter to Paul Bernays, from January 1941, he writes: "In my wretched situation I try to keep myself up by scientific work ... Here at the camp I have found two younger friends, one of them is a very competent mathematician. The other is a philosophically interested author with whom I discuss philosophical and mathematical problems." [17]. Incidentally, the "author" was the Austrian writer Jean Améry who, in his autobiography of 1971, remembered Grelling with the following statement:

> "Grelling is very seldom quoted nowadays. Rather than getting ahead professionally, he was forced into a train to Auschwitz; Laval had set the course. The logician and mathematician was taught the logic of history about which he previously did not want to know anything." [18].

Indeed: in September 1942, Grelling and his wife were deported to Auschwitz where, in all probability, they were murdered on the day of their arrival.

Emigration: the Rescue Squad Abroad

As we have already pointed out, emigration was not an easy thing to undertake. First of all, the decision to emigrate was difficult enough in most cases, for material as well as for emotional reasons, in spite of increasingly unbearable circumstances. If the decision was reached in principle, then the next step: to find a reasonable place to go, was by no means simpler. We have illustrated many of these difficulties in the above biographical sketches; for a detailed study of all aspects of the emigration of mathematicians from Germany, we refer to [19]. We now want to concentrate on one important point, namely, that for many mathematicians emigration to most countries would have been close to impossible without the constant help of other mathematicians residing in the host country.

Among those who tried to help their German colleagues to escape from a hostile and dreadful environment, we want to single out the following persons who were particularly active and helpful: Harald Bohr, Richard Courant, Stephen Duggan, Godfrey Harold Hardy, Emmy Noether, Oswald Veblen, and Hermann Weyl.

Harald Bohr (1887–1951), the brother of the famous physicist Nils Bohr, became very well known through his work on almost periodic functions. He escaped to Sweden during the Nazi occupation of Denmark (where Käte Sperling-Fenchel, the student of Issai Schur mentioned above, worked as his secretary after her emigration). One of the closest friends of Bohr was *Godfrey Harold Hardy* (1877–1947), an outstanding analyst and number theorist who supported no less than 18 emigrated German mathematicians at Cambridge; among them were Kurt Hirsch, Bernhard Neumann, and Richard Rado from Berlin. Bohr and Hardy had, like Veblen, strong ties with Göttingen. Both had been active in reintroducing German scientists on the international scene after the first World War, strongly opposing the boycott against them. They had made many friends among German mathematicians, and they took an active interest in the political developments in Germany. Bohr and Hardy had, in particular, closely collaborated for some time with Courant, Veblen, and Weyl, notably in assigning fellowships of the Rockefeller foundation to young mathematicians. Their joint efforts for the well being of mathematics and mathematicians continued naturally under the worsened conditions after 1933. Bohr had critically commented on Bieberbach's article of 1933 [5]. Bieberbach's answer was printed in the *Jahresbericht* without the consent of the other editors (which triggered the struggle in the DMV already mentioned above); Hardy added a carefully worded but adamant verdict in *Nature*. In the following years, they concentrated more and more on rescuing German mathematicians of Jewish background.

Much of the rescue work in the United States was coordinated by *Stephen Duggan* (1870–1950), a political scientist who had founded in 1919 the *Institute of International Education* in New York, with strong support from the *Carnegie Endowment for Peace*. Duggan also initiated after 1933 the *Emergency Council in Aid of Displaced German (later: Foreign)* Scholars (EC). This council organized temporary appointments for more than 300 scholars. Among them were Alfred Brauer, the student of Schur, and Hilda Geiringer, the later wife of Richard von Mises. Duggan committed suicide when he came under investigation during the McCarthy era.

Another influential American mathematician who did a lot to help foreigners was *Oswald Veblen* (1880–1960) in Princeton, one of the pioneers of algebraic topology. He had spent time in Göttingen and was quite appealing to Europeans (as Hardy used to say: "Veblen combines the best qualities of an American with the best qualities of an Englishman"). In appreciating his efforts it must be kept in mind that the economic situation in the United States was difficult in the early thirties, in particular, there was a severe shortage of jobs in academia. This made it all the more difficult to get positions for foreigners in American universities, and caused some trouble for those who tried.

Among the German mathematicians who had come early to the USA, perhaps the strongest effort to help their countrymen was launched by *Richard Courant* (1888–1972), *Hermann Weyl* (1885–1955), and *Emmy Noether* (1882–1935). Weyl, certainly one of the most influential mathematicians of this century, had become Hilbert's successor in Göttingen in 1930. In 1933, he was invited to join the Institute of Advanced Study in Princeton. At first, he refused but, concerned about the Jewish ancestry of his wife Hella, he decided to leave Göttingen and accepted the second invitation. Somewhat to his own surprise, Weyl found the scientific atmosphere at Princeton much to his liking. He continued his research work almost uninterruptedly, notwithstanding his deep concern about the situation of science in Germany under the Nazis. His new position made it possible for him to exert some influence in favour of other emigrants and he spent a lot of energy in doing so.

Richard Courant, student and coauthor with Hilbert, and an influential analyst with a distinctive sense for interesting applications, had become one of the leading figures in the Göttingen institute during the twenties. With the appearence of the Civil Service Law – which, again, did not apply to him – he was "put on leave" but not dismissed right away. In this uncertain situation, he began to explore possibilities for emigration even though he did not want to go at all. In a letter to Harald Bohr he expressed feelings which were certainly shared by many other German Jews: "I have been harder hit by the turn of events and less prepared than I should have thought. I feel so close to my work here, to the surrounding countryside, to so many people and to Germany as a whole that this 'elimination' hits me with an almost unbearable force." [20]. He turned down the offer to go to Istanbul (which Richard von Mises later accepted) and chose, after some deliberation, to join New York University at the end of 1934; at that time, this was a rather inconspicuous place for mathematical research. Half of his modest first salary was paid by the EC. But Courant soon started to work for other emigrants and, of course, also for NYU; the institute built by him and named in his honor grew into one of the leading centers in mathematics.

Emmy Noether was "in the judgement of the most competent living mathematicians,...the most significant creative mathematical genius thus far produced since the higher education of women began."(Albert Einstein [21]). She pioneered the abstract algebraic approach that became so fundamental for the modern development of mathematics. Since she was a woman, her academic career was slow. Even though she was able to get her *Habilitation* in Göttingen in 1919, in the more liberal post-war climate, she never achieved more then a non-permanent professorship. In 1933, she was dismissed on the basis of the Civil Service Law. In the fall of 1933 she came to the USA where she found a position as visiting professor at Bryn Mawr College. Soon after her arrival, she founded with Weyl the *German Mathematicians Relief Fund.* This association asked for two percent of the salary from all German emigrants who had found a job, in order to help those who had just arrived or were about to leave. Emmy Noether was, until her untimely death in 1935, one of the most generous supporters of this fund, cf. Fig. 18.

After the War

The situation of German mathematics and German mathematicians immediately after the war was difficult for many reasons: the majority of the very best had left, the international network provided little support, and the country was physically as well as economically destroyed. Courant had foreseen this correctly in a letter he wrote to Hellmuth Kneser in early 1933: "It is a pity to think what treasures are going to be destroyed in this way after more than ten years of reconstruction. It pains me most to see what senseless damage will be done to Germany...In any case, the spirit of our institute has already been destroyed. Ugly signs of opportunism have become evident. I am much afraid that apart from what has happened to me irreversible actions have been taken." [22]. The strong support of the USA initiated a period of rapid rebuilding in West-Germany, often referred to as the "Wirtschaftswunder", but this only slowly helped to further the level of mathematical education and research. For East Germany, the situation was even more difficult since the country remained under autocratic rule, thus lacking the influx of ideas, technologies, and capital the western part experienced.

The buildup of the educational system which we see in Germany today happened only after 1970. Until then, the possibilities of academic employment were small, and many promising mathematicians found their way to American universities, thus producing a "second wave of emigration". This illustrates how difficult it is to reestablish excellence once it is lost. Besides a high level of general education it seems mandatory to reach in certain centers a "critical mass" in quality, which is necessary to form a highly productive, original, and stimulating atmosphere. But there is no simple rule how to do this. The outcome depends very much on the choice of the right people, on their intellectual power, and on their abilities to interact. There are well known examples where this worked, at least for a time, and many more where it didn't. The mathematicians active in Germany today are certainly internationally very well connected, and have developed relations across the interior German boundary which was almost impenetrable for 50 years. It is fair to say, I think, that German mathematics has again a good, in certain areas a very good international standing, but it does not play a role comparable to the decades before 1933.

How did the emigrants expelled by the Nazis deal with post-war Germany? Very few of them returned as active scientists; among the Berlin group discussed here, just one out of 44: the statistician Karl Freudenberg. Some of them claimed reparation, which was granted usually only after a painful bureaucratic process (which in the case of von Mises was terminated only after his death). There were, of course, many visits from emigrants and activities to help German mathematics abroad, but the general attitude was probably close to what Bernhard Neumann, well established in Australia, wrote in a letter of August 22, 1993: "I was never approached officially about returning to Germany, though many friends asked me if I would. I have often been back to Germany, a country I love to visit because I still have family and friends there, and because I can speak the language, but where I could not envisage to live." [23].

On the German side, there was, of course, a formidable process of "reeducation", a lot of discussion about the atrocities of the Nazi regime in general and about how to prevent such disasters in the future, but also a long period of silence as far as biographical details of survivors – on both sides! – were concerned. Perhaps it needed the next generation to cope with the concrete facts. Then it would not be surprising that in Germany public interest in the details of the Nazi years culminated only in the last decade of the 20th. And it may not be accidental either that the International Congress of Mathematicians returned to Germany only in 1998, after an intermission of 94 years. It was on this occasion that the exhibition was conceived which formed the basis of this report, and it was also on this occasion that all emigrated mathematicians from Berlin who were still able to do so, accepted the invitation to visit the Congress as guests of honor of the DMV and of the city of Berlin. They formed a small group (Annice and Franz Alt, Michael Golomb with his daughter Deborah Sedwick, Rushi and Walter Ledermann, Dorothea and Bernhard Neumann, and Feodor Theilheimer's daughter Rachel) but their presence added to the atmosphere of the ICM in a very special way. More than usual, the largest business meeting of mathematicians was put in a historical perspective. The personal experience of the emigrants made it very clear again that mathematics, in spite of its eternal truth and unrivalled beauty, rests on a fragile and always endangered human basis. It showed how easily selfishness and opportunism can destroy a long tradition of intellectual culture and friendship, given the right circumstances. But it also showed that, always, we must go on and try again.

References

[1] Jochen Brüning, Dirk Ferus, and Reinhard Siegmund-Schultze: Terror and Exile. Persecution and Expulsion of Mathematicians from Berlin between 1933 and 1945. An Exhibition on the Occasion of the International Congress of Mathematicians. Technische Universität Berlin, August 19 to 27, 1998

[2] Max Dehn in: *Jahresbericht der Deutschen Mathematiker-Vereinigung* 14 (1905), 535–537

[3] Herbert Mehrtens: The Gleichschaltung of Mathematical Societies in Nazi Germany. *The Mathematical Intelligencer* 11 (1989), 48–60

[4] Herbert Mehrtens: Ludwig Bieberbach and 'Deutsche Mathematik'. In: E.R.Phillips (ed.): Studies in the History of Mathematics. Washington 1987

[5] Ludwig Bieberbach: Persönlichkeitsstruktur und mathematisches Schaffen. *Unterrichtsblätter für Mathematik und Naturwissenschaften* 40 (1934)

[6] Kurt Hirsch: Sixty Years of Mathematics. *Mathematical Medley* 14 (1986), 469–473

[7] Ludwig Bieberbach,...as in [5], p. 236

[8] Horst Tietz: Private Communication

[9] Georg Hamel: Die Mathematik im Dritten Reich. *Unterrichtsblätter für Mathematik und Naturwissenschaften* 39 (1933), 306–309

[10] Max Menahem Schiffer: Issai Schur. Some Personal Reminiscences. In: Heinrich Begehr (ed.): Mathematik in Berlin:Geschichte und Dokumentation. Aachen 1998

[11] Ansprachen anläßlich der Feier des 75. Geburtstages von Erhard Schmidt durch seine Fachgenossen. Mimeographed Notes. Berlin, 13.1.1951

[12] Rudolf Rothe: Die Aufgaben der Technischen Hochschule auf dem Gebiete der Geisteskultur. *Die Technische Hochschule* 3 (1921), 226–236

[13] Papers of Theodor von Kármán. Archives, California Institute of Technology, Pasadena

[14] Walter Ledermann: Issai Schur and his School in Berlin. *Bulletin of the London Mathematical Society* 15 (1983), 97–106

[15] Alfred Brauer: Gedenkrede auf Issai Schur. In: Issai Schur. Gesammelte Abhandlungen 1.Band, Berlin 1973

[16] Robert Remak: Kann die Volkswirtschaftslehre eine exakte Wissenschaft werden? *Jahrbücher für Nationalökonomie und Statistik* 131 (1929), 703–735
[17] V.Peckhaus: Von Nelson zu Reichenbach: Kurt Grelling in Göttingen und Berlin. In: L.Danneberg, A.Kamlah, A. und L.Schäfer (eds.): Hans Reichenbach und die Berliner Gruppe. Braunschweig 1994
[18] Jean Améry: Unmeisterliche Wanderjahre. Stuttgart 1971
[19] Reinhard Siegmund-Schultze: Mathematiker auf der Flucht vor Hitler. Quellen und Studien zur Emigration einer Wissenschaft. Braunschweig 1998
[20] Constance Reid: Courant in Göttingen und New York. The Story of an Improbable Mathematician. New York, Heidelberg, Berlin 1976, p.143
[21] Auguste Dick: Emmy Noether, 1882–1935. Basel 1970, p.37
[22] Constance Reid,...as in [20], p.145
[23] Bernhard Neumann: Letter to Reinhard Siegmund-Schultze of August 22, 1993

Mathematics and Culture in Russia

SILVANO TAGLIAGAMBE

Translated from the Italian by Emanuela Moreale

Russia and Modernity

In order to understand the general characteristics of the Russian philosophical and scientific thought, one has to take into account its peculiar character which makes it unique in the general panorama of European culture: its being placed entirely in the modern age, both chronologically and in terms of the constitutive process marking its origin. In fact, its history is brief: all scholars, whether Russian or not, agree that in Russia it took hundreds of years of "prehistory", up to the eighteenth century, for philosophy to reveal itself as an "autonomous discipline", with its own subject and problems as well as with its own specific language.

Such a long prehistory was defined as a "centuries-old silence" by one of the major scholars of Greek patristics, G. Flovorskij. Some, such as P. Ja. Caadaev, attribute it to the tight connection with the Byzantine cultural heritage, *miserable Byzance,* which was considered sterile in terms of speculation and unable to produce creative results within it. Others, such as Flovorskij himself, attribute it to the incapacity of Russian Byzantinism to take in and grasp the very rich stimuli that could be inferred from the research spirit and the intellectual curiosity typical of Byzantine culture.

What is certain, however, is that, through Byzantium, Russia was christianised but not Hellenised: the "pagan" philosophical heritage of ancient Greece, which Byzantine culture never forgot, even in its period of decline, did not significantly affect the thinking of the Russian people, who always saw the Greek-Byzantine heritage only as a tool of faith and never of creative energy.

This caesura with respect to the great tradition of western thought should be attributed also to the Paleoslavonic language or "ancient ecclesiastical Slavonic". Introduced in Moravia by Cyril and Methodius, the two saints who evangelised the Slavonic peoples, and later developed in Bulgaria, which had been converted before Kiev. It was later taken to the Russian lands, through translations of texts with a primarily moral character such as liturgical, apocryphal texts, saints' lives, anthologies, collections of famous quotes by scholars such as Socrates, Plato and Aristotle. Although a great means to evangelise people, this language ended up by separating the Russian people from Latin, the language of western culture. What resulted was a progressive consolidation of what Ivan Vasil'evic Kireevskij, one of the fathers of Slavonic studies, called "the Chinese wall" standing between

Russian and Europe: a wall in which Peter the Great managed to open a few significant breaches and openings and which, after him, has been deteriorating on a daily basis, but has nevertheless continued to exist.

Peter the Great's "Revolution from the Top"

In this situation, Peter the Great's actions constituted a moment of total interruption, which started a major and controversial process of forced "modernisation" aimed at building a bridge between Russian culture and society and Western Europe. Central to this process was the rebuilding of Saint Petersburg, that from the very moment of its foundation became not only one of the biggest metropolises in Russia, contending the political and cultural supremacy with Moscow, but above all a *symbol*, the concrete realisation of a dream, a project and an almost obsessive idea.

The dream, the project and the obsessive idea belonged to Peter I the Great. It was he who, in 1703, initiated the building of a new town in the swamp where the Neva river pours the waters of the Ladoga Lake into the Gulf of Finland and flows into the Baltic Sea. The tsar saw the new city as a "window on Europe", in both a physical and figurative sense. Physically, it had been thought out and built as a naval base and, at the same time, a commercial centre and therefore – par excellence – a place for exchanges to take place and a communication junction. Such a place was meant to be able to grasp and condense all the cosmopolitan stimuli that might turn up following the increase in communications with other countries in the European continent. Figuratively, through its very birth, the new city was meant to signify that Russian history had to start again from the beginning, regenerate itself, detach itself from all the stratified native traditions accumulated by the Russian *narod* (people) of which it was the expression. Moscow had always been the sacred symbol of purity of the blood and the land, exactly because located and deep-rooted in the heart of Russia. In order to further strengthen this symbolic function of the new city, Peter the Great decided to make it totally different from Moscow and all other Russian cities, aiming in particular to avoid disorganised and chaotic agglomerates of winding streets. The map of the new city was designed as a system of islands and canals, with the city centre placed opposite the harbour. It was the geometric and rectilinear model that had been typical of western town planning since the Renaissance: for its realisation, foreign architects and engineers were called in from Italy, France, Holland and England.

The effort of preparation, planning, organisation and construction was really immense: within ten years from the beginning of work, as many as thirty-five thousand buildings had been built in the swamps; two decades later, there were almost one hundred thousand people and Petersburg was already one of the biggest metropolises in Europe. The city, both in its town-planning and architectural structure, bore marks of the project that had led to its foundation and of its cultural and symbolic function as bridge towards the West. Large spaces, big city structure, classical symmetrical perspective, baroque monumentality, typically western facades, with no concessions to the traditional Russian styles, strictly followed ratios (of 2:1 or 4:1) of the streets width to the height of the buildings, so as to give the overall panorama the aspect of an infinite horizontal plane: everything helped to give the idea of a space use which had been accurately planned, according to an order which left little or no room to chance or improvisation.

But the new city on the Neva was not the only product of this massive effort of modernisation, which was wanted and imposed from above, by a tsar who was determined to change the destine of its country. Peter I realised that the new Petersburg he had conceived and realised only made sense and, above all, would only be able to carry out its role as a bridge towards Europe which he had wanted for it, if accompanied and supported by industrial development and a new lay culture able to impose itself on traditional culture. Otherwise, it might have remained victim to its radical heterogeneity, compared to the rest of the nation, both from a point of view of location and from a social and ideological viewpoint. This heterogeneity could have provoked such a violent resistance and reaction that people might have perceived it as a foreign body, to be isolated and rejected because of its wide and seemingly unnatural dimensions.

The Industrialisation Process in Russia

Before Peter I rose to the throne, there had been little industrialisation: the process had been hindered by the lack of qualified labourers as well as of an internal market and capital. The first factories started appearing sporadically after 1650, when tsar Alexis – inspired by the protectionist attitude adopted by western sovereigns towards industry – introduced the cultivation of cotton and mulberry. The year 1681 saw the foundation of a velvet factory that, however, soon closed down and, in 1684, a Dutchman opened a factory for the production of cloth in Moscow.

The accession to the throne of Peter the Great marked a new wave of interest in industry. The new tsar built a large number of ironworks in the Urals and Siberia: production increased by so much that, in 1716, Russia started to export iron. After the end of the war with Sweden, Peter turned his attention to increasing exports: between 1722 and 1724 five new factories were built in the Urals, of which four were destined to the production of copper. In Siberia a steelworks factory was built in Irkutsk, employing also the Swedish prisoners of war. Finally, there were 38 foundries in the governorate of Kazàn, just as many in Petersburg, 39 in Moscow and 70 on the Volga and Oka. In 1718 production was reported to have reached a global volume of just over 25,000 tons.

Although the number of installations was high, these factories lacked equipment and, above all, specialists, which is why the quality of production remained rather low. To solve this problem, foreign specialists, mainly French and Dutch, were called in from Western Europe: with their technical know-how, innovations and capital to invest, they carried out a prominent role in Russia's development and modernisation. Even though the tsar intended foreigners to bring in just their capital of technical knowledge, rather than financial capital, soon the two functions started to converge, because of the connections between technical innovation on the one hand and venture capital and management on the other.

Thanks partly to their contributions to the textile sector, large factories with up to five hundred looms were created; in 1702 a factory for the production of linen for sails was built and more factories for the production of fabric followed in 1705. By the time Peter the Great died, there existed about ten cloth factories and about as many linen ones, as well as a dozen factories for the production of silk items. Trimmings, stockings, hats and tapestry were also produced.

As to shipyards, the tsar's interest in this sector was legendary: during his reign, Petersburg became the most important dockyard in Russia. It became the place where ships were built after the Dutch models that he had been able to study during his visit to the Zaandam dockyards.

Overall, it is estimated that, during the reign of Peter the Great, at least 118 (according to some calculations as many as 233) new factories were built.

To boost and ensure the continuity of this massive process of industrialisation, not only did Peter the Great attract specialists from western and central Europe, but he also tried to create favourable conditions to the birth and development of a new scientific and technical culture in Russia: his aim was to provide a basis for the blossoming of the necessary expertise to ensure a solid development of the industrialisation process in his country.

The Monopoly of the Orthodox Clergy on School and Culture

This aspect of the situation was far from satisfactory. In Russia, the art of reading and writing had spread only in the seventeenth century, with the beginning of schools, mostly thanks to the work of the orthodox clergy. For a long time, the orthodox clergy remained at the centre of the educational system because it had a large network of elementary schools, but also owned the only secondary and specialist schools in existence at the time of Peter the Great. From this viewpoint, Kiev preceded Moscow both chronologically and qualitatively. In particular, the schools run by Polish Jesuits appeared as an exemplary model, because the Orthodox Church saw teaching as the best weapon to fight Calvinism and Catholicism that were spreading in the Polish-Lithuanian reign at the time.

The most famous Orthodox school in Kiev was modelled after colleges opened in Poland by Jesuits, who, as is well known, had a near monopoly in the teaching arena in Europe from the seventeenth to the eighteenth century. The school was founded by Pëtr Mogila (or Mohila, according to Ukrainian pronunciation) in 1632 and became a Theological Academy in 1694. It had borrowed from Jesuit schools the pedagogical method and the curriculum as well as the use of the same theological manuals for teaching. Because many lay students attended it, it was, in practice, the first university in the Orthodox Slavonic world. At the beginning, it accepted students from all social classes, so that among its students were sons of priests, nobles, Cossacks and merchants. In 1734, its fame reached beyond Kiev and Ukraine and as many as 722 students were enrolled, of whom only 380 were seminarians.

A special issue of the "Harvard Ukrainian Studies" (VIII, No. 1–2 June 1984) entitled "The Kiev Academy" was dedicated to its activity: among other things, this includes a quote by A. Kniazeff, rector of the Institut de Théologie Orthodoxe Saint-Serge in Paris – one of the main theological academies of emigration:

> The school of Kiev created a clear and disciplined thought. It forced definition and experimentation. It did not refuse in bulk all ideas coming from the West and did not assume an attitude of fear or despise before western science. On the contrary, it encouraged Orthodox theologians to utilise all the best features that Western thought might have had, thus eventually favouring the birth of academic spirit.

The end of 1687 saw the foundation of the future Slavonic-Greek-Latin Academy under the management of two Greeks, the Lichudi brothers, who had arrived in Moscow in 1685. Both had a degree in philosophy from the University of Padua. After teaching in Kiev, they wrote manuals of poetics, rhetoric, logic, (Aristotle's) physics and psychology. Their teaching however still followed the model of the old philosophy school and did not reflect the ferment of ideas of the Paduan university that they came from. Nevertheless, they had a hard time in Moscow and they soon had to leave the city. The school was in difficulty, because no suitable replacement could be found.

Academies aside, between 1721 and 1765, the Church managed to found as many as twenty-eight seminaries modelled after the seminary of the Academy of Kiev employing teachers coming from that very city (the Moscow Academy only started to provide lecturers towards the end of the eighteenth century).

An interesting comment was made by Vladimir Ivanovic Vernadskij (1863–1945), a geochemist and founder of a new and evolutionary approach in mineralogy; a sharp and versatile thinker, to whom we also owe the early detailed research into the history of Russian science. In a long essay entitled "Ocerki po istorii estesvoznanija v Rossii v XVIII stoletii" (Introduction to the history of science in Russia in the eighteenth century), published in 1914 in the journal Russkaja mysil, he pointed out that this prominent position of the clergy in the country's formative system had very negative consequences, because of its total disinterest in scientific research and its problems:

> In the long history of the Russian church, it is difficult to find someone who willingly took an interest in the environment or went deeply into mathematics. Among these very few, there were no important scientists. It was impossible for this attitude not to affect Russian culture.

Peter the Great realised that, in order to make up for the lack of a well-rooted and consolidated scientific and technical culture within the country's society, it was insufficient to send off young Russians abroad to learn experimental research and new techniques which were blossoming in varied fields, just as he himself had done in the shipbuilding sector. Likewise, it was not sufficient to call in foreign mathematicians and engineers, jurists and political theorists, managers and political economists. What was needed, instead, was a reorganisation of the whole educational system, so as to take it away from the monopoly of the Orthodox church and, above all, allow it to gravitate around a centralised, statefinanced institution of great prestige.

The Birth of the Idea to Found an Academy of the Sciences in Russia and First Contacts With Leibniz

From this need derived the tsar's idea to found a scientific academy or university in Russia. He himself talked about this idea already in 1698, during a conversation with patriarch Adrian.

From his first-hand experience of foreign academies and universities and information obtained from people who had similarly visited them, Peter the Great had persuaded himself that, so far in Europe none of these cultural institutions matched Russia's conditions and intentions. It was during this intense activity of exploration of what already existed and of planning something new and different that Leibniz came into his own and his opinions started exerting a direct and remarkable influence on the star.

When the two met, Leibniz was already a permanent member of the "Royal Society" of London, founded in 1660, and of the "Académie des sciences" in Paris, which had started its activity in 1666. Moreover, Leibniz was working on founding the "Berliner Akademie der Wissenschaften": the Berlin academy was eventually founded in 1700, mostly thanks to the efforts of princess Sophie and her daughter Sophie Charlotte, from 1684 the wife of prince Frederick, later to become king of Prussia. Leibniz was nominated first president of this new scientific institution. He was, therefore, at the time, one of the main experts in organisation of culture and a connoisseur of the still short experience of the main European academies.

It was therefore natural for Peter the Great to turn to Leibniz. The German also had a particular interest in and curiosity for Russia, in that it constituted the connecting link between East and West, the most direct way to reach China and Asian countries and its linguistic and cultural heritage strongly attracted him. Moreover, the tsar's intentions to finally establish tight links between his reign and Europe constituted a unique opportunity to have at one's disposal first-hand geographical, historical, ethnographical and linguistic information which had so far been difficult to get hold of. Flattering the German philosopher was also the attractive idea of being able to have a remarkable influence on the cultural and social development of a large nation that had decided to change direction and move towards the West, thanks to the work of an enlightened sovereign, whose fame was starting to spread with favourable echoes throughout the whole of Europe.

The commitment with which Leibniz faced this task is testified by a large quantity of letters, notes and plans he himself drew up over nearly two decades. They show his great confidence in the reformation aim and in the organisational capabilities of the Russian sovereign.

The Constitution of the Academy of Sciences and Arts in St. Petersburg

But Leibniz did not have the satisfaction of seeing the opening of the Russian Academy of Sciences for which he had fought and worked so hard. After his death, his position as advisor to the tsar was taken over by a person recommended by Leibniz himself, Christian Wolff (1679–1754), who had been professor in Halle since 1707. Wolff took active part not only in the working out of the final plan for the Academia, but he was also tasked with selecting and personally choosing the German scientists who should be called in to be a part of the new institution. Blumentrost, on behalf of the tsar, offered him the position of vice president of the Academia and coordinator of teaching in physics and mathematics in a letter of which we have a dateless copy, but which must nevertheless date back to the first quarter of 1723. The offer was not successful because of the excessive financial expectations of the Halle philosopher, who declared he would only accept the post if a deposit were paid in advance and if a lump sum of twenty thousand rubles were paid as compensation for living in Russia for five years (just to enable comparison with other amounts, in the first half of the eighteenth century, a taler was worth 70 Russian kopecs, i.e. 70 cents of a ruble). Moreover, Wolff had clearly indicated several times and during talks with different people that he was aiming for the position of president of the Academy. The tsar, being unable to accommodate these demands, decided to avail himself of the collaboration of Leibniz's pupil only for the choice of more scientists and their sending to Russia.

At the beginning of 1724, Peter the Great was finally able to ratify the act of association of the Academy of Sciences and Arts and the annexed university in Petersburg; the act was approved by the Senate, in his presence, during a meeting on 22nd January 1724. This document explicitly referred to the necessity of choosing the members of the new society from among the most illustrious foreign researchers of the time, given the lack among Russians of "natural philosophers" who could worthily join it. What is more, it contained the open invitation to foreign members to bring with them one or two of their most promising students, so as to have enough personnel capable of carrying out teaching in the gymnasium, which was part of the academy, with the aim to prepare the future generation of Russian researchers. The twenty-two permanent members of the Academia, appointed by Catherine I between 1725 and 1727, were all foreigners: of the hundred and ten academics appointed between 1725 and the end of the eighteenth century, only twenty-eight were Russian.

With the realisation of this new, big project, Peter the Great gave a further boost to his modernisation process from the top. To underline the deep connection between the Academy and the foundation of the city which housed it, both produced by the same, enormous strategic design, Peter decided that the *Kunstkammer,* the "cradle of Russian science" (as it is called in a memorial plate on the building), first headquarters of the Academy of Sciences, was to be placed in the Vasil'evskij island. This was the biggest island of the city, with an area of over one thousand hectares, located opposite the dockyard and destined to become the administrative and cultural centre of the new capital. By order of the tsar, it was here that the most important public buildings had to be located and, starting from 1716, the nobles were authorised to build their homes only in this part of the city. The *Kunstkammer,* whose construction started in 1718 and terminated in 1734, i.e. nine years after the tsar's death, housed several laboratories, a large library, the typography of the Academia and the first Russian observatory, placed in the tower above the building.

The "Modernity" of Russian Culture

From the outline picture drawn so far, a first, fundamental, distinctive trait of Russian science with respect to Western science emerges. Russian science is part of a cultural background which is doubly modern: chronologically, as mentioned earlier, as well as for its acts and the forced modernisation process forcefully started by Peter the Great. His reforms were inspired above all by a need for technical-military catching up with the most advanced European countries. As Vittorio Strada[1] observed, they also mark the conclusion of the historical phase of the "first Russia", i.e. of the developing Russia in the first seven hundreds years of the second millennium and the beginning of the "second Russia", extending until October 1917. The latter is already fundamentally different from the former: it is directed towards the West, which is seen as a civilisation to imitate. The two Russias are linked by a deep and contradictory connection, consisting of a set of values, situations and problems, which mark the history of the country rather deeply, leaving permanent traces.

> The Russian ethnic-national conscience in the first six or seven hundred years of its development identified its own "other" in two directions: East and West. Starting from an early Christian community, Russia gradually affirmed its own national Christianity of Byzantine origin that caused it to contrast with Western confessions, first Catholic and then also Protestant. A first reason for the formation of a Russian identity was therefore a religious cause, represented by the pravoslavie, the Orthodox Christianity. This offered Russia membership of the (Slavonic and non-Slavonic) Christian Orthodox community but, from the fall of Constantinople, gave Russia a chance or a claim to privilege and hegemony, as heir of the second Rome. The other direction Russia's sense of "other" developed in was the East, reaching its most intense point with the Tartar-Mongolian domination. This domination was felt by some historians to be a catastrophic detachment from the West and Europe, yet reinforced the specificity of Russia, giving it new connotations, because two hundred years of dominion left a deep trace in Russia's ethnic-national identity[2].

Peter the Great's grandiose renewal operation produced a radical upsetting of this picture just because of its detailed and pervasive nature which combined Europeisation with

> A secularisation that deprived the Russian Church of its residual autonomy and made the Russian state a special type of absolutism, in which the sovereign holds not just the political power, but also, to use an anachronistic but clear

1 *V. Strada, La questione russa. identità e destino, Marsilio, Venezia, 1991, p. 116.*
2 *ibid., pp 115–116.*

term, ideological power. The consequence of this was that to the Russians, at least for the large majority who remained faithful to the national tradition, the anti-Christ was not only the false Christian Westerner, but even Peter I himself, a false Russian, according to a legend which denied the sovereign even Russian nationality. There resulted two national self-consciences: the first restricted to the power circles, the second typical of the masses. It is appropriate to note that both national self-consciences were rather clear, although the one belonging to the high classes had a better conceptual coherence. This coherence developed itself within the new social classes, born following the Europeisation and fuelled by the contact with European culture: the intelligencija, to call this social class with a term that was to be used later. At first, the intelligencija's national self-conscience did not differentiate itself from that of the masses, but then gradually took on its own particular connotation, diverging more and more from the official one until, in the second decade of the nineteenth century, the two collided. The result of this process, which is clear to those who are familiar with the history of Russian culture, was that the national self-conscience was split in three dimensions: an official expression, a popular one and an intellectual one. What's more, as the reality in Russia got more complex, further differentiation occurred within each of these self-consciences. This is especially true of the intelligencija, which saw the formation of different and opposed groups and parties within itself. It is also true, to a certain extent, for the power circles, because their conservative or liberal tendencies were connected to different views of Russia, of its past and of its peculiarity [...]. As to the popular national selfconscience, this is the least clear and well-known, but there is no doubt that this too must have been variegated, as is shown in popular utopias, the religiosity of old believers, the very behaviour of the peasant masses who, while faithful to the myth of the tsar, at the same time had not forgotten Pugacëv's rebellion[3].

If we have insisted so much on these aspects connected to the articulation and differentiation of the routes of what has been called the Russian "national self-conscience", it is because this internal dialectic had remarkable consequences on the history of the Russian philosophical thought and therefore simply could not be ignored.

3 *ibid., pp 116–117.*

Criticism of the Idea of Progress and of the Conception of History as Linear Development

The first and most relevant of these consequences consists in Russian culture and society's peculiar attitude to time: this attitude was very dissimilar to the view that had been asserting itself in Western Europe. The main element on which to focus one's attention in this world is the very small importance of the concept of progress, or the possibility of a controlled change, based on the firm belief that it is possible to arrive at a new and better future through a process of growth, that is a continuation, evolution and capacity to reap opportunities offered by the present situation and by the tradition of the past, whose limitations are however felt. This model of history as evolutionary schema, as underlined by – among others – C. Hill[4], has had a very important role in Western history from the beginning of the seventeenth century onwards; on the contrary, in Russia it was eclipsed by the concept of change as eschatological reversal. Because of the absolute predominance of such view, the dynamic process presented peculiar aspects, which could cause one to see change exclusively as a radical rejection of the preceding phase and the new as the result of a pure and simple transformation of the old or, better said, of an operation of reversal. It was Lotman and Uspenskij[5] who clarified this characteristic trend in Russian culture and the reason for its substantial immutability across its various phases. The fundamental peculiarity of this culture, in their opinion, consists in fact in

> its fundamental polarity, which is expressed in the dual nature of its structure. The basic cultural (ideological, political, religious) values in mediaeval Russia were arranged in a bipolar value field divided by a sharp line and without any neutral axiological zone".[6]

A typical example of this situation is the fact that, in Russia during the Middle Ages, there was not perceived to be a middle ground (purgatory) between the two extreme concepts of Hell and Heaven. As an immediate result, it was impossible to identify, in our lives on Earth, a type of behaviour which could be characterised as neutral, not saintly but not even sinful, such as to act as a neutral axiological zone in this bipolar value field and as a reserve from which to draw those elements which, exactly because not involved in an extreme judgement (whether of exaltation or condemnation), could form a mediation buffer between the two different development phases and thus grant a passage from the one to the other without too many jolts and fractures.

4 *Hill C (1965) Intellectual Origins of the English Revolution, Clarendon Press, Oxford - Italian edition: Le origini intellettuali della rivoluzione inglese, Il Mulino, Bologna, 1976.*

5 *Lotman Ju M, Uspenskij B A (1977) Rol' dual'nyh modelei v dinamike russkoj kul'tury do konca XVIII veka (The role of bipolar models in the dynamics of Russian culture until the end of the eighteenth century), Trudy po russkoj i slavianskoj filologii, XXVII. See also: Lotman JM, Uspenskij BA (1984) Semiotics of Russian culture, (A. Shukman Ed.) Ann Arbor, Michigan: Department of Slavic Languages & Literatures, The University of Michigan.*

6 *Ibid, p 4.*

In the Western world, as Lotman and Uspenskij point out, the presence and availability of a wide variety of behaviours considered neutral and social institutions considered neutral allowed society's contemporary critics to derive their ideals from well-defined circles part of the (social non-ecclesiastical system, lower middle class family) reality around them. Thus, their fight became an attempt to eat into and turn upside-down the current value hierarchy, ensuring that elements taken from the neutral zone became standard values, i.e. the norm. There resulted the possibility to establish a real continuity between the today that was negated and the long-awaited and hoped for future. It is just by virtue of the recognition of this possibility that a new view of life gradually emerged and consolidated itself. This view accepted the challenge of fear, anxiety and anguish that affect human existence without however giving in to the temptation of shunning reality.

In Russian culture, on the contrary, the absence of an idea of progress, seen as an opportunity to derive from elements of the present the conditions for its transformation into new forms, resulted in the prevailing of mechanisms that fatally reproduce aspects of the past within this culture. This peculiarity is not limited to the Middle Ages or to the period preceding the end of the eighteenth century. In fact, it can be found in various phases of its development and in seemingly heterogeneous aspects: for instance, in the wave of the religious popular turmoil, usually known as *raskol* (schism, split), which rose in the second half of the seventeenth century. As A.I. Klibanov, an attentive historian of the *raskol,* notes:

> the insufficient development of social relationships, the fact that the new economic phenomena, in the seventeenth century, were still at the beginning of their history and had affected only the insignificant strata of the peasant classes, had as a consequence the fact that, in the raskol's view, the predominant motive was the idealisation of old patriarchal habits. This finds its expression in both the contraposition of the old faith to Nikon's "new" one and the demand to return to the old habits.[7]

However, the influence and predominance of such type of attitudes were not limited to the seventeenth century and, in fact, can be found even as late as the end of the nineteenth century, just three decades before the October Revolution, within a trend of the *postnicestvo* (fasting) which had formed within the sect of *christovoverie* (Faith of Christ). At that time, V.F. Moksin attempted to introduce a new form of christovoverie devoid of any form of prejudice, superstition and ignorance and founded on an idea of "progress" of middle class origin. But this effort to respond to a changed reality through a reform of the movement accord-

[7] *Klibanov A I (1980) Storia delle sette religiose in Russia, La Nuova Italia, Firenze, p 67. History of Religious Sectarianism in Russia (1860s–1917), Molokan Heritage Collection, translated by Ethel Dunn, 1982.*

ing to the middle class interests and values which gave it meaning, found a huge obstacle in the competition of other, better-known exponents of this trend, as well as in the resistance of the lower social classes. The latter were still attached to the old and believed that a real renewal could only take place within the boundaries given by the traditions of the original *christovoverie.*

On the other hand, this very tendency to counter the existing with preceding forms of organisation and to think that a renewal can be the result of a return to the old has permeated the whole history of the Russian revolutionary movement. Already Aleksandr Ivanovic Herzen (1812–1870) had noticed that Slavophiles had tended to fill with real meaning that *narodnost'* (term derived from narod, "people" and "nation") which was one of the official passwords at the time of Nicholas I and to uncritically extol popular traditions and patriarchal ways of life, while negating the more modern and less native ones. Their adherence to the Russian mediaeval tradition, of which they declared themselves heirs and which they wanted to continue, caused them to fully condemn Peter the Great, because he created a state that openly followed the idea of renewal and modernisation. Hating their contemporary world, they extolled the most ancient forms of possession and distribution of land in rural communities: hating the state, they wanted to feel close to the Russian people and peasants and rekindle with feeling the Church. Theirs was therefore an idolization of their origins, a myth of a Russia outside time which supports Lotman and Uspenskij's claim that, in Russian history, change does not normally happen through the working out of alternative and, as far as possible, new models, but rather through an axiological exchange, so that what was positive becomes negative and viceversa. The most immediate consequence of this is that the very concept of "new", in general, turns out to be the realisation and the putting forward again of concepts whose roots go back to the ancient past.

The constant and massive presence of this peculiar trait is also confirmed by many scholars of rather different background. So, for instance, in a paper published in the American Historical Review in October 1953, A. Gerschenkron observes that populists had correctly started from a solid awareness of the economic backwardness of their country, but had then rapidly distorted their intuition, ending up by paradoxically affirming that

> the preservation of the *old* rather than the easy adoption of the *new* constituted the advantage of backwardness. The result was a tragic surrender of realism to utopia. Here is perhaps the main reason for the decline of populism. When the rate of industrial growth leapt upward in the middle of the eighties, after the government had committed itself to a policy of rapid industrialization, the divorce between the populist utopia and the economic reality became too great, and the movement proved unable to survive the repressions that followed Alexander III's advent to the throne.[8]

[8] *Gerschenkron A (1968) Continuity in History and Other Essays, Harvard University Press, Cambridge, pp 55–456.*

Even Venturi, in the Introduction to his classical work Russian Populism, after remembering Slavophiles' decisive opposition to revolutions and despotisms and, in general, all barbarian methods to combat barbarianisms, observes:

> It is exactly this attitude that seems to be the deepest root of the interest, which is currently resurfacing in today's Russia, for these figures so far away from today's Russia, for these nineteenth century romantics who seem to have been covered in scorn and hate for decades.

Referring to this interest, he makes a further comment that is of particular relevance to our analysis:

> A deep movement supporting a return to old Russia and the religion of the forefathers invite us to look at the past differently, to again consider and appreciate values that seemed destroyed and buried (one can easily be persuaded of this simply by seeing how Russian mediaeval art is considered or by reading Pasternak's and Solzenicyn's work or even simply by watching Tarkovskij's film on Andrej Rublëv). What is most important is to see how this deep and varied movement now present in the Soviet Union ends up attacking an adversary and an enemy that is both feared and hated: the despotic and bureaucratic state. This is just what happened in the 1830s and at the time of the rise of Slavophilia.[9]

Looking at most recent Russian news, one could add to Venturi's many other interesting and significant examples of this deep movement of return to old Russia.

Additional proof of the little effect of the idea of progress based on a gradual transformation and renewal and seen as controlled change, is the predominance of programs of *integral* transformation of Russian society. The theme of regeneration, rebirth and palingenesis is constant in Russian culture, both before and after the revolution. As mentioned earlier, this idea of change as total reversal triggers a mechanism that inevitably brings us to look to the past and to reconsider with interest forms of life and culture that were already tested out and consumed at the appropriate time. Herzen had already understood this when he noticed how the juxtaposition between Slavophile and Westernising factions was the fight of two alternative models which were totally distinct, apart from their common tendency to look intensely to the past, be it mediaeval Russia (the former) or Peter the Great (the latter). The exhortation he addresses to both factions is significant:

[9] *Venturi F (1972) Il Populismo Russo, vol I, pp LIV–LV.*

> It is time for humanity to forget what is not necessary in its past, or rather, to remember all, but at the same time to remember that it is past and no longer existing.[10]

A further boost to this small or non-existent inclination to ideas of progress, growth and modernisation came from the widespread distrust in industrialisation and the development of certain economic sectors. As Gerschenkron notices:

> the creation of large industrial centres threatened to infect Russia with the "cancer of proletariat", as they would have put it at the time. The government was anxious to ward off the menace of peasant revolts and had no wish to evoke the menace of urban revolutions. The traditionalism implicit in an economic structure based on agriculture appeared to be a much better guarantee of political stability than the restless changeability of modern industrialism. Among the forces that, in the Russia of the second half of the 1850s, were capable to make themselves heard, there was none that could push the government towards a decisive policy more favourable to industrialisation. The upper and lower nobility, considered as a group, did not wish at all to see widespread urban growth, because this would have threatened their preeminence within the social structure of the Russian state. The *intelligencija* was mostly radical and political stability simply did not figure in its ideals: it hindered the aristocracy and supported a type of peasant emancipation that was well beyond the limits of acceptability for the government. However, in its hostility towards industrialisation and in its making rural society's (actual or alleged) values its own, it was remarkably close to the government's position, although for very different reasons.[11]

On a structural level, this distrust towards industrialisation, combined with a view of change based on the principle of juxtaposition and alternation of opposites, i.e. of radically different social models in constant competition among themselves, caused the process of development of all sectors of Russian society to have a characteristic, slow and stuttering course. This process can be easily documented and illustrated. While the Russia of Catherine II placed itself among the greatest economic powers of the eighteenth century, for number of industrial establishments, production volume and its part in European commerce, already in the midnineteenth century, industrial Russia was going through a slump and was not participating in the general movement which had been transforming western economies. Moreover, while the last decade of the nineteenth century was characterised by impetuous progress which fixed the traits of a new industrial geography for the next thirty years, it was soon fol-

10 *Ibid., vol I, p.4.*

11 *Gerschenkron A (1974) Politica agraria e industrializzazione in Russia, 1861–1917, in: Storia Economica Cambridge, vol VI, II, Einaudi, Torino, p 771.*

lowed by a prolonged period of stagnation, during which (especially between 1901 and 1903) the country's economy, and in particular its metallurgic industry, was hit by a serious crisis. There was a continuous wavering between the two poles around which were arranged the ideas relating to the model of society to be aimed for: on the one hand, opening to the West and choice of a process of modernisation following in the steps of what had been done mainly in England, Germany and France; on the other, allout defence of the specificity of genuine Russian tradition. The initiatives of the state were often characterised by a bitter struggle against what had been laboriously built over several decades. This situation could not but determine a periodic destruction of what had been achieved by preceding generations, with the consequent lack of that process of accumulation of results and experiences that is an indispensable condition for a stable and lasting development.

This concept was grasped with great shrewdness by Pëtr Caadaev (1794–1856) in the first of his eight *Filosofskie pis'ma,* dated 1st December 1829, in which he critically listed the evils of the Russian past, which appeared to him to be substantially lacking a history:

> One of the most reproachable aspects of this odd civilisation of ours is that we still have to discover the most trivial and obvious truths even among peoples that are certainly less advanced than us. The fact that we have never walked together with other peoples, that we do not belong to any of the big families of the human race, neither to the West, nor to the East, and we do not have the tradition of either. It is as if we were placed outside time, so that the universal lessons of humankind never spread among us. The remarkable concatenation of human ideas in the sequence of generations and the history of the human spirit, which has enabled it to reach the levels at which it is today in the rest of the world, have had no effect on us. What is elsewhere the very foundation of society and life is for us simply theory and speculation".[12]

The "Reaction" Against Philosophy

An aspect of remarkable importance in our discourse – and one that characterises the cultural situation in Russia in the second half of the nineteenth century and first years of the twentieth century – is the widespread hostility and distrust towards philosophy that steadily came to affect progressive thinkers and revolutionaries. The latter, in general, had grown up with a deep suspicion towards the "official" culture that never failed to provide ideological support to the ruling classes and the monarchy. "Academic" philosophy could not but be

[12] *Caadev P Ja (1991) Polnoe sobranie socinenij i izbrannye pis'ma (Complete Works and Selected Letters), vol 1, Nauka. Moskva, pp 323*

affected in this globally negative opinion, particularly since idealistic and spiritualistic trends and approaches clearly prevailed within it. Next to these approaches, a positivist trend emerged and progressively became stronger, as capitalism developed. Positivism was represented above all by thinkers such as K.D. Kavelin (1818–1885), Vladimir Viktorovic Lesevic (1837–1905), Evgenij Valentinovic De Roberti (1843–1915), Grigorij Nikolaevic Vyrudov (1843–1913), Nikolaj Ivanovic Kareev (1850–1931), just to quote the most important ones. Most of them had been educated at the school of revolutionary intellectuals such as Herzen (1812–1870) and Petr Lavrovic Lavrov (1823–1900) and, with the worsening of social and class tension, they sided with the bourgeoisie. The necessity to fight the materialistic tendencies prevailing within revolutionary organisations caused this positivist movement to come to agree with neo-Kantian trends that had emerged within German positivism. Another consequence was that the positivist movement adopted the conclusion of Otto Liebmann, who, in his work "Kant und die Epigonen" (Kant and his followers) published in 1865, had terminated his examination of the four main trends in German post-Kantian philosophy – the idealism of Fichte, Schelling and Hegel; Hebart's realism; Fries's empiricism and Schopenauer's transcendentalism – with the invitation "We therefore must go back to Kant".

In these conditions, a sense of impatience towards philosophy grew in progressive and revolutionary circles. The most radical expressions of this trend were to be found in the work of two of the most active exponents of "Zemlja i Volja": Aleksandr Aleksandrovic Serno-Solov'evic (1838–1869) and Nikolaj Isaakovic Utin (1840–1883). They considered philosophical thought to be the heritage of preceding generations, imprisoned in its utopian dream of a free and unconditioned human personality, and therefore able to develop and make progress without boundaries. By this time, however, philosophers and romantics had been replaced by a bigger and bigger crowd of revolutionary socialists who, as Utin wrote, were moved

> not by abstract ideas, but by a rigorous adherence to harsh reality, which has always the power to draw to itself and constantly remind of its own presence". Therefore, philosophy needed to be replaced by a new, authentic science, capable of "filling with loathing for the current social regime" and to indicate the "nondeferrable and irrefutable necessity of a new order based on maximum freedom" (Narodnoe delo, 2–3, 1868, p. 40).

Petr Nikitic Tkacev (1844–1885), one of the main ideologists of revolutionary populism, also came to be hostile to the most recent western philosophical trends (in particular Mach and Avenarius and their empirio-criticism, which found supporters even in Russia[13]). So much so that he theorised the necessity of

[13] *Particularly important, in this context, V.V. Lesevic, author of Ot Konte k Avenariusu (From Comte to Avenarius), which gives evidence on and explains the evolution of traditional positivism into the "second positivism" of Mach and Avenarius.*

overcoming philosophy, which he considered a corruptor of youth, because it made them lose sight of the practical tasks which life put in front of them. While scientific investigations and generalisations taking place within these refer to real and actual phenomena, according to Tkacev, philosophical investigations "refer to a set of phenomena of purely speculative nature and are totally detached from reality" (Tkacev PN (1877) O pol'ze filosofii (On the utility of philosophy), *Delo,* 5, p.83).

The characteristic aspect of these statements about philosophical thought is its clear tendency to consider the latter as an ideological weapon, a fighting instrument in the ideal and cultural, but also political and social, battle that was taking place in the country. In other words, it was the object of a dispute which went beyond the specific topic of its assumptions and conclusions and which involved problems concerning the historical situation at the time. The positions of the various authors were thus considered with reference to a "general climate" in force in Russia at that time: they were judged on the basis of the supporting role to certain political tendencies, which – more or less occasionally and instrumentally – they might end up carrying out. The strict connection that was made between the spreading of philosophies of a speculative type, characterised by a large use of abstractive methods on the one hand and the strengthening of reactionary and conservative tendencies in politics on the other, led the exponents of revolutionary organisations and movements to put into discussion most of the available philosophical heritage.

The effect of this anti-philosophical attitude at the hands of anarchists and populists is testified by the persistence of similar positions also in the debate following the October Revolution. One of the first issues of *Pod znamenem marksizma* (Under the flag of Marxism) – the new review described as an "organ of militant materialism", which had started being printed in January 1922 – contained an article significantly entitled Filosofiju za bort! (Throw Philosophy overboard!) by S.K Minin. The author, an expert on religious criticism, presented the whole of philosophical thought as a variant of religious ideology, since both lack any cognitive [conoscitivo] value. Its most authentic meaning, according to Minin, was instead that of constructing the indirect and mediated, but not less genuine, expression of the interests of the middle classes and other ruling classes. The essay started in a characteristic way, by immediately communicating its author's intentions and aspirations as follows:

> In the last few months, the fight on the front of abstract thought has been rekindled: the guns of reviews have again started to crackle and the heavy artillery, composed of treatises and volumes, has restarted to shell. What a comforting symptom! But in this impetuous attack, we show quite a bit of disorder and at times, unfortunately, this happens exactly with reference to the fundamental problems. Examples? Here is a wonderful and surprising example: our being busy around a "Marxist philosophy". (Minin S K (1922), Filosofiju za bort! (Throw Philosophy overboard!) *Pod znamenem marksizma,* n. 5–6) [25]

In a successive work entitled Osnovnye voprosy marksizma (the Fundamental Questions of Marxism), Minin further explains his point of view, identifying with clarity the aims of his argument:

> But, in spite of everything, we continue to speak of this very 'philosophy' and to proclaim its importance. So Plehanov often uses this non-Marxist expression, 'Marxist philosophy' or 'philosophical aspect of Marxism'. In his preface to the second edition of *Materialism and Empirio-Criticism,* Lenin himself went as far as to write: 'I hope that, independently of the polemic with Russian Machists, it [this work] will not be useless as support to the knowledge of Marxist philosophy, dialectic materialism, as well as the philosophical conclusions drawn from the most recent discoveries of natural sciences'. And the editorial staff of the new review *Pod znamenem marksizma* commits sins which are more serious than venial sins in this respect, starting with the preface 'From the editor', published in the first issue". Minin S K, *Osnovnye voprosy marksizma* (The fundamental questions of Marxism, Moscow, p.10).

Minin's conclusion was peremptory:
In completing the building of our scientific vessel and in equipping it, we must make sure we throw overboard, together with religion, the whole of philosophy.

Stalin's "diamat"

In the few months following the revolution, when he was still active effectively in command, Lenin undertook to gradually building what could be called a *critical culture* of modernisation and change. He went back to Marx's lesson, as explained rather clearly in the preface to *A Contribution to the Critique of Political Economy,* where it is indicated that the distinctive trait of the superstructures is the fact that they are subject to a far slower revolution than its structural basis. Lenin was persuaded that one could develop ? [prodotti conoscitivi oggettivabili] that could be stored and reused so as to take on a character of permanent acquisition, and be cumulative, even in the case in which historical circumstances forced their temporary shelving. In fact, in his *Proekt rezoljucii o proletarskoj kul'ture* (Outline of a resolution on proletarian culture) of the 8th October 1920, he wrote explicitly that "a really proletarian culture" is the result not of any kind of *ex novo* "invention", but of a "development" process, from a Marxist viewpoint, of the "most precious conquests of the bourgeois period". Far from rejecting the latter, the "proletarian culture" had instead, according to Lenin, "assimilated and adopted the most valid areas in the over bimillenary development of human culture and thought" and must continue to work "on this basis and in this direction"[14].

[14] *Lenin, Opere Scelte in Sei Volumi, Editori Riuniti – Edizioni Progress, Roma, 1973–1975, vol VI, pp 187–88.*

Lenin's attempt was brutally censored by the new generations of thinkers and researchers to whom Stalin gave the task of bringing about the "big turn" on the philosophy and science fronts when his power had definitively consolidated, i.e. starting from the beginning of the Thirties. M.B. Mitin and P.F. Judin, who ran the communist unit of the Institute of Red Professors, acted as standard bearers of this radical line of cultural renewal. Others were philosophers P.M. Fedoseev, F.V. Konstantinov, M.D. Kammari, physicists V. Egorsin and A.A. Maksimov, mathematicians E. Kol'man and S. Janovskaja and biologist B. Tokin. The obvious and glaring fracture characterising the passage from the generation which had Lenin as a background and origin to the one imposed by Stalin revealed itself in 1931 with a radical change of direction of the two main reviews in the country. Both reviews, *Pod znamenem marksizma* (Under the flag of Marxism) and *Estest-voznanie i marksizm* (Natural Sciences and Marxism), had been given the task of going in more depth into the problem of the relationship between philosophy and science and the cultural significance of the new theories and views which had been emerging in the scientific field. Their editorial offices were completely revolutionised: to explicitly cancel any traces of continuity with the past, the latter, "organ of the exact sciences and natural sciences section of the Communist Academy", took on the new name of *Za marksistsko-leninskoe estestvoznanie* (For Marxist-Leninist natural sciences). In fact, already the 1930 issue number 2–3 included an editorial by the new editorial staff entitled *For the party line in philosophy and in natural sciences,* preceded by a note reading "The editorial staff points out that this issue, organised and realised by the old editorial team, was published late due to technical reasons and its content does not yet reflect the turn discussed in the opening article. The problems relating to such a turn will therefore be dealt with in the next issue of this review". The first thing that emerges from reading this editorial, apart from the keen claim of the nonneutral and biased character of any result achieved within the theoretical sphere, is the tight bond which is postulated between theory and practice. This bond gave away the increasing importance of the tendency to judge the value of scientific hypotheses on the basis of the success (or lack of success) of their concrete applications.

What these practical aims were which the development of science was considered destined to submit to, can be clearly deduced from the resolution entitled *Balance and new tasks on the philosophical front* approved at the end of a joint meeting of the sections of the Institute of Philosophy of the Communist Academy and the Moscow organisation of the society of militant dialectical materialists of 24th April 1930. It was, so to speak, a compromise document, still straddling two phases, the one that was already setting and the one, which was just starting, of Soviet cultural debate. The latter phase therefore reflected contrasting requirements and trends: in any case, it already clearly showed the aim towards which all the "healthy forces" in the country had to converge:

> The future activity on the philosophical front of the fight for Marxism-Leninism must be affected first of all by those general tasks that nowadays face the country of proletarian dictatorship. The proletarian revolution, in its current stage of evolution, that is to say in the conditions of fierce class struggle

> and victorious construction of socialism, has reached the fundamental task of the socialist revolution in our country: the eradication of the roots of capitalism. On the basis of the successes met in the industrialisation of the country, the full collectivisation and the removal of the *kulaki* as a class mean laying the foundation for the collective running of the national economy. As long as socialistically collectivised industry continues to be contrasted with the wide margin of discretionality of small rural properties, spontaneousness will continue to have an important role in social relations.
> The current stage of revolution has placed in front of us tasks from the solution of which depends the transformation from "blind" necessity into the chance of a conscious choice on the part of the producers themselves. This is a trend which is emerging in the class struggle and which leads to the creation of the premises for the destruction of classes.[15]

The "political-social" objective of reaching a swift and full collectivisation of the countryside and the economic aim to reach a noticeable and prompt increase in agricultural production are therefore two of the most important motives behind the fight against science and bourgeois scientists. In fact, it was thought that a new, collectivist structure of the agricultural production should be matched by a totally new agronomic technique derived from scientific principles and totally different from the traditional ones. It is hardly surprising that these movements ended up leading to a very distorted view of the complex problems concerning the relationship between political reasons, scientific requirements and economic-social needs. This is particularly true if we remember that the institution and the consolidation of the socialist production chain in the countryside was one of the "theoretical" nodes that proved the most difficult for Marxism to solve. Alternatively, one could analyse the tribulations of the agricultural policy of the Soviet Communist Party in the years immediately following Lenin's death.

Having set out the question in the terms described so far, the participation of science to the construction of socialism might be considered really useful only in the case of a reformulation implying above all a review of the work methods and times followed until then. This review, aiming first of all to carry out planning, that is the systematic management and organisation of experimental and scientific activities, had to be carried out on the basis of principles determined by the party line. The reference to dialectics, considered as universal scientific method, was above all meant to underline this latter need. In fact, since the indissoluble link between theory and practice, in the reductionist sense explained earlier, had been singled out as the fundamental principle of dialectical materialism, the reference to that philosophical doctrine was meant to be conceived essentially as a

[15] *Tagliagambe S (1978) Scienza, Filosophia, Politica in Unione Sovietica (Science, Phi-losophy and Politics in the Soviet Union), 1924-1939, Feltrinelli, Milano, p 285.*

reference to a constant link to concrete problems which the party was having to face and with the directives which it issued in the attempt to solve them.

In fact, it was this link that constituted the *leitmotiv* of all the following cultural debates. The most important among those which took place after the events described so far, was the debate held in March 1931 on the occasion of the joint meeting of the direction of the association of natural sciences of the Communist Academy, the managing committee of the natural sciences section of the Institute of Red Professors and the general editorial team of the *Great Soviet Encyclopaedia*. The meeting was open by an introductory report by O.J. Smidt, as chief editor of the Encyclopaedia. While admitting that inadequacies and mistakes were present in the work carried out in the field of natural sciences, he also claimed its substantial validity and effectiveness. There followed a speech by A.A. Maksimov, who was very critical towards the criteria and choices that had been followed during the working out and realisation of the work plan for the publication of the Encyclopaedia itself. After analysing the content of some scientific entries, he declared:

> We must conclude that, as far as natural sciences are concerned, we do not have a Marxist view: we have attentively sought Marxist entries and we must frankly say that there are none. At most, some of them need not be criticised that much and can therefore be left alone and not commented upon: if, however, one must refer to the large majority of the entries, it has to be said that, not only do they not adhere to a Marxist view, but they even express views which are antithetical to those of Marxism. Our conclusion, which was already formulated in the Theses that we developed and expressed in the headquarters of the Presidium of the Communist Academy, is that the natural sciences section of the Great Soviet Encyclopaedia, headed by comrade Smidt, must be labelled as anti-Marxist. This conclusion, as we have said, is undoubtedly correct. The ideological errors signalled so far are matched by errors in the choice of authors[16].

During the discussion, many specialists advanced a punctilious examination of the "ideological" inadequacies of each scientific section of the encyclopaedia. As to mathematics in particular, it was S. Janovskaja who spoke and pointed to the inadequacies and errors that, in her opinion, affected the section dedicated to this discipline.

She observed:

> Mathematics occupies a peculiar position among scientific disciplines, if nothing else because the hopes of idealism are strictly connected to it. It was not for

[16] *Za marksistko-leninskoe estestvoznanie, 1931, 1:73.*

> nothing that, in *Materialism and Empirio-Criticism,* when quoting the *History of Materialism* by F.A. Lange, Lenin notes that 'Hermann Cohen is, as we have seen, enthusiastic of the idealistic spirit of modern physics and goes as far as to preach the teaching of superior mathematics in schools, so as to infuse in secondary-school pupils the idealistic spirit driven back in our materialistic time'[17]. In fact, mathematics, being the most abstract science, constitutes a fertile ground for idealism and, just because of this, even if it is not yet possible to supply a positive elaboration of it at the level of dialectic materialism methodology, criticism of idealism must be clear and precise. Now, if we examine the content of the mathematics section of the Encyclopaedia, we must note that, not only is there no trace of any criticism of idealism, but on the contrary, as comrade Maksimov has pointed out, idealism is extolled.
> For instance, the entry "Arithmetic" exudes a pretty utilitarian spirit: it would be useless to look for any reference to Marxism, to class or ideology struggle in it (...) And this is nothing: the point is that this is a direct recommendation of the most ingrained idealism in mathematics: 'The problem of the principles of arithmetic', the author wrote, 'has found a satisfactory solution in the works of Robert Grassmann, Karl Weierstrass, Julius W.R. Dedekind, Georg Cantor and others'. Later on, it is true, he himself recognised that the problem of the principles of arithmetic could not be considered solved once and for all. However, his hopes of a solution were obviously placed in the formalism of David Hilbert, that is to say in the work of an author who, without any embarrassment, had turned mathematics into a game, with the result of preventing the assignment of any concrete, real content to its basic concepts. Idealism under the mask of utilitarism: this is the – scientifically very mediocre – general methodological formulation of this entry.[18]

Janovskaja similarly criticises the entries for "Algebra" and "Axiom". Talking about the latter, she observes that its fundamental flaw is its

> separation of logic from history, which, in turn, is the result of the attempt to treat logic not as an awareness and understanding of historical necessity, but rather as something which can be explained or put forward in a totally formalistic way, without any reference whatsoever to history, reality or practice. Thus, this is not simply a separation of logic from history, but also of the inevitably connected separation of form from content and of theory from practice. Even the author of this entry, just as O.J. Smidt himself, limits himself to talking of "specially prepared structures and methods" which we *"gather, choose, build and adapt* at first in a not totally conscious fashion, but then reach a phase of full awareness and consciousness" and "are improved thanks

[17] *Lenin, Materialismo ed empiriocriticismo (Materialism and Empirio-Criticism), Editori Riuniti, Roma, 1970, p 302.*

[18] *Za marksistko-leninskoe estestvoznanie, 1931, 1:74–75.*

to a slow adaptation". 'gather', 'choose', 'build' and 'adapt': we do everything but *reflect*. In the latter case, in fact, there would be no conventionality whatsoever and mathematics could not be considered, as it is by all Machists, not a science referring to material reality, but rather a simple store of ready-made confections, from which physicists can choose as they please, guided only by considerations of 'opportunity and convenience' and 'economy of thought'. This is a point of view that constitutes a complete deformation of historical reality, a real distortion of what really happens in the relationship between mathematics and physics.

Here materialism is clearly confused with idealism, with feeling and experience; here there is no talk of experiment, but rather of experience, a term which is known to lend itself to the most varied interpretations. The author keeps to the empiricist view and, in all conviction and sincerity, confuses this empiricism of his with materialism. He would like to stick to our point of view, but does not know and cannot grasp that empiricism and materialism are not the same thing.

I would also like to dwell on another point. The author is right when he claims, or nearly claims, that an axiom is simultaneously a definition and a judgement. This statement highlights its synthetic character, since it can be dealt neither as a simple definition, nor as a simple judgement. The author, instead, starts from the fact that an axiom is also a definition and that in each definition there is an aspect of conventionality to draw the conclusion that axioms are a kind of convention. This is not true: this is not the case at all. The difference between our logic and formal logic lies exactly in the fact that, while for the latter talking about something being true or false is always and only expressing a judgement, we could also be talking about a correct definition. This is because a correct definition reflects exactly the essential and peculiar traits of the object (thing, process or relationship) under discussion, while an incorrect definition cannot do that. Games, empty definitions and purely nominal determinations were never admitted by Marxism: in fact, Lenin fought against them with all his might. I would be wandering too far if I continued talking of this problem, but I still feel I have to stress that for us, just like judgements, even definitions can be true or false, correct or incorrect, because they too reflect reality.

Finally, the entry should have mentioned that one cannot talk of a single axiom or judgement, but only of a *whole system* of judgements and that the system itself can only make sense if it is considered as part of a totality, of a system.

As to the fact that, with change and development in science, axioms change and develop too, it would have been necessary to examine how they change and show that this development *determined by practice* brings us ever closer to absolute truth. The problem is that many idealists confuse these two facts, that is the compactness of the axiom system, which should be considered as a totality, and its variability, to the only end of distorting the reality of things. Thus, to idealists, a certain axiom system allows us to define real numbers:

these, however, could be defined differently, that is following the socalled genetic method which fully corresponds to historical development and is linked to progressive change, to the 'extension' of the concept of number and operations that can take place on it. From this point of view, though, it is not the genetic approach which is fundamental and decisive: everything is turned upside down and the axiom system ends up being considered as something which, just like Minerva, is fully complete on exiting Jove's head, that is it exits the head of the modern mathematicians in an already complete form. It is exactly this system that becomes the demiurge of mathematics, the person whose task it is to establish the course and limits of the next genetic extension of the concept of number. The discussions on the genetic and the axiomatic methods continue to shake the philosophy of bourgeois mathematics, but their traces cannot be found anywhere in the entry. Yet, it is very important for us to explain the subordinate role of axiomatics in science and to clarify that it does not constitute the real starting point, but it is instead the result of an analysis that can be carried out starting from an already high level of scientific development. This attribution of the fundamental function to axiomatics is useful to idealism, in that what the latter does in mathematics is take relations which are found in reality and turn them upside down, so as to be able to assert that the trunk has originated from the head and not viceversa: this is the formulation that we must fight. This is why I believe, in conclusion, that the 'Axiom' entry is to be considered as idealistic and not Marxist[19].

For its linearity and the clarity of its position, Janovskaja's critical analysis might well summarise and be exemplary of the overall debate on natural sciences, and on mathematics in particular, which developed starting from 1931 and continued for the whole of the Stalin era.

Conclusion

We have proposed two moments that we consider to be particularly significant in the debate on mathematics in Russian scientific thought and tried to place them in the social and cultural climate of their respective historical phases. These "probings" certainly do not aspire to summarise and express the complexity of a long and tormented historical event, particularly its stages in the first half of the twentieth century. They can however give an idea of the consequences that inevitably derive from a certain type of attitude towards scientific research; consequences which Julian Huxley admirably summarises in his 1949 work dedicated to the analysis of the "Lysenko case":

[19] *Ibid., pp 79–81.*

Science cannot flourish and cannot be fully fruitful save in certain material conditions and in a certain moral and intellectual atmosphere. As Muller (1949) well puts it, "It has taken thousands of years to build the basis of that freedom of inquiry and of criticism which science requires. It has been possible only through the growth of democratic practices, and through the associated progress in physical techniques, in living standards, and in education, applied on a grand scale. Only in modern times have all these conditions advanced sufficiently to permit the widespread, organized, objective search for truth which we today think of when we use the word Science."
But the atmosphere necessary for science to flourish can be all too readily destroyed or poisoned, whether by ignorance and mental laziness, by prejudice and vested interests, or by authoritarian power[20].

[20] J. Huxley, La genetica sovietica e la scienza, Longanesi&C., Milano, 1977, pp. 193-194.

Mathematics and Economy

Economists and Mathematics from 1494 to 1969 Beyond the Art of Accounting

Marco Li Calzi, Achille Basile

Translated from the Italian by Emanuela Moreale

The master economist must possess a rare combination of gifts.
... He must be mathematician, historian, statesman,
philosopher to some degree.
J.M. Keynes (1936)

Introduction

The first link between mathematics and economics dates back to calculations of a commercial nature, which merchants had to carry out daily in ancient times. This relationship was later enriched by the financial calculations needed by usurers and bankers: it developed into *a corpus* of applications of arithmetic that is nowadays called accounting. Its practical relevance is at the basis of the importance attributed to the ability of getting sums right, or numeracy, by anyone investing their savings, carrying out a breakdown of tax in an invoice or filling in their tax returns declaration.

During the High Middle Ages, when numeracy was an art limited to few, merchants developed a method for "balancing the accounts" of their businesses and keeping track of income and expenditure. It was in Italy, probably in the second half of the thirteenth century, that double-entry book-keeping was developed: this system is still being used and is still at the heart of the book-keeping activity in any business today. A first description of its methodology was published in Venice in 1494 within a compendium of economics and mathematics by Luca Pacioli (1445 ca. –1514 ca.), *Summa de Arithmetica, Geometria Proportioni et Proportionalitate.* In homage to the merchants' lagoon city that hosted the third convention on "Mathematics and Culture", Venice will be our starting point in telling the story of the evolution of the relationship between mathematics and economics.

Our thesis is simple: as long as the practical needs of commerce or finance were satisfied by the arithmetic collected in book-keeping manuals, the relationship between mathematics and economics was constant but superficial. Merchants needed to keep their accounts tidy and mathematicians possessed the art necessary to do this. The progressive transformation of secular economic precepts into a social science caused the surfacing of both a deeper relationship between economics and mathematics and a new profession – that of economist – entitled to manage it.

Modern economists often have not had the opportunity to use double-entry bookkeeping. However, before becoming economists, they had to study differential calculus, linear algebra, probability theory and statistics. Later, depending on their specialisation, they are required to achieve quite a deep knowledge of those areas of mathematics that they need to build, understand and use quantitative models. Their activity, nowadays, resembles that of an engineer who strives to realise the precepts of an art that has reached the *status* of a science.

Since 1969, a symbol of this status has been the Nobel Prize in Economic Sciences (or, more precisely, the Bank of Sweden Prize in Economic Sciences in memory of Alfred Nobel). Our story of the evolution of the relationship between mathematics and economic sciences starts from the year of the publication of the first work on mathematics and book-keeping (1494) and ends with the international recognition of the figure of the economist as a scientist (1969). During this period of almost five hundred years[1], the relationship between mathematics and economics radically changes.

Mathematics, seen initially as a way to express economic concepts, has increasingly become one of the languages in which economic theory can be formulated and communicated. Having reached the status of language, all economists must face it. In the best of cases, it becomes a genuine knowledge tool. More often, though, it is used as a technique to prove the logical coherence of a theory. At times, its use degrades into rhetoric abuse whose only aim is to give a scientific appearance to an argument. In any case, by the time that the Nobel Prize in Economic Sciences starts being awarded, knowledge of mathematics is considered a requisite for this profession.

This article summarises the main phases of the story of this transformation through the voice (or, rather, the pen) of many of its protagonists, by offering a small anthology with comments quoted from their works. Some of the quotes in this article have been inspired by a reading of [2]. Further details on the role of mathematics in the construction of economic theory can be found in [3] and [4].

Mathematics as a Source of Examples

The mathematical competences of the first economists are probably limited to the arithmetic necessary to book-keeping and accountancy. Since economics deals with prices and values, theoretical discussions cannot avoid the use of numbers altogether. But this use is mostly an example to persuade the reader that "the accounts balance", that is that what has been stated derives from the internal logic of economics.

1 *The three-decade period between 1969 to 1999 is dealt with in the article "Who said that a mathematician cannot win the Nobel Prize?", also published in this volume.*

In fact, the reflection on economics is not yet a science. Researchers are trying to uncover the "natural behavioural laws" of economic systems and arithmetic is contributing decisive help in identifying the economic phenomena that need to be explained. In his *Essay on the Nature of Commerce,* written around 1720 but published in 1755, Richard Cantillon (1680–1734) illustrates how the same amount of work carried out by a man can correspond to a different buying power:

> If, for example, one man earn an ounce of silver every day by his work, and another in the same place earn only half an ounce, one can conclude that the first has as much again of the produce of the land to dispose of as the second.

In his *An Inquiry into the Nature and Causes of the Wealth of Nations* (1776), Adam Smith (1723–1790), unanimously considered the father of economic science, does not use proportions – but rather multiplication and division – to clarify how specialisation can increase a worker's productivity:

> a workman not educated to this business [...] certainly could not make twenty [pins a day]. [... But] where ten men only were employed, and where some of them consequently performed two or three distinct operations [...] they could [...] make among them upwards of forty-eight thousand pins in a day. Each person, therefore, [...] might be considered as making four thousand eight hundred pins in a day.

Only a few years later, Thomas Malthus (1766–1834) introduces arithmetic and geometric progressions in his *An Essay on Population* (1789) to explain the risk that population growth might drastically reduce the quality of life:

> Taking the population of the world at any number, a thousand millions, for instance, the human species would increase in the ratio of – 1, 2, 4, 8, 16, 32, 64, 128, 256, 512, etc. and subsistence as – 1, 2, 3, 4, 5, 6, 7, 8, 9, 10, etc. In two centuries and a quarter, the population would be to the means of subsistence as 512 to 10.

Cantillon claims that a different salary can correspond to the same quantity of work. Smith explains that a different organization of work can increase a worker's productivity. Malthus shows that the rate of population increase can drastically diverge from the rate of development of means of subsistence. Each of these propositions can be expressed without using mathematics. However, using an arithmetic example causes the argument to stand out with greater clarity. The "accounts balance", because what is claimed can be verified numerically, which reassures the reader as to the coherence of the discourse.

Mathematics as a Language

Slowly – and mostly unwittingly – the use of mathematics as source of clarifying examples will lead economists to introduce a more explicitly mathematical language in their writing. In this process of cultural enrichment, three trends emerge.

On the one hand, there is a sincere effort to formulate the laws of economics rigorously, by introducing precise definitions and mathematically-proven propositions. An example of this is the way in which, in his *Recherches sur les Principes Mathématiques de la Théorie des Richesses* (1838), Antoine A. Cournot (1801–1877) proves the existence of a price that maximizes revenue:

> Since the [demand function] $F(p)$ is continuous, the function $pF(p)$, representing the value of the yearly sold quantity, must also be continuous. [...] Since $pF(p)$ initially rises and then dips in p, there must exist a value of p that maximises this function. This value is given by the equation $F(p) + pF'(p) = 0$.

The existence of this price – optimum price because it leads to the highest possible revenue – justifies economists' efforts to supply precepts that help businessmen determine the best price at which to sell their products. The proof is a simple but genuine application of mathematics to economics. Not for nothing, Cournot is often considered the father of mathematical economics. However, scholars like Cournot remain rare until 1870 and Cournot himself will not write about economics after 1838.

More common, instead, is the tendency to introduce a more mathematical language as a natural complement to a search for rigour, which however is not identified with mathematical rules. The main communication means used by scholars in economics remains mainly non-mathematical, with occasional intrusion of mathematical terminology.

As a representative of this prevailing trend, we can consider the manual *Principles of Political Economy* (1848, 1st ed.; 1871, 7th ed.) by John Stuart Mill (1806–1873) – the reference text in the formation of economists until at least the second half of the nineteenth century:

> The equation of international demand [...] can be concisely stated as follows. The produce of a country exchanges for the produce of other countries, at such values as are required in order that the whole of her exports may exactly pay for the whole of her imports.

Mill states that the exchange rate of a currency must be such that the value of imports is equal to the value of exports. Expressed as "equation of international demand", the same proposition can be given in a mathematical form. Thus, mathematics can be used as a language to discuss economics.

One hundred years later, we find full awareness of this in an anecdote about physicist Gibbs reported by Paul A. Samuelson (born in 1915 and winner of Nobel Prize in economics in 1970) in a passionate defence of the importance of mathematics to economic theory [5]:

"The great Willard Gibbs was supposed ever to have made [one single speech] before the Yale Faculty. [...While professors] were hotly arguing the question of required subjects: Should certain students be required to take languages or mathematics? Each man had his opinion of the relative worth of these disparate subjects. Finally Gibbs, who was not a loquacious man, got up and made a four-word speech: 'Mathematics is a language'."

The spreading among economists of the awareness that mathematics can be a language with which one can communicate economic theory and build applications is one of the important causes of all the successive transformations in the relationship between mathematics and economics. For simplicity, we will gather them in three categories.

For most, following Mill, the assimilation of the axiomatic method and demonstration techniques borrowed from mathematics will provide the basis for the transformation of a collection of vague and contradictory economic concepts into a systematic and logically coherent corpus. In Samuelson's words [5]:

> The problems of economic theory – such as the incidence of taxation, the effects of devaluation – are by their nature quantitative questions. [...] When we tackle them by words, we are solving the same equations as when we write out those equations. [...] Where the really big mistakes are is in the formulation of premises. [...] One of the advantages of the mathematical medium – or, strictly speaking, of the mathematician's customary canons of exposition of proof, whether in words or symbols – is that we are forced to lay our cards on the table so that all can see our premises.

Some great scholars will follow Cournot's example. Going beyond the use of mathematics as a mere technique, they will turn it into a genuine knowledge tool indispensable for the advancement of economic theory. Again, Samuelson writes [5]:

> Euler's Theorem is absolutely basic to the simplest neoclassical theory of imputation. Yet without mathematics, you simply cannot give a rigorous proof of Euler's theorem.

Finally, for others, the paraphrase of mathematical language will only serve to attribute more authority to their theses. Leaving out the examples of Smith and Malthus, since these are simple and understandable by anyone, in this case mathematics is not used for cultural enrichment, but to obscure laymen's understanding of economic laws. John Maynard Keynes (1883–1946), possibly the greatest economist ever, warns against this degenerative temptation in an anecdote told by his friend Roy F. Harrod in his biography *The Life of John Maynard Keynes* (1951):

> When I asked him in 1922 how much mathematics it was needful for an economist to know, he replied that Johnson, in his article in the Economic Journal, had carried the application of mathematical analysis to economic theory about as far as it was likely to be useful to carry it.

In order to better evaluate this claim, it should be pointed out that the article by Johnson that Keynes refers to was published in 1913: it used differential calculus and determinants. It is, therefore, a mathematical background similar to that which (in 1922!) is expected from an engineer.

The three following sections will show how, in its relationship with economics, mathematics has been seen and used as a technique, a knowledge tool and a rhetorical tool.

Mathematics as a Technique

Let us first of all consider the role of mathematics in the development of economic analysis techniques. An example will help us in understanding the terms of reference. Let us read again the passage by Chantillon – quoted earlier – on the problem of establishing which "natural law" gives the first man's working day a higher value than the second man's working day. Presumably, the work carried out by the first man is paid more because it is more productive than the work carried out by the second man. Let us now formulate two hypotheses whose concatenation can explain the phenomenon at hand: 1) a higher productivity creates a higher value; 2) a higher value results in a higher salary. How can we turn these hypotheses into knowledge?

One could proceed by empirical induction, by observing whether a higher productivity corresponds to a higher value and whether a higher value corresponds to a higher salary. However, since the value of a working day is not a measurable quantity, we can only ascertain whether a higher productivity results in a higher salary. The result of this experiment is not conclusive: a positive correlation between productivity and salary is a necessary but not sufficient condition for the existence of the two positive correlations we hypothesised. Vice versa, if we find a negative correlation, we cannot determine whether one of the two hypotheses is correct.

In general, this is a typical problem encountered in all social sciences: the consequence is that it is difficult to devise conclusive experiments for even the simplest propositions. Because this prevents us from employing empirical induction, many scholars have been forced to proceed deductively. If we define the concepts of salary, value and productivity appropriately, we can write a theorem and show that better productivity creates higher value, even if not necessarily a higher salary.

The first scholar to consciously stress the importance of the deductive method in mathematics as a basis technique for economic science was William S. Jevons (1835–1882) in his *opus magnum* on *The Theory of Political Economy* (1871): "Economy, if it is to be a science at all, must be a mathematical science."

The characteristics of his method are evident in his letter dated 1st June 1860 to his brother Herbert:

> I obtain from the mathematical principles all the chief laws at which political economists have previously arrived, only arranged in a series of definitions,

axioms, and theories almost as rigorous and connected as if they were so many geometrical problems.

Jevons was followed by a radical methodological change – known as marginalist revolution – which revolutionised economics. Marginalists' introduction of the deductive method is at the basis of the neoclassical theory, which even today constitutes the paradigm of economic science. For many years, the historians of economic thought have been debating the relationship between the introduction of the marginalist method and progress in economic science. Here we will simply observe that it has two concomitant effects.

The first effect is that of "mathematising" part of the economic discourse, opening the discipline to the contributions of other scientists, particularly physicists and engineers. They, in turn, concentrate on and study the problems that best fit the deductive method. Economic science gains in depth in some areas, but at the same time, risks losing track of other – equally important – areas. In fact, the second effect is that of pushing into the background all historical and institutional aspects that are difficult to treat "mathematically".

The most characteristic example of the first effect is the introduction of a utility function to describe consumers' behaviour. As Tjalling Koopmans (1910–1985, Nobel Prize in economics in 1975) explains in his *Three Essays on the State of Economic Science* (1957):

> A utility function of a consumer looks quite similar to a potential function in the theory of gravitation.

Engineers familiar with rational mechanics – among these Italian Antonelli, Pareto and Boninsegni – or physicists used to the principle of conservation of energy only need to transpose their knowledge into the economic field to generate a myriad of new results and enlightening analogies. The economist who best represents this process is, again, Samuelson. In his article *How "Foundations" Came to Be* (1998), he writes:

> I was vaccinated early to understand that economics and physics could share the same formal mathematical theorems (Euler's theorem on homogeneous functions, Weierstrass's theorems on constrainted maxima, Jacobi determinant identities underlying LeChatelier reactions, etc.), while still not resting on the same empirical foundations and certainties.

Between 1870 and 1930, the success of these exchanges is accompanied by a methodology for the use of mathematics in economic research. This methodology is quite close to what was in use in rational mechanics at the end of the nineteenth century. Not for nothing, the greatest success of neoclassical economics is the theory of general economic equilibrium conceived by Leon Walras (1834–1910) and consolidated by Vilfredo Pareto (1848–1923) who sees economics as a system of contrasting forces in search of an equilibrium.

Nowadays, more modern mathematical methods have been introduced: yet, this formulation survives in the most mathematically accessible – and therefore more widespread – formulations of neoclassical theory. This has two main effects. The first is that economists often continue to adopt those hypotheses that can be more easily taken back to this analogy, instead of using the most realistic ones. In fact, in this respect, very little progress has been made since October 1901, when Henri Poincaré wrote Walras a letter criticising his axioms of economic behaviour:

> You regard men as infinitely selfish and infinitely farsighted. The first hypothesis may perhaps be admitted in a first approximation, the second may call for some reservation.

The second effect is the little attention that economists usually dedicate to alternative paradigms (such as evolutionary theories from biology) or to the most recent theory in physics (such as relativity or quantum theories). Yet, in some cases, mathematical theories have been suggested by economic problems and later rediscovered in physics. But, in these cases, more than a technique, mathematics is being used as a knowledge tool and this is the topic of the next section.

Mathematics as a Knowledge Tool

The most successful applications of mathematics to economics have extolled its role as a knowledge tool. Numerous economic theories have been set out and perfected thanks to the systematic and enlightened use of mathematics. A thorough review of such theories can be found in [6].

Among these applications, we will mention seven: a) the theory of general economic equilibrium, based on a mathematics similar to that of rational mechanics and later on the methods of differential and algebraic topology; b) the theory of rational expectations, founded on statistical inference and on dynamic programming; c) game theory, which has even generated its *ad hoc* mathematics; d) economics of uncertainty, based on the same theory as game theory as well as on probability calculus; e) the theory of social choice, founded on algebraic methodology; f) mathematical finance, which has enriched the theory of continuous time stochastic processes; g) the theory of optimal resource allocation, from which linear programming – and, more generally, operational research – developed.

Some of the main authors of these theories have received the Nobel Prize in economics. Further details can be found in the article "Who said that a mathematician cannot win the Nobel Prize?", published in this volume, which also deals with Debreu and economic equilibrium, Nash and game theory, Arrow and social choice and Kantorovich and optimal allocation.

We will mention three mathematicians who died before 1969: Ramsey, von Neumann and Bachelier. At least in the case of the first two, we are certain that the only reason they did not receive a Nobel Prize in economics was its late institution. In fact, in a brief autobiographical piece written in 1990, Robert T. Solow (born in 1924 and Nobel Prize in economics in 1987), claims that the three non-economists who have been most important to economics have been Ramsey, von Neumann and the mathematical statistician Harold Hotelling (1895–1973).

The first author is Frank P. Ramsey (1903–1930), who died at a very young age, but authored absolutely original contributions on three different problems: providing a general criteria for decision-making under conditions of uncertainty (1926, published posthumously in 1931); determining the best income taxation system (1927) and establishing the best way to accumulate national savings (1928). The latter work introduced the use of calculus of variation in economics, bringing to economists' attention the necessity to introduce techniques to solve dynamic optimisation problems.

The second author is John von Neumann (1901–1957): versatile genius, with sporadic interest in economics, he nevertheless provided absolutely original contributions to this discipline. In fact, he found the first general solution to purely antagonistic games between two people (1928); provided the first model of economic growth with conditions for the permanence in time of the equilibrium (1937); together with Oskar Morgenstern (1902–1977), he placed into a system framework some previous intuitions by founding game theory and decision theory (1944).

The third author, still unknown to most economists even today, is Louis Bachelier (1870–1946). In his doctoral thesis (1900) on speculation problems, he represented price movements of financial activities through random walks, anticipating by a few years Einstein's Brownian motion (1905) and by seventy years the use of martingales in the mathematical representation of an efficient market.

The contribution of mathematics to economics, however, is not limited to its role as a knowledge tool with respect to theories. As said earlier, starting from 1870, the marginalist revolution consciously transforms the use of mathematics from a language into an analysis technique. In this process, economic theory is more and more characterised by the systematic use of the axiomatic formalism typical of the deductive method. The limitation of this approach is that it allows the development of theories of general importance, but does not provide predictive models bound in space and time.

Starting from 1930, under the pressure of statisticians, a second – and parallel – process of formalisation concerning empirical observations affirms itself. To understand the world and foresee what is going to happen, a quantitative science is needed which can manipulate statistical data, from which it needs to derive descriptive and predictive models, whose validity is based on formal but empirical criteria. In a sense, it is the revenge of the inductive method over the deductive method. Next to mathematical economics, which reduces economic phenomena to theorems, econometrics is born, with its aim to measure economic phenomena.

The birth certificate of this new discipline is the foundation of the Econometric Society, *"an international society for the advancement of economic theory in its relation* to statistics and mathematics" by Irving Fisher (1895–1973) and Ragnar Frisch (1895–1973, Nobel Prize in economics in 1969). Here is how the latter describes econometrics in the article *On a Problem in Pure Economics* (1926):

> Half way between mathematics, statistics and economics, we find a new discipline that, for want of a better name, we can call *econometrics*. The purpose of econometrics is to subject the abstract laws of economic theory or "pure" economics to experimental and numerical tests, so as to turn pure economics, as far as possible, into a science in a strict sense.

After some understandable initial difficulty, the Econometric Society, also thanks to a series of favourable financial circumstances, has an unexpected success and econometrics establishes itself rapidly, distinguishing itself from traditional economics. Frisch is usually credited with the invention of the terms "microeconomics", "macroeconomics" and "econometrics" which even today still designate the three main subjects in the first year economics undergraduate curriculum.

Before closing this section on the successes of mathematical economics, it may be useful to compare Frisch's full-of-hope words with an anecdote told by Keynes in his *Essays in Biography* (1933):

> Planck [...], the famous originator of Quantum Theory, once remarked to me that in early life he had thought of studying economics, but had found it too difficult! [...] Planck could easily master the whole corpus of mathematical economics in a few days. [...] But the amalgam of logic and intuition and the wide knowledge of facts, most of which are not precise, which is required for economic interpretation in its highest form is, quite truly, overwhelmingly difficult for those whose gift mainly consists in the power to imagine and pursue to their furthest points the implications and prior conditions of comparatively simple facts which are known with a high degree of precision.

Even if econometrics and mathematical economics have contributed to transforming economics into a science, their task is not over. Economics and society are constantly changing and their genuine interpretation requires yet more mathematical instruments and knowledge.

Mathematics as a Rhetorical Tool

A consequence of the successes in the application of mathematical methods and instruments to economics is the spreading of the mathematical language as a rhetorical tool. Banal propositions, when duly made up to look like a theorem, can look like scientific arguments (particularly to lay people). In these cases, unfortunately, mathematics contribute neither to the analysis nor to the development of economic science, but is reduced to a tool used to dress up and confer greater dignity to studies which – evidently – have little merit in themselves.

The thesis that economics suffers from an excessive "mathematisation" is often based on this widespread rhetorical artifice. Very many economists, and even some Nobel Prize winners such as M. Allais and W. Leontief, have attacked – sometimes violently – the practice of passing mathematical theorems as good economics.

One of the best-known criticisms to the use of mathematics as an economic analysis tool was made by Alfred Marshall (1842–1924) in a letter to Bowley dated 27th November 1906:

> A good mathematical theorem dealing with economic hypotheses [is] very unlikely to be good economics: [...] the rules – (1) Use mathematics as a shorthand language, rather than as an engine of inquiry. (2) Keep to them till you are done. (3) Translate into English. (4) Then illustrate by examples that are important in real life. (5) Burn the mathematics. (6) If you can't succeed in (4), burn (3). This last I did often.

Coherently, in his *Principles of Economics* (1890, 1st ed.; 1920, 8th ed.), the most widely-used economics manual across the nineteenth and twentieth century, Marshall relegated his formal system to the appendix. But, as his pupil Keynes explains in his *Essays in Biography* (1933), he did so to avoid giving the impression that mathematics provides answers to real life problems just by itself.

A necessary condition for a fertile application of mathematics to economics is that it must contribute to the understanding of important phenomena. Otherwise, it is better to be silent. The risk that mathematical formalism can take over the substance of problems, on the other hand, is the same risk described in general by John von Neumann in [7]:

> As a mathematical discipline travels far from its empirical source [...], it is beset with very grave dangers. It becomes more and more purely aestheticizing, more and more purely l'art pour l'art. This need not be bad, if the field is surrounded by correlated subjects, or if the discipline is under the influence of men with exceptionally well-developed taste. But there is a grave danger that the subject will develop along the line of least resistance, that the stream, so far from its source, will separate into a multitude of insignificant branches, and that the discipline will become a disorganised mass of details and complexities.

Mathematics as a Professional Requisite

The versatility of the roles that mathematics carries out in economics – such as technique, analysis and rhetorical tool – implies that it is to be considered by all means a professional requisite indispensable to the modern economist.

Here is Samuelson's advice to a youth with modest mathematical background who wishes to study economic analysis in depth [5]:

> It happens to be empirically true that if you examine the training and background of all the past great economic theorists, a surprisingly high percentage had, or acquired, at least an intermediate mathematical training. [...] Moreover, without mathematics you run grave psychological risks. As you grow older, you are sure to resent the method increasingly. Either you will get an inferiority complex [...] or you will get an inferiority complex and become aggressive about your dislike of it. [...] The danger is almost greater that you will overrate the method's power for good or evil.

A good mathematical awareness must provide economists with a capacity to tell a good theory from a bad one and prevent them from being taken in by rhetorical sirens. After all, mathematics is a tool that should be judged on the basis of the use that is made of it. He goes on to advise as follows:

> Mathematics is neither a necessary nor a sufficient condition for a fruitful career in economic theory. It can be a help. It can certainly be a hindrance, since it is only too easy to convert a good literary economist into a mediocre mathematical economist.

There is no mathematical *via regia* to economics. But it is a fact that the importance of mathematics in professional economic communication is undisputed. For instance, a research of G.J. Stigler (1911–1991, Nobel Prize in economics in 1982) and other two collaborators noted that, taking into account the whole of the five main economic reviews, the percentage of articles making use of differential calculus or other advanced techniques has risen from 2% in 1932–33 to 31% in 1952–53 and from 46% in 1962–63 to 56% in 1989–90 [8].

Perhaps more surprisingly, another study by T. Morgan [9] notices how in 1982–83 the "mathematical models without empirical data" were 42% in the *American Economic Review* (a reference review for economists), 18% in the *American Political Science Review* (reference review for political scientists), 1% in the *American Sociological Review* (reference review for sociologists), 0% in the *Journal of the American Chemical Society* (reference review for chemists) and only 12% in the *Physical Review* (reference review for physicists).

Apparently, therefore, there is (by far!) more mathematics in economics than in any of the other social sciences and even than in more traditional scientific disciplines. It is therefore all the more important for economists to have a solid mathematical background, so as to avoid suffering from any inferiority complexes and to be able to distinguish good from bad economics autonomously. A last reason is not to betray that quarter of mathematical nobility that, as quoted in the epigraph, Keynes attributes to a true economist.

Bibliography

[1] Keynes J M (1936) The General Theory of Employment, Interest and Money, Macmillan, London
[2] Zakha W J (1992) The Nobel Prize Economics Lectures: A Cross-Section of Current Thinking, Avebury, Aldershot
[3] Ingrao B, Israel G (1987) La Mano Invisibile: L'equilibrio Economico nella Storia della Scienza, Laterza, Bari. English Edition: Ingrao B, Israel G (1990) The Invisible Hand: Economic Equilibrium in the History of Science, The MIT Press, Cambridge (Mass)
[4] Mirowski P (1991) The When, the How and the Why of Mathematical Expression in the History of Economic Analysis, Journal of Economic Perspectives 5:145–157
[5] Samuelson P A (1952) Economic Theory and Mathematics – An Appraisal, American Economic Review, Papers and Proceedings 42:55–66
[6] Various Authors (1981–1991) Handbook of Mathematical Economics, 4 vols, North-Holland, Amsterdam
[7] Von Neumann J (1947) The Mathematician, in: The Works of the Mind, R.B. Heywood (ed), University of Chicago Press, Chicago, pp 180–196
[8] Stigler G J, Stigler S M, Friedland C (1995) The Journals of Economics, Journal of Political Economy 103:331–359
[9] Morgan T (1989) Theory versus Empiricism in Academic Economics: Update and Comparisons, Journal of Economic Perspectives 2:159–164

Who Said that a Mathematician Cannot Win the Nobel Prize?

Achille Basile, Marco Li Calzi

Translated from the Italian by Emanuela Moreale

Is Mathematics Supportable or Is It to Be Supported?

It is well known that mathematicians do not win Nobel Prizes. A Nobel Prize is awarded in Chemistry, Physics, Literature, Medicine and, since 1969, a Prize in Economic Sciences has also been given. (In this context, we will obviously disregard the Nobel Prize for Peace). The Nobel Prize is awarded to people who, in the preceding years, *"have conferred the greatest benefit on mankind"* [1].

There simply is no Nobel Prize in Mathematics. The implication that mathematics does not provide services to humanity is logically inconsistent and false. Nevertheless, the idea hovers, even if only subconsciously, in the thoughts of most people and certainly also in those of so called educated people.

It is understandable if, at the dawn of the century, economics was not considered to be as important as more traditional and then more prestigious disciplines. Even today, in fact, the Nobel Prize in Economics causes controversy; see for instance [2, pp.345–346]. Nevertheless, it is difficult to accept that, a century later, there still does not exist a recognition of mathematical studies that is equivalent in all respects to the Nobel Prize.

Leaving aside anecdotes, jokes and conjectures about the omission of mathematics from Alfred Nobel's will (see articles [3] and [4]), we mathematicians, with our daily actions, are all at least partly responsible for the absence of an equivalent prize in mathematics.

Of course, the Fields Medal has been conferred since 1936. Yet, in terms of celebrity outside the world of mathematics, in terms of monetary value and overall social impact, this prize is nearly negligible. There are also other prestigious scientific awards that can also be conferred to mathematicians, even if not specifically dedicated to mathematics. Some of these even involve considerable amounts of money; an example, geographically close to the authors, is the "Premio Balzan" [5]. However, many of these awards are often neglected by the mathematical community.

As to the prize which mathematicians consider their "Nobel", that is the Fields Medal, we would like to dwell on the issue of its social impact, for we believe that this reflects the opinion that our society has of mathematics.

In order to evaluate its fame, we invite our readers to reflect on the two following issues and to reconsider them in the contexts they find most appropriate:

- Given a group of educated people, how many among them are aware of the existence of the Fields Medal? On the other hand, are there really any among them who have never heard of the Nobel Prize?
- If a person has heard of the Fields Medal, how many Fields Medallists can she list? On the other hand, how many Nobel Prize winners can she name?

It is unsurprising that our experience in this respect has been rather disappointing, often even among graduates in mathematics or mathematicians by profession.

Moreover, Nobel Prize announcements are immediately released on radio, television and in the press. This certainly does not happen because of the immediate relevance that the rewarded work has in everyday life, for in fact contemporaries almost inevitably fail to grasp such relevance. By contrast:

- How many journalists could immediately say or write something meaningful about the Fields Medal?

Moving on to the monetary value of the Fields Medal, it is evident that, in itself, it is unimportant whether the monetary value of a scientific prize accrues to one Euro or to one million Euros. However, if the amount is considerably high, this is bound to impress the collective imagination. Moreover, when an institution such as the Bank of Sweden decides to finance a Nobel Prize in Economic Sciences, this implies recognition of the social value of the work of thousands of economists the world over.

The fact that mathematics has (nearly) nothing similar suggests that society puts up with but does not appreciate mathematicians. An odd coincidence provides a further argument in support of our claim: while researching in preparation for the present article, we came across the Nobel Prize entry in an old encyclopaedia published in the mid-Seventies.

Quite understandably, the table illustrating the various Nobel Prize winners does not contain a column for economics, as this would have been extremely shorter than the others. In the text, there is a reference to the prize in economic sciences having been awarded since 1969 together with a list of the few award winners until then. However, quite strangely, the year 1972 is missing: this happens to be the year when the Nobel Prize in economics was awarded to Arrow, the first "mathematician" to have been given such an award in the 30 or so years.

Mathematicians win Nobel Prizes in Economics

But let us go back to the main topic and to our initial claim which we would now like to modify as follows: mathematicians do not win Nobel Prizes meant specifically for them. Luckily, since intelligence can hardly be restrained, it so happens that mathematicians end up winning, so to speak, Nobel prizes meant for others. In particular, mathematicians seem rather skilled at winning prizes reserved for economists.

To this day, this has definitely happened in at least 10% of cases and, if the term "mathematician" is taken in a somewhat less strict sense, the percentage rises to 17% of cases. We are certain that, even in the future, mathematicians will win

Nobel Prizes or that these prizes will be won by economists who have made contributions of an essentially mathematical character[1].

Going back to the past, we will now turn our attention to the Nobel Prizes in Economic Sciences that have so far been awarded to "mathematicians":

- Gerard Debreu (1983);
- Kenneth J. Arrow (1972, together with John R. Hicks);
- John F. Nash (1994, together with John C. Harsanyi and Reinhard Selten);
- Leonid V. Kantorovich (1975, together with Tjalling C. Koopmans).

It seems appropriate to justify our rather uncontroversial, yet not quite univocal, selection. Moreover, we feel obliged to clarify the link we see existing between mathematics and economics.

Economics is a science endowed with a profound social connotation and therefore the relevance of the themes it studies is to be evaluated as a function of their repercussions onto society. It is exactly the complexity of the most central of these themes that is often best tackled with the aid of logic and mathematics. This is true both in terms of processing capabilities and in terms of analysis of the various propositions. The risk that the development of sophisticated models disconnected from reality might be passed for good economics research, when in fact they are a good intellectual exercise at best, can only be overcome by promoting the diffusion of greater mathematical culture among economists. In this fashion, the discrimination of what is relevant for the development of economic sciences from what is only a formal exercise will not suffer from difficulties in deciphering and recognising the terms in which the questions are posed.

Within the mathematical community, the traditional criteria used to determine the relevance of a piece of work are based on its mathematical profoundness, its capacity to introduce new ideas or new methods or the fact that it solves a problem which had long been unsolved. These criteria, however, are not appropriate for the evaluation of the contribution of mathematics to economics (or to any other discipline). It is instead necessary to take into consideration how much the use of the tool of mathematics contributes to the development of the discipline to which it has been applied and therefore to its comprehension of the real world[2].

We think that our choice of Kantorovich and Nash can be hardly objected to: both of them are definitely mathematicians. Even if neither of them had ever met economics and had therefore never won the Nobel Prize in this discipline, they would still be remembered for the unforgettable traces they left in the mathematics of this century. Even the choice of Debreu is hardly surprising. As we will see, both because of his background and of the specificity of his contribution to the Theory of Economics, it is difficult not to see him as a mathematician.

[1] *While this article concentrates on economics, we could not refrain from mentioning that the last Nobel Prize in Chemistry was co-awarded to the British mathematician J.A. Pople in 1998 for developing calculation methods that made possible the treatment of mathematical equations that are at the basis of the application of quantum mechanics to chemistry problems.*

[2] *In terms of solving problems which could not otherwise have been tackled or in terms of clarifying the fundamental concepts.*

On the other hand, our choice of Arrow may turn out to be more controversial. However, we believe that there is good reason to list him under the "mathematicians club". This has mostly to do with his background, the relationship between his work and Debreu's and the explicit reference – made in the press statement [6] announcing his Nobel Prize – to a "theorem" (theorems evoke Mathematics) as one of his most important contributions to social welfare theory. The main reason is, however, his exemplary intellectual attitude in facing questions of clear political and social relevance with an intimate need for logic coherence which is satisfied by introducing and promoting the use of all the necessary formal instruments, even very abstract ones.

Gerard Debreu

Debreu was born in Calais in 1912 and completed his university studies at the Ecole Normale Supérieure during the Nazi occupation of Paris. He studied mathematics and physics. As Debreu himself declared, the lecturer who had the most influence on him was Henri Cartan, but in general it was the whole bourbakist school of thought that influenced the development of his mathematical taste. Subsequently, between 1946 and 1948, he converted to economics under the influence of the volume *A la Recherche d'une Discipline Economique* that had recently been published by Maurice Allais (who too was to be awarded the Nobel Prize in 1988).

Later on, Debreu applied himself exclusively to economics (or mathematical economics) by taking more and more prestigious positions at important European and American scientific institutions. He was awarded the Nobel Prize in 1983 for the following reason: "having incorporated new analytical methods into economic theory and for his rigorous reformulation of the theory of general equilibrium".

This implicitly refers to the introduction into economic theory of mathematical analysis involving multifunctions, of convex set theory and of convex analysis. Today these tools, which go far beyond differential calculus, are part of the many tools in the "toolbox" of a good economic theorist. There is also an explicit reference to his main work [7] on general economic equilibrium.

The theory of general economic equilibrium extends the partial analysis of equilibria in which a single goods market is typically studied and which involves the simplifying (but unrealistic) assumption that it is not influenced by the markets of the other goods. Instead, to be more realistic, a general approach is needed in which all the markets influence one another and the economy is studied as a whole, with the aim to determine the simultaneous prices and equilibrium quantities for all the goods.

The first formulation of this theory goes back to Leon Walras in 1874, but it was only in the 1950s, thanks to a famous work by Arrow and Debreu [8], that it received its first formal confirmation of logical consistency through the proof of the existence of an equilibrium. In 1959 Debreu published [7], in which he summarised all his work from the 1950s and thus he also inevitably referenced Arrow's work on the same issues.

Having embraced the axiomatic approach, Debreu adds to it a fairly definitive theory known as general economic theory in the case of perfect competition. This is at least true of what was the most developed area at the time: in fact, issues related to the uniqueness and stability of equilibria are not touched upon. The universality, as well as the elegance, allowed by the axiomatic approach are explicitly acknowledged in the extensive motivation of the Nobel Prize [9] which lists several contexts in which this theory is applied. To us mathematicians, this is no news (on the contrary, it is our strong suit): good formal abstract theories are characterised by freedom from particular world interpretations that suggested them in the first place and are therefore applicable in different contexts. But, for the world of economic theory of the 1950s, this was a real innovation, if not even a revolution, which was to strongly influence future developments.

We will now very succinctly hint at the content of Debreu's monograph [7], limiting ourselves to the case of exchange economies. Most of this monograph was written with the aim of arriving at appropriate definitions for the concepts the theory deals with and that of studying the formal relations among them. At the end of this effort, its fruits can be gathered in the form of term definitions such as that of *abstract economy* and *equilibrium,* in the proof of the existence of at least one equilibrium allocation and in the exhibition of its proprieties of optimality.

An economy consists of:

- a finite set A of agents;
- a vector space V in which a vector represents a goods basket;
- a vector subspace P of the algebraic pair of V in which vectors p (prices) allow a comparison between the various goods baskets x through the numeric value $<x,p>$;
- for each agent, a triplet $\{V(a), e(a), \precsim(a)\}$ where: $V(a) \subseteq V$ is the set of the possible purchases of A, containing the goods basket which a may be interested in acquiring; $e(a)$ represents the initial endowment of a; $\gtrsim(a)$ is the preference relation over the baskets in $V(a)$.

Aiming to satisfy its own preferences, agent a takes into consideration prices p and demands the goods basket $d(a,p)$ which is maximal with respect to the relation $\gtrsim(a)$ in the set $B(a,p)=\{x \in V(a): <x,p> \leq <e(a),p>\}$. For simplicity, we will assume that the demand consists of only one vector.

Given an overall goods offer of e (sum of all the initial amounts $e(a)$), a price p is said to be of equilibrium if the total demand $d(p)$ (the sum of all the individual demands $d(a,p)$) is equal to e.

The three main points in [7] are: the proof of the existence of at least one functional equilibrium price, the proof of Pareto-Optimality of $d(a,p)$, when p is an equilibrium price and, finally, the acknowledgement that every Pareto-Optimal resource allocation $x(a)$ coincides with the system of individual demands $d(a,p)$, for an appropriate equilibrium price p in an economy in all respects similar to the initial economy, apart from the initial amounts, which are now equal to x(a).

After [7] there was another famous result by Debreu (and Scarf). This was the demonstration of Edgeworth's conjecture that, in economies with a large number of agents, when dealing with equilibrium allocations, it is possible to use a stronger optimality criterion than Pareto's: the core criterion. The correct formulation of the equivalence theorem is rather involved and therefore is beyond the scope of this book (although it is all but insurmountable). The reference to this work is justified by its importance as well as by the fact that it will be taken up again in the section dedicated to Kantorovich.

Kenneth J. Arrow

Just like Debreu, Arrow too was born in 1921. In New York, his native town, he studied mathematics (with, among others, Alfred Tarski) at Columbia University, obtaining a Masters degree in 1941. Later, under the influence of courses by Harold Hotelling, he started a Ph.D. in economics, which he was awarded in 1951. Part of the research he carried out for his Ph.D. thesis and for other early work took place at the Cowles Commission, where he met Debreu, and at the Rand Corporation. The latter is an environment which now appears magical due to the involvement of researchers such as R.Bellman, D.Blackwell, H.F.Bohnenblust, J.Milnor, J.Nash, P.Samuelson, L.S Shapley all the way up to the very special presence of J.von Neumann.

Arrow is now Emeritus Professor at Stanford University where he has taught economics, statistics and operational research since 1949 (with a ten year break at Harvard). He was awarded the Nobel prize in 1972 with the following justification: "for ... pioneering contributions to general economic equilibrium theory and welfare theory".

Of course, Arrow's contribution to the theory of equilibrium has largely a common basis with that of Debreu: although the two started from similar motivations, they did not quite develop the same approach. In particular, the American scholar is constantly motivated by applying his research. A comparison between the two personalities is beyond the scope of this contribution, which is meant to be a review rather than a thorough examination. Readers interested in the thorough examination of this worthwhile subject should read chapters IX and X in [10]. Here we will only point to a major contribution that Arrow has made to the development of a theory of uncertainty within the general economic equilibrium. We will also deal in more detail with his other most fundamental line of research, welfare theory, through his discussion of the impossibility theory that is believed to be *"perhaps the most important of Arrow's many contributions to welfare theory"* [6].

Although not technically difficult, in terms of technical tools required, Arrow's impossibility theory possesses great mathematical elegance and, we dare to say, sublime interpretive contribution. The field in which it operates is the study of acceptable mechanisms for the aggregation of individual preferences with the aim of constructing a social preference relation.

Given a set X of possible alternatives, let us suppose that each individual *a* of society A possesses an individual preference relation $>(a)$. A mechanism of preference aggregation is an application that transforms the preferences $\{\succ(a): a \in A\}$ into a single collective preference $>(A)$. The technical term usually employed in place of the aggregation mechanism is *social welfare function.*

Term aside, a social welfare function designates a very general concept whose acceptability depends on the realisation of particular rationality requirements. To begin with, we expect a "good" social welfare function to obey *unanimity*: if each individual prefers alternative x to y, then society too must prefer x to y. We can express this requirement with a very simple formula:

$$\cap_{a \in A} \succ(a) \subseteq \succ(A), \text{ for every } \{\succ(a): a \in A\}.$$

Slightly more technical, but of easy interpretation, is the requirement of *independence of the irrelevant alternatives.* This requirement is met by a social welfare function

$$\{\succ(a): a \in A\} \rightarrow \succ(A)$$

when, given two arbitrary alternatives x, y and two arbitrary individual preference relations $\{\succ(a): a \in A\}$ and $\{\rhd(a): a \in A\}$, if it is the case that, for each $a \in A$, the relations $\succ(a)$ and $\rhd(a)$ coincide on $\{x,y\}$, then we have $>(A) = \rhd(A)$, on $\{x,y\}$.

Social welfare functions independent of irrelevant alternatives and respecting unanimity constitute an acceptable model of individual preferences aggregation. Of course, the question is whether they exist. The answer to this question is simply affirmative, although social welfare functions display a disturbing behaviour that we have just termed "acceptable".

Given an agent b, we consider the application

$$\{\succ(a): a \in A\} \rightarrow \succ(A) := \succ(b). \quad (1)$$

It is obvious that this paradoxical aggregation mechanism (according to which the opinion of the whole collective A is that of individual b) is both independent of irrelevant alternatives and respectful of unanimity. However it is also obvious that, psychologically, we would like to have social welfare functions that do not smell of dictatorship. Arrow's impossibility theorem tackles the question of existence of good social welfare functions by establishing that these non-dictatorial functions exist only if a society is composed of an infinite number of individuals, or if this number is finite and society must decide only between two alternatives. In the socially much more significant case in which a society has a finite number of individuals and at least three alternatives from which to choose, there are no acceptable social welfare functions other than (1).

John F. Nash

If we leave out a few isolated publications, Nash's scientific contribution is concentrated between 1950, the year of his Ph.D. thesis on non-cooperative games in Princeton, and 1958, when he published the paper "Continuity of Solutions of Parabolic and Elliptic Equations" in the *American Journal of Mathematics.*

These are two very important papers in Nash's life (and for science in general). In fact, his brief doctoral thesis gained him the Nobel Prize in Economics, while the 1958 paper brought him very close to winning the Fields Medal and gave him the definitive recognition among his colleagues as a star of the first order in contemporary mathematics.

The Fields Medal was never granted to Nash who, in all truth, never managed to obtain a teaching chair. Nash's opinion [11] was that the Fields Medal was not conferred to him because of the 1957 publication by Ennio De Giorgi (a contemporary: both were born in 1928) in which the same results were obtained for the elliptical case. But Nash's results also covered the parabolic case and introduced techniques which were different from those introduced by De Giorgi, so there were no doubts as to the independence of the two works. It is much more likely that the Fields Medal was not granted to Nash because of his great difficulties in having ordinary relationships (whatever this may mean in the case of mathematicians) within his work environment [2].

Nash's production consists of seven papers in Economics and Game Theory (written in his early Princeton years) and of seven papers in pure Mathematics. In spite of this relative low number of publications and leaving aside the considerations that lead to him winning a Nobel Prize, he is considered one of the most genial mathematicians in the 20th century. A meaningful proof of the consideration he enjoyed from his colleagues is the *Duke Mathematical Journal* publication of two special volumes [12] bringing together essays by revered scientists written in Nash's honour which mainly testify the influence of his thought on all post -1958 research.

The 1995 publication can be misleading: in fact, as one of the curators, H. Kuhn, reminds us, the idea and the work for these volumes were started in 1993 and thus one year before the Nobel Prize. Therefore, the motivation for this work was not the Nobel Prize itself. Moreover, Nash's motivation, difficulties with relationships and the fact that he had been absent from active mathematics for 30 years should all be taken into consideration.

In other words, we would like to stress that the Duke volumes were not the tribute of friends (of whom he has very few) or students (of whom he had none): they were the homage of true fans appreciating his innovative ideas. Today, his ideas still have a key role in at least three fields: Game Theory, Geometry and Mathematical Analysis.

Nash was given the Nobel Prize in 1994 for his *"pioneering analysis of equilibria in the theory of non-cooperative games"* and in particular [13] for developing a concept of equilibrium known as the Nash equilibrium which has laid the foundations for all successive analysis and refinements.

Here too we notice the curious phenomenon that is at the centre of our argument. Just like in Debreu's case, we can say that the Nobel Prize in economics was really awarded to Nash in recognition of his work as a mathematician. In fact, Nash's contribution is purely mathematical in the traditional sense. If we simplify things further, it consists in having come up with an appropriate definition together with an appropriate theorem of existence. The fruitfulness of his idea is borne out by the work of the many scientists who later examined its most varied application.

A *game* (without the possibility of cooperation and with complete information for all of the *n* players) is a function:

$$f: S_1 \times \ldots \times S_n \to R^n.$$

Each player *i* simultaneously chooses a strategy s_i from the set of available strategies S_i. In this way, we obtain the vector of played strategies $(s_1, \ldots s_n)$ and its corresponding payment vector for each player. In particular, each player I receives the amount $f_i(s_1, \ldots, s_n)$ which, of course, the player aims to maximise.

The basic question is simple: is it possible to foresee the result of the game? In other words, is it possible to foresee what the combination $(s_1, \ldots s_n)$ of strategies played will be? With simple examples, which we will not list here for reasons of brevity, it is possible to highlight that the attempt to isolate a privileged behaviour of the players, on the basis of naïve criteria of rationality, is generally destined to fail. Even the use of so called mixed strategies, which allowed von Neumann to solve two-person zero-sum games, turns out to be insufficient. Nash's idea was to base his concept of solution on the criterion of a rational game described below.

A result $(s_1, \ldots s_n)$ is rational if, for each player *i*, the following property holds: as long as the other players do not change the strategies declared in $(s_1, \ldots s_n)$, player *i* has no interest in changing his or her strategy. It is intuitive that, if during the execution of a game, a rational result (in the sense just described) was obtained, players would have no reason to regret their choices and therefore, should the game be played again, each player would confirm his or her choice. It is this property of permanence of choices that leads to calling rational results "(Nash) equilibria".

In our opinion, this is not an elementary concept. Certainly, the proof of the existence of Nash equilibria is simpler. On the other hand, he did not receive the Nobel Prize for the proof of the theorem of existence (which only provided consistence to his idea of equilibrium), but rather for conceiving this notion of equilibrium.

Leonid V. Kantorovich

A rather different scientist (to Nash), but an equally powerful figure, is that of Kantorovich. Those who have studied functional analysis no earlier than the end of the 1970s will probably have come across him (although not his activity as an economist) by reading one of the several translations of his thick book on Functional Analysis, written together with Akilov.

While Nash's formation occurred in a tranquil provincial context [2], Kantorovich was born in Saint Petersburg in 1912 (died in 1986) and was thrown into the vortex of history and social unrest. As he himself remembers, one of his first clear childhood memories is the October Revolution.

While Nash met with difficulties in seeing his own talent formally recognised, Kantorovich met with brilliant success. He also appeared to be perfectly at ease in the social context of his time. In Kantorovich's case, the numbers speak for themselves:

- he enrolled at university at the age of fourteen;
- he left university at the age of eighteen and at twenty-two he already held a Chair in Mathematics in the same university;
- he wrote 300 scientific articles;
- he created a functional analysis school which is still active and influential today, even if it has spread across the world;
- he founded and managed a laboratory for the application of mathematics to economics within the prestigious Soviet Academy of Sciences;
- in 1965 he won the Lenin Prize;
- he was given several government posts, including managing the research office of the Institute for National Economic Planning, thus influencing the upper ranks of soviet bureaucracy in their work of managing and governing the country's economy.

The Nobel Prize in Economics was awarded to Kantorovich in 1975 for his *"contributions to the theory of optimum allocation of resources"*. This is a theory dealing with establishing the most efficient allocation of specified resources with respect to:

- the need to use them in a certain process
- their scarcity
- the possibility to employ them in alternative ways

Kantorovich came across this theory in 1938, but his passion for political economy and modern history (obviously due to living in his time) date back to a far earlier time. In USSR, the 1930s saw the consolidation of the soviet regime and of the collective effort aimed at the success of the socialist idea.

In mathematics, the general question of the utility of intellectual work produced a strong emphasis on applications. Kantorovich was given the following problem: a firm in the timber industry organises its production across several plants, each of which has its own technological (production and market) characteristics. The requirement is to find the optimal way to distribute the timber across the various production units so as to maximise the overall production, given some constraints.

Mathematically, this problem becomes a problem of maximisation of a linear function on a convex polyhedron. The main difficulty in arriving at the solution is not conceptual, but rather operational, due to the great number of variables involved. Having recognised the common mathematical formulation of a great variety of economic problems, Kantorovich tackled and managed to solve the problem of conceiving efficient techniques for solving the problem. His results led to a mathematical theory that we now call Linear Programming. This all happened several years before the same results were obtained by Dantzig, independently, in the United States.

In this case, too, therefore, the Nobel Prize is the recognition of an activity which is mainly that of a mathematician, although, obviously, of exceptional relevance because of its applications in the economic field.

To conclude, going back to H. Kuhn's paper on Linear Programming included in this volume, we would like to highlight how another mathematical idea closely linked to the name of Kantorovich plays a very important role in the most recent economic theory. We are referring to the notion of vector lattice associated with the theory of positive operators.

These lattices are vector spaces that are partially ordered and have the property of possessing a supremum for each pair of their elements, whereby they are also lattices themselves. Their usual name in the Western world is that of Riesz spaces, named after the Hungarian mathematician F. Riesz, who, during a World Congress of Mathematicians, in Bologna in 1928, was the first to add to the study of the interaction between algebraic and topological structures in function spaces also their interaction with ordered structures. Starting from the mid1930s, the theory was developed independently in the West (Birkhoff, Freudenthal) and mostly in the Soviet Union, thanks to Kantorovich's decisive contribution. Around this theory, Kantorovich created a real school of functional analysts. In USSR, vector lattices are known as K-spaces.

Today, vector lattices are a consolidated topic in pure mathematics, with dozens of monographs, a couple thousand papers and even a couple of magazines always well disposed towards it. Moreover, in economic theory, from the second half of the 1970s onwards, one comes across more and more models that make use of vector lattices. For instance, vector lattices provide an appropriate way to describe the deriving of sophisticated financial products derived from simpler, more basic tools:

$$(\text{F-kI})^{+} = (\text{F-kI})\text{v}0$$

The above refers to a European call option on a basic activity F with strike price k (where I is a non-risky activity which always pays, independently of what happens tomorrow).

The structure of a vector lattice can be tightly linked to deep results in Economic Theory. For instance, it was recently shown that Debreu-Scarf's equivalence theorem mentioned earlier is equivalent to saying that the price space P is a vector lattice. More specifically, given appropriate hypotheses, it is possible to prove that (even if V is infinite) equilibrium allocations coincide with Edgeworth's allocations, if and only if P is a vector lattice [14].

Bibliography

[1] http://www.nobel.se

[2] [2] Nasar S (1999) *A Beautiful Mind:* A Biography of John Forbes Nash, Jr, Winner of the Nobel Prize in Economics, 1994, Touchstone, Simon and Schuster

[3] Garding L and Hormander L (1985) Why is there no Nobel Prize in Mathematics?, *The Mathematical Intelligencer* 7:73–74

[4] Morrill JE (1995) A Nobel Prize in Mathematics, *Amer Math Monthly*, pp 888–891

[5] http://www.balzan.it, http://www.balzan.ch

[6] http://www.nobel.se/economics/laureates/1972/press.html

[7] Debreu G (1959) *Theory of Value*, Yale University Press, New Haven, CT

[8] Arrow KJ and Debreu G (1954) Existence of an Equilibrium for a Competitive Economy, *Econometrica* 22:265–290

[9] http://www.nobel.se/economics/laureates/1983/press.html

[10] Ingrao B and Israel G (1987) *La Mano Invisibile*, Gius. Laterza et Figli Spa, Rome 1987; English translation: *The Invisible Hand*, Economic Equilibrium in the History of Science, MIT Press, London 1990

[11] http://www.nobel.se/economics/laureates/1994/nash-autobio.html

[12] Kuhn HW, Nirenberg L, Sarnak P and Weisfeld M (editors) (1995) A Celebration of John F Nash Jr, Duke Mathematical Journal 81

[13] http://ww.nobel.se/economics/laureates/1994/press.html

[14] Aliprantis CD and Burkinshaw O (1991) When is the Core Equivalence Theorem Valid?, *Economic Theory* 1:169–182

Mathematics, Arts and Aesthetics

From Space as Container to Space as Web

Capi Corrales Rodriganez

Space & Time

Terms used in philosophy to describe the structure of nature. They are sometimes described as containers in which all natural events and processes occur, and sometimes, as relations which connect such events.
Collier's Encyclopedia

The two words "container" and "relations" describe, respectively, the idea of space held in eighteenth century and in contemporary mathematics. Two centuries ago, mathematicians associated space and the Physical Universe, and they believed – as many people still do today – that the model which describes the space as an inmense empty box is an exact reproduction of the universe around us. In fact, this space assumed to be a replica of the physical world was the only one considered in mainstream mathematics, and it could be described by the following properties: a *cubical* receptacle (i.e., with three dimensions, namely width, depth and height), *infinite* and *homogeneous* (which means that it has the same properties in all of its parts); it opposes *no resistance to motion*, and it is endowed with an *analytical* model (a system of coordinates) which gives mathematicians a reference to take measures from.

Nowadays, mathematicians describe a space (notice that it is no more "space" but "a space") as an abstract entity consisting of two things: a *set of objects* – any objects –, and a *web of relations* among these objects. As an example, a space could be formed by certain curves and relations among them; another by triangles and relations between triangles. We could say that the objects will be the bricks and the relations the cement to be used in the construction of mathematical spaces.

These two radically different notions of space reflect and give testimony of the change that has taken place in the way mathematicians think of space since, say, Newton (1642–1727), one of the first mathematicians to mention explicitly the word "space", up to Hausdorff (1868–1942). But, as Hegel so beautifully explained, the spiritual structure is such that no will, distance or special education is sufficient to obtain the isolation necessary to avoid being seduced by the reflection growing in the culture around us. This being so, the evolution of the notion of space in mathematics depicts a cultural change in the way of looking at space, in the "glance", we could say, that is found also in the different representations and models of space along time of many western painters, musicians or writers, to name a few. As example, the two paintings *Las Meninas* by Velázquez (1599–1660) and *Las Meninas* by Picasso (1881–1973), illustrate the definition of the Collier´s dictionary given above.

The process that took mathematicians from a space thought of as container to a space thought of as a web of relations has two parts. The first part consisted in getting rid of the constraint that supposed the use of the objects of Euclidean Geometry – lines, points, etc. – as only possible bricks of a mathematical space. This part of the process took place mainly during the nineteenth century, and by the beginning of the twentieth century, any collection of objetcs could be thought of as adequate set of bricks or "abstract points" with which to construct a mathematical space. Once we allow any set of objects to form the basic elements or points of a mathematical space, the second part of the work starts: we need to construct the tools that will allow us to build spaces using these elements as bricks. We need the cement, so to speak, the way of linking, of relating these objects among themselves so we, in fact, obtain a structure, a space out of them. A process we could call of arithmetization and which characterizes much of the mathematics of the twentieth century.

In these pages we will briefly describe the first part of this process, which takes us from the cubical container-space to the notion of a space as any byproduct of relations among objects as defined by Hausdorff in 1914. And we will illustrate the main changes taking place with the works of a few mathematicians and a few painters. This process could be described as taking place in several phases:

Precursors: Constructions in mathematics and representations in painting which challenge the notion of space as cubical container. Examples: cartographers and Velázquez.

Phase 1: Emergence of geometries other than Euclidean Geometry. Examples: Gauss and Goya.

Phase 2: Theoretical permission to conceive spaces other than Euclidean Space. Example: Riemann.

Phase 3: Development of the tools necessary for the construction of spaces whose elements are not necessary lines and points, or three dimensional, or reproductions of Physical Space. Examples: Cantor, Seurat, Cézanne.

Phase 4: Construction of the first abstract spaces. Examples: Ascoli, Volterra, Fréchet, Monet.

Phase 5: Final formalization of abstract spaces. Examples: Hausdorff, Kandinsky.

Precursors in Mathematics: From Newton to Euler

The first notion of a mathematical space (which, as has been mentioned, was supposed to reproduce the physical universe and had the shape of an inmense empty box), was weaved by the seventeenth and eitghteenth century mathematicians using essentially four threads: euclidean geometry, the philosophical and theological discussions on space, the arabian process of arithmetization of geometry, and the real numbers. As examples, Isaac Newton discusses in some of his letters written in the seventeenth century the properties of space (in the context of his philosophical debates and relating them with God's nature; the reader can find more on this in [Gr] and [Gry] in the bibliography), and, in 1748, Leonard Euler gives in the second volume of his *Introductio* two different systematic descriptions of three-dimensional polar coordinates. In fact, the appendix of this book is considered by many mathematicians as the first treatise of three dimensional analytical geometry written in the form of a text book.

This cubical space-container was the only space explicitly or implicitly assumed in mainstream mathematics for many centuries. But there were several contexts in which this choice had been challenged for a long time. Perhaps the most interesting one is that of the navigators and cartographers. All possible map representations pose the same mathematical difficulty: how to depict geographical accidents so that navigators, for example, can rely on the distances and directions appearing in the map. Besides their lack of knowledge about many countries, cartographers faced the difficulty of having to transfer the measurements taken on the surface of the Earth to their corresponding point on the map. They could find no way of doing it without great distortions. Navigators needed two features: that the "north" route from any point on Earth would appear as the straight up direction on the map, and that all compass directions would be correctly depicted with respect to the north direction. For example, a road going east-west should appear in the map as a horizontal line, while one going north-east should appear with a 45° angle. The first map respecting these two principles was drawn in 1569 by Mercator (Gerhard Kremer), a Flemish cartographer. Figure 3: Mercator's map.

Mercator's map has a drawback: in order to respect the angles, the further we are from the Equator, the bigger the distortion is. Could one draw a map of the Earth that would respect Mercator's map features, but without the distortion? The answer to this question, proven by mathematicians in the eighteenth century, is no. It is a geometric fact that if we do not want to sacrifice either feature of Mercator's map (a vertical north and appropiate directions with respect to such north), then our only possibility is to accept distortion.

Since non-distortion seemed to be incompatible with a fixed northern direction, the cartographers decided to turn around the question: could it be possible to draw a map without distortion and with a variable northern direction? For centuries cartographers tried, unsucessfully, to answer this question.

The problem was finally solved in the XVIIIth century by the mathematician Leonard Euler. In his article *De repraesentatione superficiei sphaericae super plano*, published in 1778, Euler proved that no portion of the Earth can be reproduced in a flat paper surface without distortion. Euler's theorem tells us that the perfect map does not exist. Depending on our purpose, one type of map reproduction will be more adequate than another. Euler's theorem forced cartographers and mathematicians to study spherical geometry and trigonometry as a subject of its own, independent of Euclidean geometry.

In Painting: Velázquez

As happened with the works of mathematicians, most of the paintings done by the beginning of the nineteenth century depicted space as a cubical container of the objects described in the scene. But, as we have just seen happened in mathematics, pictorial spaces which parted from, and challenged this, conventional huge box which contains but is unrelated to the objects, had started to emerge along the years.

During many years of its evolution, the european art had received instructions first from theology – didactic art – and then from politics – art which follows a normalized idea of beauty. This situation starts to change at the end of the XVIIth century – the period in which the mathematical idea of space starts to break its close relation with God's nature. At the begining of the XVIIth century we still find the artist "sustainer of the State" (and thus sustained by this), and it is in the XVIIIth century when, once artistic creation has been depatronized, a process of reflection starts within art. It is only natural that this reflection would take the artist, sooner or later, to analyze space, which is precisely, in our opinion, what characterizes the work of Velázquez: his profound and spectacular analysis of space.

Let us look back at his piece *Las Meninas*, a painting so much about space that we find represented in it the space which is not in the painting. The mastery of the standard rules – perspective rules, not mathematical rules – and tricks with which to construct an Euclidean Space is complete in this painting. It is impossible to stand in front of this painting without asking oneself where the eye describing the scene is placed. In the painter? In the girl? In the visitor appearing through the door at the back of the scene? Taking any of these eyes as origin of a coordinate system, the geometrical construction is impeccable. As it is, if the eyes looking are those of the king and queen reflected in the mirror at the back. A mirror that brings into the painting the space not in the painting, the space where we, the audience, stand.

Yes, this painting of Velázquez shows the perfection with which Euclidean Space could be graphically constructed in Newton's time. But it also shows something else: a relation between the space and what inhabits in it. Velázquez creates this relation in two ways.

On the one hand, he establishes a double relation between the object (scene to be represented) and the space around it (the place where the scene takes place), by introducing both his studio and the audience in the painting. The space around the scene when it was painted, and the space around the canvas each time it is looked at, are depicted in it. On the other hand,

> "Velázquez searches for the impression of things. Impression is formless, and accentuates the matter – silk, velvet, canvas, wood, organic protoplasm – with which things are made".
> "Is it licit to be surprised on hearing people of a certain formality calling Velázquez a Realist or Naturalist? With wonderful inconsequence they supress this way all the merits of Velázquez. Because if Velázquez had been mainly concerned with things, the *res* or nature, he would have been no more than a

disciple of the Flemish and Italians of the quattrocento. Those were the conquerors of things, of the *naturas* of things.
There is nothing more opposed to Realism than Impressionism. For this, there are no things, there is no *res*, there are no bodies, space is not an immense cubic receptacle. The world is a surface of luminous values. The things, starting here and ending there, are melted in a marvellous crucible, and start to flow into each other's pores.
Who is able to pick an object in a painting from the last period of Velázquez? Who is able to determine where a hand in *Las Meninas* starts and ends? We could perhaps dream of having between our arms the ivory and languid body of Mona Lisa; but that maid who hands the vase to the cesarean girl, is like a shadow, and if we tried to fetch her only an impression would be left in our hands". (José Ortega y Gasset; [OG] in the bibliography).

The neat distinction between space and the bodies inhabiting this space established with such precision by Aristotle in his definition of *topos* of a body – we recall it: "interior frontier of that which contains it" –, starts to disappear in Velázquez. His paintings, like those of the impressionists after him, introduce us into a space of things reflecting onto each other, flowing into each other and relating with each other. As the mathematicians of his time – Newton, for example –, who were already considering space as a geometric object itself.

Phase 1. Emergence of Geometries other than Euclidean Geometry.

In mathematics: Gauss

Up to the XIXth. century, the shape of the Earth had been deduced from the study of the sun and the stars. If, like Venus, the Earth happened to be perpetually covered by clouds, what would we do? This question was answered by Carl Friedrich Gauss, director of the astronomical observatory of Göttingen since 1807. In 1818 he carried on a study of the land covered by the kingdom of Hannover. Reflecting on the data obtained from direct measures on the Earth, Gauss noticed, and proved, that it would be enough to carry on certain specific measurements on the surface of the Earth to find out its shape, and that a similar process could be followed with any surface. Said differently, it is not necessary to "step out of" a surface – like observing the Sun and stars in the case of the Earth – to determine its shape. This is what the expression *intrinsic geometry of a surface* means: the geometry – shape – of a surface not only characterizes it, but it can also be described from the surface itself, without leaving it. Let us see this with the example of triangles.

It is a fact that in flat geometry the angles of any triangle add up to 180°. In triangles with their sides drawn directly on the surface of a sphere, the sum of the angles adds up always to more than 180°, while on a surface shaped like a saddle the angles of a triangle will add up to less than 180°. Now, let us suppose we want to determine the shape of a surface. We can triangulate it, and measure the size of the angles in each triangle. Where all of the triangles have angles which add up to 180°, our surface must have a flat or a cylindrical shape; in the regions of the surface where the triangles' angles add up to less than 180°, the shape will be similar to that of a saddle; and where the angles of the triangle add up to more than 180°, we will have a shape similar to a piece of sphere or ellipse. (Fig. 1).

Another fundamental property of flat or Euclidean geometry, known as the Parallel Postulate: if the angles α and β in figure 2 add up to less than 180°, then lines l and m should meet somewhere in the same side of the plane with respect to line n (or equivalently, given a line and a point not in the line, we can construct only one line which is parallel to the given one and goes through the given point). (Fig. 2).

In Spherical Geometry, lines are the maximal circles perpendicular to the equator, and they all meet at the north pole; hence, no parallel to any one of them can be constructed through a point external to it, and so, in Spherical Geometry the Parallel Postulate does not hold true. This fact rang a very particular bell for mathematicians. The basic constructions of Euclidean Geometry are the unlimited extension of lengths and the construction of parallels. If up to Riemann (1854), no one in mathematics ever questioned the possibility of extending a length beyond any limit, the debates on the construction of parallels started

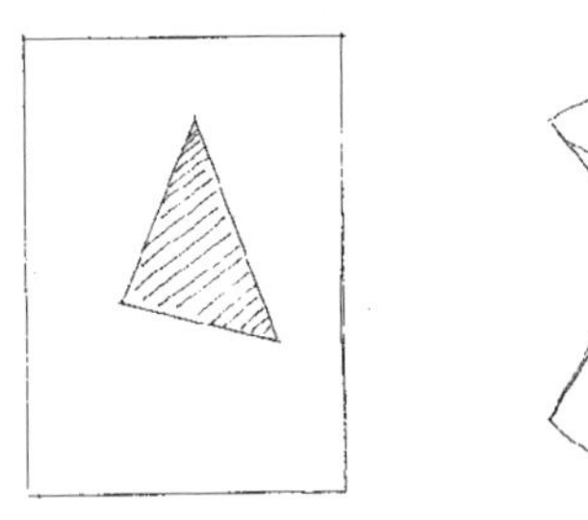
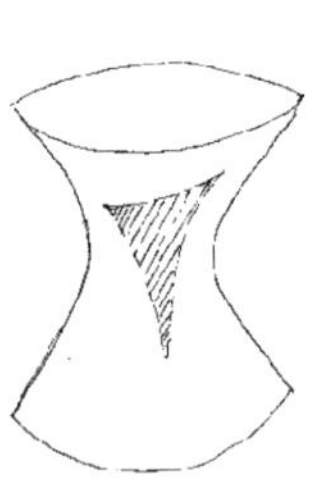
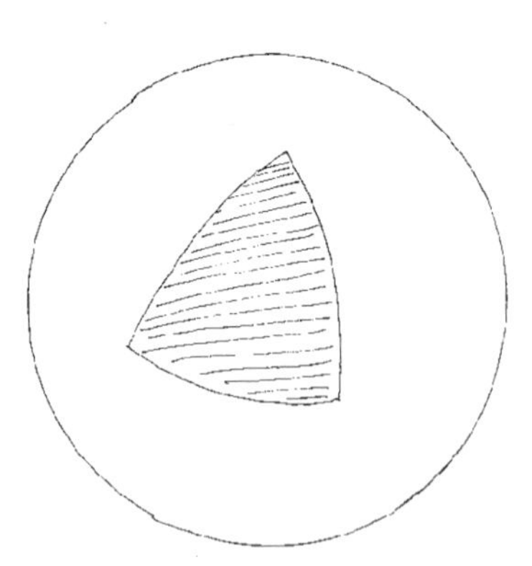

Fig. 1. Triangles on different surfaces

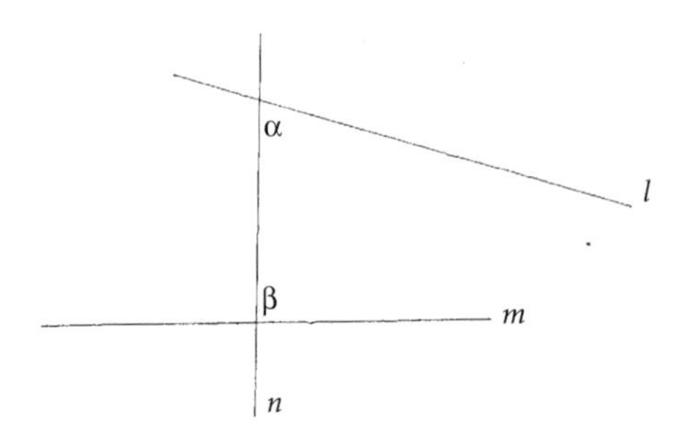

Fig. 2. Euclid's Parallel Postulate

Fig. 3. Asymptote lines

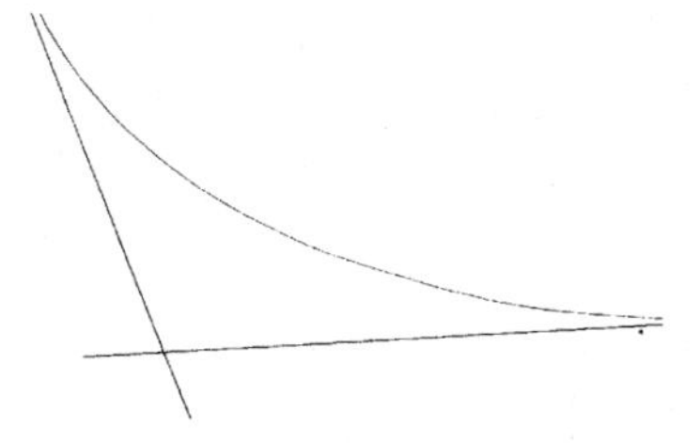

already in Euclid's time, because it assumes, without proof, that we know what happens to two lines in places infinitely far away from us. Far away, they argued, lines l and m in the previous figure could bend and behave like asymptotes, getting closer and closer to each other without ever touching. Why not? (Fig. 3)

At the beginning of the nineteenth century, the mathematicians Gauss, Bolyai and Lobachevsky, independently had the same idea. Let us suppose we have a geometry which is exactly like Euclid's geometry except for the Parallel Postulate; say that we accept the construction through a point not in a given line of more that one line parallel to the given line. After all, in Spherical Geometry there exists no such parallel line; why not play with the theoretical possibility of a geometry with more than one? When they developed their idea, they found that theoretically, mathematically, such a geometry was perfectly valid, and it was named *hyperbolic geometry*. With the geometries of Gauss, Bolyai and Lobachevsky, Saint Euclid fell, and the door was left open for the new spaces and geometries that soon started to visit the mathematical world.

Before we look at the work of Goya, an example of a painter contemporary to Gauss, let us see a plane graphic model of a "world" with a geometry that is hyperbolic, and which was developed many years later by Poincaré. This "world" consists of a half plane Ω limited by a horizontal line $\partial\Omega$ which is considered the "edge" of the world, or "the line at infinity". The characteristic feature of Poincaré´s half plane is that everything in it gets smaller as it approaches $\partial\Omega$, and larger as it moves away from $\partial\Omega$.

As a consequence of this (metrical) property, a person living in Poincare´s world would never be able to reach its edge $\partial\Omega$, because if she starts, say, one kilometer away with a certain size, by the time she is half a kilometer away from $\partial\Omega$ she will be half of her previous size, and as she gets closer and closer to $\partial\Omega$ she will be smaller and smaller, and no matter how small she might get, she will always be able to be half as small, so she will never touch $\partial\Omega$.

There is another graphic model of the Poincare´s half plane Ω. It is obtained by "closing" the edge line $\partial\Omega$ of the previous model into a circle (it is not difficult to think of a straight line like $\partial\Omega$ as a huge circle, with a radius so big that its graph looks to us as a straight line); Poincaré´s world will be the section enclosed by the disc. The metric of Ω can be easily understood by looking at Escher's drawings of different tilings in its second model. (Fig. 4).

Let us study the lines in Ω. We define a "line" as the shortest path between two points. With a little thinking one can deduce that the "lines" in Ω will look like

Fig. 4. Poincaré's model of the Hiperbolic Plane

semicircles perpendicular to $\partial\Omega$ (in the plane model they will also look like vertical lines, i.e., huge semicircles). How will parallel lines look like? They will be lines that do not meet in ?, and so two lines will be parallel both when they do not meet at all, or when they meet "at the edge of the world", at the infinity line $\partial\Omega$. Hence, both lines through the point P in the two models of next figure are parallel to line l, and so, Euclid's Parallel Postulate does not hold true in Poincaré's Hyperbolic Plane. (Fig. 5).

Non-euclidean geometries, like the one illustrated by Poincare´s half plane Ω (a more extense description of this plane can be found in [De] in the bibliography), remained for a long time as marginal constructions in mathematics. They were not integrated in the mainstream of research until after Riemann carried on his studies on space in the 1850's. Nevertheless, the first step had been taken: geometries that parted from and challenged Euclidean Geometry and its conventional cubical container had emerged in mathematics, and they would stay for good.

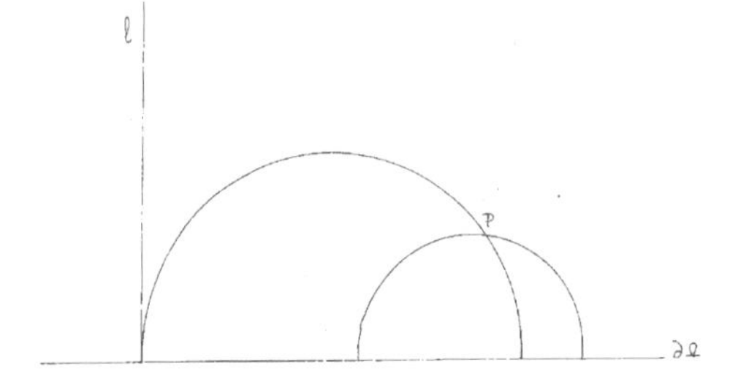

Fig. 5. Counterexample to Euclid's Parallel Postulate in Poincaré's Hiperbolic Plane

In Painting: Goya

"With the strokes of Goya, starts the evolution in painting that culminates with the dematerialization and decorporalization of the impressionistic textures, which Van Gogh would later describe as more music and less volume", writes Hofmann (see [Hof] in the bibliography). We will not repeat here what was already explained when describing the pre-Impressionistic space we find in Velázquez. Goya's paintings take Velazquez's challenges to the cubical container several steps further, and it is in two of these new steps that we will focuss:

We start to find paintings that represent spaces which are neither "fantastic" nor "real"; by "fantastic" we mean non plausible spaces, like, for example, those represented in the paintings of Hyeronimus Bosch; by "real" we mean those that reproduce a recognizable piece of the world around us.

We also find paintings in which no references extrinsic to the objects depicted are used to help the painter describe the action.

Let us explain these two aspects on concrete paintings. Among all the paintings of Goya there is one which is specially striking: *The buried dog* (1819).

Let us recall for a moment the paintings of Hyeronimus Bosch; in them, they consciously represent something which does not exist, and could not exist in the world as we know it. But we recognize the water, the trees, the grass. Goya´s painting is different. When we first see it, it does not strike us as something fantastic. We inmediately see there is a buried dog. But buried where? What type of space has painted Goya in *The dog?* Euclidean and with perspective? Euclidean and flat? Elliptical? In any of these spaces could this sand exist. If we decide that it is sand. It could also be a tree, or a piece of cloth. There is nothing in the texture of the matter burying the dog that allows us to identify it with exactitude. It is an imaginary matter, but not an unreal one. We could not say it is not something existing in nature and still we can not identify what exactly it is. The space the dog lives in is possibly not Euclidean nor spherical, but it is plausible that it is either; the matter burying the dog cannot be identified with precision, yet it is a plausible existing matter. This painting is not what we would call real, yet it is not fantastic either. It does not represent an identifiable piece of physical matter, but it can neither be considered a piece of fiction.

As we described before, Gauss' great idea was that of geometry as intrinsic to a surface: the environment surrounding the surface is not necessary to describe the surface. We find something similar in Goya's black paintings (1820–23).

In this series of paintings describing the war that was taking place around him, Goya places us directly before the human bodies, our only reference in his brutal depiction of the effect and desperation of war. In these paintings there are no soldiers, no weapons, no corpses. No "tomato". Like Gauss, Goya does not need the environment in order to describe clearly and precisely the brutality and tragedy we read in his paintings.

Phase 2. Theoretical Permission to Conceive Spaces Other than Euclidean Space.

Riemann

The contribution of Bernard Riemann to the process we are describing is essential and theoretical. The emergence of geometries different from the Euclidean one in the mathematical context, took him to analyse the restrictions under which mathematicians had worked so far and, as we will see, to pose a number of theoretical questions. As a consequence, he freed mathematics from the limitations imposed by the exclusive use of Euclidean objects as constituents of a mathematical space, and he introduced the theoretical possibility of using arbitrary mathematical objects instead. Let us describe his reflections in some detail.

Once geometries other than the Euclidean one are accepted as valid by mathematicians, the next natural step, taken by Riemann, is to accept also spaces different from Euclidean Space as valid ones. But once we have several mathematical spaces at our disposal, it is no more inmediate to decide which of them fits more adequately with the Physical Space. As Riemann points out, this decision depends both on our pre-conceptions about the Physical Space and about the experimental data at our hand. Why should Physical Space be a huge cubical container?, he argues. Why couldn't it be a huge sphere or a huge ellipsoid? If such were the case, given the size of the Universe we would perceive it unbending, as we basically do.

As Riemann noticed, if we think carefully about it, there is no evidence of the Physical Universe being actually unbending nor infinite. It could be spherical, and thus finite even if unlimited. This can be easily understood if we think of an ant walking on the surface of a sphere; it could go on walking forever and never reach a boundary, and still, it would be covering a finite region. Lines on an spherical Universe would be circles, and hence unlimited paths but finite.

Reflections similar to these led Riemann to take several fundamental steps, which he beautifully explained (with only one little formula!) in the memoire with which he entered the School of Philosophy in Göttingen (see [Ri] in the bibliography).

In the aspects that are of interest to our story, Riemann's contributions can be summarized as follows:

1. He differenciates between Physical Space and a geometrical space.
2. He realises that in geometry, line and point are not important notions; he suggests as essential, instead, the idea of position given through a metric.
3. He defines a general mathematical space to be an n-dimensional space, which will have as "points" arbitrary elements, and will be measured, if possible, with a metric (example: n-tuples). In this way, with Riemann, metric spaces are born.
4. Physical Space is only a particular case, an example, of a three-dimensional space. And it needs not be flat. It could be spherical or elliptical, for example. In such cases it would be unlimited, but finite.

The questions that Riemann posed made mathematicians realise that there was much work still to be done. For example, the notion of "dimension" needed to be defined. Also, in those spaces where we have a metric we know how to proceed. But, what about those cases where we know of no metric? What about the Abstract Spaces suggested by Riemann, whose elements are no more points or lines but objects of any type, integrals, functions and such? It was necessary to find a way of defining what position is, and comparing the position between two elements, when there is no metric. Riemann's reflections had suggested wonderful possibilities, but the tools to carry them on had still not been built.

Phase 3. Construction of Tools.

In Mathematics: Cantor – the Dimension Problem and Sets of Points –

In order to carry on Riemann's ideas, mathematicians needed, among other things, to define with precision the notion of "dimension". At the time of Riemann's death (1866), when mathematicians used the expression "the dimension of a space", they meant – intuitively – the number of independent coordinates necessary to determine the position of a point in such a space. For example, in a spherical space we only need two coordinates to determine the position of a point (we can call them longitude and latitude, as in the geometry on the surface of the Earth); hence, a spherical space has dimension 2. But this definition had never been made precise, and now it was necessary to do so.

In their correspondence of 1874, G. Cantor and R. Dedekind reflected on this intuitive idea of dimension, and both agreed that a set of dimension 2 should be bigger, in some sense, than a set of the same nature but dimension 1. The easiest way to check whether or not two sets have the same size is to pair up all of their elements in couples: if after doing so there are no elements left in either set, both sets have the same size.

Thus, Cantor and Dedekind assumed that, for example, one should not be able to establish a one-to-one correspondence between, say, a line (space of dimension 1) and a surface (space of dimension 2).

> But to their surprise – and of every one else's in mathematics –, in 1877 Cantor was able to construct a one to one correspondence between a segment and a square, a segment and a cube, and, in general, a segment (with one dimension) and a p-dimesional cube with p arbitrary. And that went against the intuitive idea of dimension.

After much thought, Dedekind ended up with the following conjecture (known now as *the theorem of the invariance of the dimension*): "if we are able to establish a one-to-one correspondence between the points of two sets with different dimensions, then the correspondence must for sure be discontinuous" (a correspondence is discontinuous if its graph is not smooth, but has holes or jumps, for example). This theorem was not proved by either Dedekind, Cantor or any of their contemporaries. After many years and many people trying it, it was L.E.J. Brouwer and H. Lebesgue who, independently and following different methods, succeeded in doing so in 1911.

In the different unsuccessful proofs of the invariance theorem he constructed, Cantor had to decide whether certain functions were continuous in all of its points. This requires one to determine whether it has or not "exceptional" points, points where the function has a behaviour exceptional for whichever reason. And, in order to characterize these exceptional points, Cantor had to define, with rigor and precision, notions such as continuous set, and interior, exterior, accumulation, isolated or limit point in a set, among others.

In Painting:

A. Sets of Points: Seurat.

In some painters of this period we find graphic examples of attempts to reduce the relations of depth (three dimensions) to relations in a surface, a line or a set of points (2, 1 or 0 dimensions) by means of points. A notorious case is that of Seurat. He chooses points, which have dimension 0, and – as in the case of Cantor – relations among points (color, density) to describe objects in three-dimensional space.

Let us look at Seurat's painting Poseuse de face (1886) and let us take any of the yellow points in the head of the girl. If we want to describe what characterizes this head point and makes it different from any of the yellow points in, say, the shoulder, we need to consider this point in relation to the points around it. For example, we could say that a yellow point in the head has mostly dark points around it, while a yellow point in the shoulder will have mostly yellow and white points around it. The precise formalization of descriptions like this one is an example of the type of questions Cantor was addressing on mathematical sets of points.

B. The Dimensions Problem: Cézanne

A graphic attempt to reduce the relations of depth to relations in a surface is found in Cézanne. He sees a three-dimensional object, and he wants to construct it on a two-dimensional canvas without illusions (perspective rules or color grading), and to do it without deformations. Of course, Cézanne fails: conformal representation in a two-dimensional surface of a three-dimensional object is impossible, as Euler had already proved. Nevertheless, in his attempts Cézanne succeeded in setting himself (and the painters that came after him) free from the tyranny of three-dimensionality. Cézanne achieves this dance between the bi-dimensional and three-dimensional, by using lines, colors and planes, and by following simultaneously several strategies which can be seen in most of his paintings like his series on *Mont Saint Victoire* (1904–1906) and *Château Noir* (1900–1904);

i) We rarely find continuous lines drawn around an object. With his strong colors and strong black lines, Cézanne gives depth to the painting, and he obtains volumes, three-dimensionality. But then there are many places in which the lines get lost, dissolved in the background or simply are drawn in pieces. These dissolving and disappearing lines contribute an element of bi-dimensionality. Both aspects are constantly combined in the space of the painting.

ii) Vertical planes, moving towards the end of the scene, take us back to flat paintings. With lines he constructs planes, with planes volumes. A strategy we already find in early painters like Giotto.

iii) Additionally, Cézanne never tries to describe the matter the objects are made of. He gives them all the same flat texture, and sometimes, as we see in *Mont Sante Victoire,* even the same type of strokes and colors (he distinguishes the mountain and the sky by a few discontinuous black lines). Prior to these times, painting tried to adapt itself to the touch and texture of the different objects, mantaining the hard as hard, the soft as soft, the untouchable as untouchable. This is very related to establishing the diference between object and background, between space as something exterior and object as that which lives in space. Gradually, in painting, like in mathematics, "bodies" and "intermediate spaces between bodies" are considered equivalent.

iv) We also observe in Cézanne's paintings that sizes never vary according to perspective rules. Depth, three-dimensionality, is achieved by compensating volumes in such a way that it respects the bi-dimensionality of the painting: varying the distance between vertical planes, Cézanne creates depth, tension and rhythm, always in relation to the plane, the bi-dimensional.

Phase 4. Construction of Abstract Spaces.

In Mathematics

The works of Cantor had many deep consequences. Among other things, they took a definite step towards effecting a divorce between mathematical space and spacial intuition. In his works on sets of points he developed the necessary tools to handle ¨abstract points¨, and this allowed the emergence of spaces whose "points" were elements of quite diverse nature: curves, functions, integrals,... As long as one can define a notion of distance or proximity among the elements of a set, one can construct a space with them.

Among the first mathematical examples of spaces whose points are not arithmetic points we find those constructed by Ascoli and Volterra. In 1883, G. Ascoli considers spaces whose "points" are curves (see [As] in the bibliography). In 1887, V. Volterra constructs spaces whose "points" are functions ([Vo] in the bibliography).

With Volterra, the modern notion of mathematical space (suggested by Riemann), reaches its correct definition: "space" is any set of objects and relations among these objects (the relations will tell us how to measure distances or compare positions between the elements). And the mathematician can freely choose both the objects and the relations among these objects that he wants to study. Volterra uses a correct definition of abstract space, but he doesn't formalize it, he doesn't give it explicitly. The first axiomatic definition of an abstract space (as a set of arbitrary elements among which we establish a proximity relation) is established by Fréchet ([Fr] in the bibliography); the tool he uses to establish proximity among elements of a space is the notion of limit. A couple of years later, in 1908, F. Riesz gives a new axiomatic definition of abstract space, with the idea of condensation point as basic tool ([Ris]).

At this stage of the process, all the proximity relations that mathematicians define in their spaces are still quite attached to the notion of "distance", and hence to an Euclidean vision of the structure within the space.

In Painting: Monet

If the mathematician can freely choose the objects and relations he wants to study, the same happens with the painter. Thus, we find, for example, the *Impressionists*, studying the relations among colors, among strokes, among lights.

If the mathematical spaces of this period are still linked to an Euclidean structure, the pictorial spaces are still linked to nature, to the sensorial world around us. The Impressionistic paintings represent still landscapes, people, flowers, animals. But they are spaces like Cézanne's, spaces which make no distinction between the objects and the setting of the objects. The same strokes, the same textures are used for air, water, matter and bodies.

If the mathematician can freely choose the objects and relations he wants to study, the same happens with the painter. Thus, we find, for example, the Impressionists, studying the relations among colors, among strokes, among lights.

If the mathematical spaces of this period are still linked to an Euclidean structure, the pictorial spaces are still linked to nature, to the sensorial world around us. The Impressionistic paintings represent still landscapes, people, flowers, animals. But they are spaces like Cézanne's, spaces which make no distinction between the objects and the setting of the objects. The same strokes, the same textures are used for air, water, matter and bodies.

Abstractly speaking, Ruskin's discovery refers to giving up that shape that painted or drawn imitation of reality must portray in order to keep its immediacy. The Impressionists carried this giving up to the extreme: they try to take to the canvas only that which is perceived by the spontaneous visual act, guided by no knowledge. An impression, a primordial and original image with as much innocence and immediacy as possible. And they do it by scratching the painting with color strokes. And each stroke is nothing but a stroke, a spot of pure color associated to no shape.

Phase 5. Final Formalization of Abstract Spaces

In Mathematics: Hausdorff

With Volterra, Riesz and Fréchet, abstract spaces were defined in all rigour. Unfortunately, we already mentioned that all the definitions given were still attached to an Euclidean idea of distance or proximity. The final step was taken by F. Hausdorff in 1914 (see [Ha] in the bibliography):

> A *topological space* – or abstract space – is a set E formed by elements x, together with certain subsets U_x of E associated to each x; the subsets U_x are called *neighbourhoods of* x, and they are subject to the following conditions:
> a) Each point x is associated with at least one neighbourhood Ux. Each neighbourhood Ux contains the point x.
> b) The intersection between two neighbourhoods of a point x contains a third neigbourhood of x.
> c) If y is an element in U_x, there exists a neighbourhood U_Y of y contained in U_x.
> d) If x and y are two different elements in E, then there exist respective neighbourhoods U_x and U_Y without common elements".

The notion of neighbourhood is no more based on an Euclidean intuition of space. We can certainly define the neighbourhood of a point by means of the usual Euclidean distance, but we can also do it by means of any other mathematical relation. Hausdorff's definition is the starting point for the theory of abstract spaces – called also topological spaces – a theory which in mathematics is know as Set Topology or General Topology.

In Painting

At the time when Hausdorff was working on abstract spaces, Kandinsky was painting his watercolors entitled "Abstractions".

Kandinsky's watercolor, as was the artist's intention, represents an abstract space, since there are in it no forms nor objects linked to a representation of the material world of nature. But it is not, viewed with the eyes of mathematician, an abstract painting, since it is not a piece which is the fruit of a process of abstraction on material reality, as are the paintings of Cézanne or Mondrian. This difference is easily described in spanish (or italian), a languages which has two different expressions for these activities: "hacer abstracción a partir de la realidad" (to abstract from reality), and "abstraerse de la realidad" (leave reality aside).

Perhaps we could introduce a new word for paintings like those of Kandinsky: *Abstractism*. We would then say that an abstract painting is a painting which is the product of a process of abstraction done on a piece of the world around us, and painters like Cézanne, Mondrian, Matisse or Picasso, would be called abstract painters. On the other hand, the Kandinsky of this watercolor and painters like

Pollock, Malevitch or Tapies would be Abstractists. The reason why we make a point of stating clearly the difference between abstractists and abstract paintings, is that the intellectual challenges posed by the abstractists are not useful as illustrations of the ideas about space we have reflected on in these pages.

Bibliography

[As] Ascoli, G., "Le Curve Limite di una Varieta data di Curve", Atti della Reale Accademia dei Lincei, Roma 1883.

[Co] Corrales Rodrigáñez, C., "From container space to web space in mathematics and painting", Ediciones Mob Coop, Madrid 1999 (to appear).

[De] Dedò, M., "Transformazioni Geometriche", Decibel editrice 1996.

[Euc] Euclid, "The Elements", english edition by Heath, Dover Publications.

[Eul 1] Euler, L., "Introductio"

[Eul 2] Euler, L., " De repraesentatione superficiei sphaericae super plano", 1778. Complete Works Volume XXVIII.

[Fr] Fréchet, "Sur quelques points du Calcul Fonctionnel", Rendiconti del Circolo Matematico de Palermo 22 (1906). 1–74.

[Gold] Goldstein, C., Gray J., Ritter J., editores, "L'Europe Mathématique", Editions de la Maison de sciences de l'homme, Paris 1996.

[Gr] Grant, E., "Much ado about nothing: Theories of space and vacuum from the Middle Ages to the Scientific Revolution", Cambridge Univ. Press, 1981.

[Gry] Gray, J., "Idea of Space", Oxford University Press, 1992.

[Ha] Hausdorff, F., *Grundzüge der Mengenlehre* , 1914,

[Ho] Hofmann, W., "Grunlagen der Modernen Kunst", Alfred Krönen Verlag, 1987.

[Lo] Loran, E., "Cézanne's Composition" (1943), Univ. of California Press 1963, 3ª ed.

[OG] Ortega y Gasset, J., "Papeles sobre Velazquez y Goya" (1950), Alianza Ed., 1987.

[Pro] Proclo, "A commentary on the first book of Euclid's elements", translated by G.R. Morrow, 1970, Princeton U. P., New Jersey.

[Ra] Rasched, R., "Entre arithmétique et algèbre; recherches sur l'histoire des methématiques arabes", Les Belles Lettres, Paris 1984.

[Ri] Riemann, B., "Ueber die Hypothesen, welche der Geometrie zu Grunde liegen" (Habilitationschrift, Göttingen 1854), translated to french by J. Hoüel (1870), in *Oeuvres mathématiques de Riemann*, Blanchard, Paris 1968.

[Rs] Riesz, F., "Stetigkeit und abstrakte Mengenlehre", IV International Congress of Mathematics, Roma 1908.

[Vo] Volterra, V., "Sopra le Funzioni che Dipendono da altre Funzioni" and "Sopra le Funzioni da Linee", Atti della Reale Accademia dei Lincei, Roma 1887.

[We] Weil A., "Introduction to volume I of the complete works of E.E. Kummer: Contributions to Number Theory", Springer-Verlag.

Irrational Geometry

ACHILLE PERILLI

Translated from the Italian by Emanuela Moreale

My relationship with mathematics – that world that somehow controls rationality as its supporting structure – has always been a bit paradoxical. This relationship developed a bit more when – more or less thirty years ago – I decided to start a creative exploration concentrating on perspective. By analysing the reasons that regulate it, I was exposing its own essence, its being a sensational falsification of reality imposed by a constructive system of knowledge. "Investigation into perspective" was my first manifesto, published in July 1969. In it I stated: "For centuries, perspective has been a classification of the world carried out with tools that are not real, not matching and not true", to which I added, "Perspective is the most repressive form of imagination that a ruling class can come up with". This announced a new creative method for my new work: "and it is this artificial category – which we can more or less call perspective – which my analysis concentrates on: it attempts to absorb from optics those elements that are thought to be certain, but which I falsify through a series of interferences at the hands of other values (colour, tone, sign, structure)".

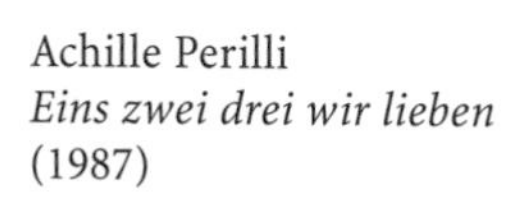

Achille Perilli
Eins zwei drei wir lieben
(1987)

Achille Perilli
Que viva Villa
(1988)

Mondrian wrote: "Art is a game and games have rules". Whenever artists open a route to visual knowledge, they start a process that is not based on the values of logic, but obeys the laws of imagination. Because of this, having understood that I was entering a paradoxical world through experimentation and being tempted to let the two "faces of the moon" – rational and unconscious world – live side by side, I tried to define what I called "the theory of geometric irrationality".

This theory originated from the thought that it is possible to think of a geometric form that is no longer determined by the laws of calculus or optics, but by the slight slipping and sliding that memory produces on visual perception data. The concept of geometric irrationality was starting to take shape. It was the start of my work on the dissolution of the geometric form, on the loss of weight, on the escape from pictorial bidimensionality with the shift from visual perception to memory, so as to produce a fundamental change. I tried to clarify this concept in the 1982 manifesto "Theory of geometric irrationality":

> If this happens in a concentrated and taut space such as geometric space, tension ends up replacing rigour, certainty and confidence. The shape turns into a field of rapid movements, furious fights and incredible deformations that define brand new, complex structures regulated by "other" laws which I had defined – in an earlier manifesto in 1975 – as "Machinerie, ma chère machine". From the relationship between two geometric modules – at times in conflict, at times similar but slightly different – a sequence is born that tends to shift in space until it starts growing from painting to painting: it starts developing and involving more elements in the search … even the reading of these sequences proposes journeys that shatter the idea of central space, surface space or light space.

Achille Perilli
Il morso del cristallo
(1994)

Achille Perilli
La solenne trincata
(1997)

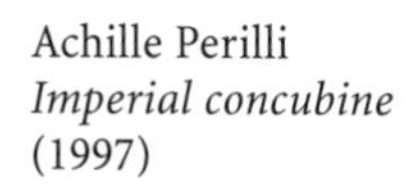

Achille Perilli
Imperial concubine
(1997)

It is an operation that tends towards expansion, not reduction, and continuously shifts the search from perception to mind, refusing any minimisation of problems related to vision: on the contrary, it keeps dilating them until it intervenes on still-unknown places between code and code, involving alien linguistic structures. Therefore, the work takes place in those territories that are under the influence of a law, but are not subject to it: they feel its presence and reality which are perceived with the utmost ambiguity of meaning, such that any possible deformation and transformation is possible. Only then can creative imagination have the chance to build a new utopia: the non-geometric form.

This has given me the chance to interfere with a three-dimensional space, by devising various and different theatrical experiences and, in the last few years, by realising sculptures that integrate geometrical elements in an organic nature.

Achille Perilli
Apollo e Dafne (1997)

Achille Perilli
La voglia di Giano Bifronte

The latest experience I have devised in the last few years is called "the trees": by operating on the surface of an organic structure like a tree or a wood fragment, by following its relief, avoiding the knots, highlighting the grain, I can insert a structure of complex geometries by moving them between full and empty spaces to arrive at that complex communication that is the fundamental theme of my work. The bet consists in bringing into three-dimensional space what I have managed to accumulate as images on the painting surface, so as to penetrate the deepest mystery of our visual world.

I was able to gather all these materials, fruit of thirty years of research, in a show entitled "De Insana Geometria" that took place at the Mole Vanvitelliana of Ancona in December 1998.

A last consideration: all this work to discover the new could find a further confirmation of the numerous possibilities it offers if it were possible to use all that modern technologies offer. But it is a field that can be made accessible to experimentation only by inserting it in an advanced didactics of a future school that can go back over the route opened by the "Bauhaus" in the Thirties and advanced by the Ulm "Hochschule für Gestaltung" in the Fifties.

On the occasion of the "Matematica e Cultura 1999" conference, the volume "Achille Perilli – 12 Acqueforti e Acquetinte" (Achille Perilli – 12 Etchings and Aquatints) – was published by the Centro Internazionale della Grafica, Venice, 1999.

Mathematics and Cinema

Moebius Strip: From Art to Cinema

Michele Emmer

Translated from the Italian by Emanuela Moreale

Moebius's Subway

In 1958 Clifton Fadiman edited the collection of mathematical stories *Fantasia Mathematica* [1]. One of the stories it contained – written by A.J. Deutsch – was entitled A *Subway named Moebius.*

August Ferdinand Moebius (1790–1860) was a great German mathematician who lived in the first half of the nineteenth century. It was he who, in 1858, first described a new surface in three-dimensional space that is now known as *Moebius Strip* (Figure 1). In his work, he explained a very simple method for constructing the surface named after him: take a rectangular strip of paper and proceed as follows: while holding one end of the strip still (for instance AB), operate on the other side (CD), by giving it half a twist (180°) along the horizontal axis of the strip, ensuring that A meets D and B meets C.

This surface has interesting properties. One consists in the fact that, if one follows the surface along the longest axis with a finger, one notices that one gets right back to the starting point, without having to traverse the border of the strip. In other words, the Moebius Strip has only one side, and not two, an external and an internal one, such as a cylindrical surface, for instance. Should one want to paint the surface of the strip proceeding along the horizontal axis, it is possible to colour the whole surface without the brush ever leaving the surface and without it crossing the surface border.

The idea of the story contained in Fadiman's collection was that the subway system in Boston had become so complex that it was possible for a train to get lost in the subway meanders, ending up in an infinite loop, in a Moebius ring. The mathematician in the story formulates the hypothesis that the subway system is

Fig. 1. Moebius AF (1886) *Zur Theorie der Polyeder und der Elementarverwandtschaft,* in *Gesammelte Werke,* vol 2, Leipzig, p 515.

a network of great topological complexity, with an extremely high number of connections. This is why train number 86 Cambridge-Dorchester had disappeared on March 4th!

Moebius: the Film

In the mid-Nineties, the young Argentine director Gustavo Mosquera R. had the idea of utilising Deutsch's story for a film to be realised by the final year students of the *Universidad del Cine of Buenos Aires* (Figure 2). The film was completed in 1996 and was an immediate success. It was awarded prizes and was invited to many cinema festivals.

Mosquera told me that he had wanted to make a film in which many different things converged [2]: first of all the subway train which disappears with all its passengers in Buenos Aires is easily linked to the terrible stories of the desaparecidos during the military dictatorship. But the exposition does not want to be too realistic: the audience might well not grasp this level of meaning. Secondarily, the great complexity of the subway network is remindful of meanders, labyrinths and therefore Borges. (The evening when the film was shown in Venice, a congress on Borges was being held in another Venice University building!)

The film features a meeting with a blind elder as well as an underground station named *Borges.* At the same time, the protagonist – a mathematician who is an expert in topology and wants to solve the mystery of the disappearance – is called Pratt. Is the reference casual? Not at all! Hugo Pratt had spent many years

Fig. 2. Mosquera RG (1996) *Moebius*, film, Buenos Aires; film poster.

in Argentina; the protagonist bears his name, has his sense of adventure, his desire to solve enigmas and mysteries. Even Pratt – in the book *Corto Maltese: Tango,* which is set in Buenos Aires [3], a thriller that Corto attempts to solve – calls *Borges* one of the train stations that the protagonist uses.

The film was made on a tight budget and, like so many other times in the history of cinema, the director has made up for the lack of funds with his imagination and ability. In order to produce the effect of the ghost train moving along the tunnels, Mosquera adapted an old movie camera dating from 1926, which was used to take one frame at a time of the real images of the passing of the train. This created a very spectacular effect, never before seen at the cinema.

A mysterious story, in which mathematics is mysterious, has its own life and does not allow itself to be understood even by mathematicians. A film without an end, with that train dashing in the tunnels, with no beginning and no end, like a Moebius strip.

Moebius and Art: Max Bill

There is no doubt that one of the clearest expositions of the possibility of a mathematical approach to art was formulated by artist Max Bill in a 1949 article. Bill was not satisfied with the way the title of his article had been translated from German into English. So in 1993, on the occasion of the publication of a collection of essays on the relationship between mathematics and art, when he republished the article as an introduction to the volume, he gave it a new title that seemed closer to his thought in those years: "The Mathematical Way of Thinking in the Visual Art of Our Time" [4,5] (Figure 3).

The breakthrough happens at the beginning of the Twentieth century: "The starting point for a new view is probably due to Kandinsky, who in his 1912 book *Ueber das Geistige in der Kunst,* laid the foundations for an art in which the artist's imagination was to be replaced by mathematical view" [6]. Then it was Mondrian who went the farthest from the traditional view of art.

He wrote

> that neo-plasticism has its roots in cubism and can also be called abstract-real painting, because the abstract (like mathematical sciences, but – unlike them – without reaching the absolute) can be expressed in painting by a plastic reality. He believed that it is a composition of coloured rectangular planes that expresses the most profound reality, which it reaches through the plastic expression of the relationships and not through its natural appearance... He also added that the new plastic art poses its problems in aesthetic equilibrium and expresses the new harmony in this way [7].

It is Bill's opinion that Mondrian exhausted the last possibilities that were left in painting. Which roads should be followed to enable a future evolution of art? The return to the old and already known, or rather the exploration of a new basic theme and, in this case, what could this new theme be?

Fig. 3. Emmer M (1993) (ed) *The Visual Mind*, MIT Press. On the cover, *Variation 12* by Max Bill, 1937.

> I am convinced it is possible to evolve a new form of art in which the artist's work could be founded to quite a substantial degree on a mathematical line of approach to its content. ... It is objected that art has nothing to do with mathematics; that mathematics, beside being by its very nature as dry as dust and as unemotional, is a branch of speculative thought and as such in direct antithesis to those emotive values inherent in aesthetics ... yet art plainly calls for both feeling and reasoning.

He went on to say that thought allows ordering of emotional values so that they can let art emerge from them. How can mathematics be useful to an artist then? Bill replies:

> Just as mathematics provides us with a primary method of cognition, and can therefore enable us to apprehend our physical surroundings, so, too, some of its basic elements will furnish us with laws to appraise the interactions of separate objects, or groups of objects, one to another. And again, since it is mathematics which lends significance to these relationships, it is only a natural step from having perceived them to desiring to portray them.

Here is a definition of what a mathematical view of art must be:

> It must not be supposed that an art based on the principles of mathematics, such as I have just adumbrated, is in any sense the same thing as a plastic or

pictorial interpretation of the latter. Indeed, it employs virtually none of the resources implicit in the term pure mathematics. The art in question can, perhaps, best be defined as the building up of significant patterns from the ever-changing relations, rhythms and proportions of abstract forms, each one of which, having its own causality, is tantamount to a law unto itself. As such, it presents some analogy to mathematics itself where every fresh advance had its Immaculate Conception in the brain of one or other of the great pioneers.

Max Bill has called his own sculptures in the shape of a Moebius strip *Endless Ribbons* and, in his article *Come Cominciai a Fare Superfici a Faccia Unica* (How I started to make singleside surfaces), he told of the occasion on which he had discovered Moebius surfaces. He said that

> it was Marcel Breuer, his old Bauhaus friend, who was really responsible for his singleside sculptured. It happened in 1935 in Zurich where, together with Emil and Alfred Roth, he was building those Doldertal houses that were so famous at the time. One day, Marcel told Max Bill that he had been asked to build a model for a show in London: this was to be a model of a house in which everything, even the fireplace, was electric. It was clear that an electric fireplace that produces light but has no fire was not the most attractive object. Marcel asked Max Bill if he wanted to make a sculpture to put on it and Max started looking for a solution, a structure that could be placed over the fireplace and that possibly could turn in the ascending draft and, thanks to the shape of its movement, act as a flame replacement. Art instead of fire! After a long time experimenting, Max Bill found a solution, which he thought reasonable

The most interesting thing to notice here is that Bill thought he had found a completely new form. Even more interestingly, he had found it by playing with a strip of paper, just how the astronomer and mathematician Moebius had discovered it many years earlier! In the film we made together in his studio in

Fig. 4. Bill M (1986) *Kontinuität,* granite.

Zurich in 1978, Bill repeats the hand gesture in the air that is meant to give an idea of a surface moving in space. It is a gesture that expresses the artist's creativity much better than many words. Unfortunately, Max Bill did not allow me to film the small cupboard that he had in his studio, in which he kept scores of paper strips which he played with and studied in order to realise great topological sculptures which he then realised in stone or metal. He feared that other artists, having seen his models, would copy his ideas!

Bill had the idea of setting up a hall with his topological sculptures for the topology section of the permanent mathematical exhibition of the London Science Museum. Unfortunately, this project was never carried out.

In 1947, Bill created a ten-foot tall version of the sculpture named *Kontinuität:* made of steel and covered in cement, it was placed in the park along the shores of the lake in Zurich. In 1948, it was destroyed by vandals. *Kontinuität* appeared in the press all over the world where it was described as a one-side surface sculpture.

In 1947–48, Max Bill discovered that *Kontinuität* did not have a single-side surface, so in 1986 he made it again in Frankfurt [9] (Figure 4). This time, it was placed opposite the Department of Mathematics, a more appropriate place for the surface that was made out of Carrara marble. In a volume of the 1986 *Kontinuität. Granit-Monolit von Max Bill,* all the phases of the creation are documented photographically: from the choice of the stone, cut, transport to the realisation proper. Although he continued to live in Zurich, Bill never rebuilt his Zurich sculpture. Is it possible that it was the topological structure that disturbed the sleep of the Zurich population?

Bibliography

[1] C. Fadiman (1958) *Fantasia Mathematica*, Simon & Schuster, New York

[2] G. Mosquera R. (1996) *Moebius*, film, Buenos Aires

[3] H. Pratt, *Corto Maltese: tango* (1998) Lizard edizioni, Roma

[4] M. Bill (1993) *The Mathematical Way of Thinking*, in: M. Emmer (ed) *The Visual Mind: art and Mathematics*, MIT Press (5th ed), Cambridge, USA. The image on the front cover of the volume is a reproduction of a work by Max Bill

[5] *Die mathematische Denkweise in der Kunst unserer Zeit*, Werk, n 3, 1949. Reprinted in English in Emmer M (ed) (1993), *The Visual Mind: Art and Mathematics,* MIT Press (5th ed), Cambridge, USA

[6] W. Kandinsky (1968) *Concerning the Spiritual in Art,* translation, original German edition by R. Piper & Co., Munich 1912

[7] P. Mondrian (1975) *Le Néoplasticisme: Principe Général de l'Équivalence Plastique*, Editions de l'Effort Moderne, Paris, 1920. English version : "Neo-Plasticism: the General Principle of Plastic Equivalence, Plastic art and Pure Plastic Art" in: Harrison C (ed), Wood P (ed), *Art in Theory,* pp 287–290 and 368–374

[8] M. Bill (1977) *Come cominciai a fare le superfici a faccia unica*, 1972, Italian edition Quintavalle AC (ed), Max Bill, Catalogo della Mostra, Parma, pp 23–25

[9] W. Spies (1986) *Kontinuität. Granit-Monolit von Max Bill*, Busche Verlagsgesellschaft m.b.H., Dortmund

A Few Reflections on the Creation of "Moebius"

GUSTAVO MOSQUERA R.

The film is drawn from a short story, entitled "A subway called Moebius" published in 1950 in a magazine of the Boston University, in which the author, A. J. Deutsch proposes Moebius' topology as the apparent cause of the disappearance of an underground train, filled with people, that was travelling in the intricate network of galleries in the underground of the North American city. This disappearance provokes endless search operations by the authorities that are in charge of the underground trains, who do not have adequate economic resources, and therefore entrust a young topologyexpert with the matter, as they believe that he has the mathematical know-how that is necessary to find thesolution. In fact, at the end of the story the young protagonist really meets the train that had disappeared and informs the authorities, but this takes place just as a second train disappears leaving the reader with an open ending and an unanswered question.

When I read the story, the initial idea fascinated me immediately, even if I was well aware of the fact that the story was too short for me to be able to use it in it's original form, to produce a 90 minute film. And I felt it was quite obvious that the construction of a similar intrigue, without a satisfactory final solution could be suited in the case of a short story, but not for a film because it would leave the public with the feeling of a film without a conclusion – which was the first problem that had to be solved.

The second difficulty instead, consisted in the fact that this story, which is only twelve pages long, was insufficient even as only a plot for a full-length film: even if it already contained an extraordinary message. Many situations and characters had to be invented, on which the story of the film had to be woven. Once the interest was aroused, it would then be necessary to support the tension of at least one and a half hours of film.

The third matter to handle was the need to establish a new geographical setting in which the story would be developed, as it had to have the colours, the structure and the characters of the city in which the film would be shot in reality. The original story was set in the city in which the author was born – Boston in the Fifties – but I did not feel there was any reason for me to adapt to this choice. From the very beginning, I thought of moving the story to Buenos Aires.

I started to rewrite the text, replacing the names of the underground stations with the names of the stations that already existed in Buenos Aires, and I invented some others in order to suggest the idea of a future underground railway with a larger extension than the present one. In this way, however, everything began to fill with a special significance, as I imagined the progress of this story. Changing the places was quite simple, when compared with the deep changes in the meaning that arose when one even only imagined the possible dialogues about the disappearance of a train filled with people – specially in a country in which so many people had disappeared for political reasons.

In this way, a new element began to add another atmosphere to the film, which was much more intense, and to offer a strong reason to adapt the text to the situation of the Republic of Argentina. And in the attempt to imagine the tone of the dialogues of the authorities of a society which had emerged from a military dictatorship, the most interesting elements emerged on this special disappearing train/object . This is where the metaphor began.

The challenge was to try to adapt, in a new and complex space characterised by a number of dimensions, the reasons that initially moved the characters of the story to play a double game between the abstract mathematical notion and the hidden story. The "political implications" on the other hand had to be dosed as the dialogues of the film unravelled, and the intention was never to expose them directly but to wait for the right moment to make them emerge.

During the preparation of the screen play that I obtained from the short story mentioned before, the characters began to "dance"on a complex line of dialogues which had a double sense.

Out of the fifteen pupils who offered to collaborate in this operation, only five of them completed the work. After three months, with our executive producer, Maria Angeles Mira, and naturally myself, there were only seven of us left with the task to prepare the proposals for the dialogues of the film. At the end I had to decide whether to accept them or not, because that enormous ocean of situations and texts had to be reduced to a unified aesthetical composition without any fractures.

It was not easy to decide up to which point it was possible to play with the dialogues without falling in the bureaucrat's cliché, without breaking the magic of the double intention of the line of the dialogue of the film, which had to play with the starting datum of a train that had disappeared and hide the new meaning of the concept of "infinity" up to the very end, a concept that was proposed originally in the short story and was now intentionally given a political significance.

Fig. 1. Film director Gustavo Mosquera R. during shooting of the film

Fig. 2. Left: Extra playing Borges's spirit, with a "bandoneon". Right: Actor Guillermo Angelelli inside a train

At that stage, the substitution of the parts started to become a delicate operation. The young spirit of the investigator topologist had to face the rough characters that enclosed another notion of "infinity", that was marked more with atrocities to hide, rather than truths to be discovered. The notion of "infinity" was starting to take its place, as a result of this new effort, in an interesting conflict of absurd elements that were inexorably connected to the same matter.

The research of the code to convey this double concept of "infinity" forced me to think of a suitable atmosphere, and I realised that the stories told by Jorge Luis Borges had a good resonance with the requirements of the film. However, to create a Borges-style atmosphere is not exactly a simple matter. I allowed myself the use of some aesthetical resources. One of these was the management of the movie camera starting from precise movements (which I would define very "ascetic") of a mathematically exact nature, so as to travel across the space without letting the movie camera disclose the presence of a human being behind it. But also the agreement of cold tone characterising the photography and the actors' reciting, to create characters who feel lost, and therefore small, when faced with the enormous labyrinth.

I felt that a visual key full of intensity was fundamentally important to enable the story to play on and to connect the fantasy of a train that had disappeared, and the cruel historical truth of the "desaparecidos" in Argentina.

Even in this manner, it was still difficult to identify the tone of the interpretation of the young topologist, because his inquisitive eyes had to contain everyone's questions. The role is played – with great talent – by the young actor Guillermo Angelelli, who discovers this new world through the circumspect eyes of the young investigator Daniel Pratt.

Finally, a disconcerting atmosphere was created with the use of real names, various numbers and/or references that would enable the spectator to find a double association so that the fiction "is" what the reality "was". Is Daniel Pratt the real name or a fictitious representation of Hugo Pratt? Is it by chance that the bureaucrats always participate in their meetings in three, or is it a chance that

Fig. 3. Actor Guillermo Angelelli during rehearsals

the representation of number three is associated with the grotesque generals of the military juntas, authors of the "putsch" in our countries?

The hope of finding a new view in this scenario without outlets, was born with the creation of the figure of the adolescent Abril, who is sufficiently young and external to the corruption that the "adults" of this labyrinth are subjected to. She faces the beginning of a process of change in her life, and her innocence, that is still intact, permits her to view the situation with an ingenuity that differentiates her from the other protagonists. During the film she plays, she does not look for anything. However, it is this tranquillity that enables her to find the answers more rapidly. With her innocence she reaches the point, that she does not even try to conceptualise. And so she becomes the almost natural passport for a hope that is still possible. For me, she is the one that represents the generation which is filled with the anxiety of knowing what happened, where all these people are, and mainly "why" it happened.

I was sure that she had to speak very little because her function was to look and play with her eyes. Annabella Levy carried out this task without ever having worked in any film or play at the theatre. Her fresh and simple young attitude was fundamental for this part.

What we learn from this trip with no return, as I perceived this initiation trip, transforms into a lesson for all the characters of the film, a lesson that is comforted by the serene gaze of the old professor who accompanies the entrance of the young Pratt in a new field of knowledge. In front of them the concept of infinity materializes, and the much feared truth, that the eyes were unable to see, is disclosed. Therefore the learning of knowledge takes place , during the same journey, with the unique experience of the moving train, that actually does not disappear, but enters a new space. Therefore the end of the film is not pessimistic, if not only a little, because it opens the doors to a new perspective, the perspective of possibility.

The presence of those who were absent, represented in the train travelling at a great speed, who had left on a no-return trip towards an inexorable experience, was one of my greatest worries. And it guided the choice of the musical theme of

Fig. 4. Film director Gustavo Mosquera R. and his students get ready for shooting in a tunnel

the final sequence, just before the epilogue, and for this I paid particular attention to the selection of the choruses which would have integrated the musical band at the end. My intention was always to insert real human voices that would give consistency to all those who had departed. It was necessary to dedicate time to listening to a number of Gregorian chants, a task that I was able to carry out one year before actually starting to rewrite the screenplay and the production of the film. The research led to a meeting with the musician who was suited for the film. The work of the young composer Mariano Nuez West was essential and at the same time brilliant.

"Moebius" – which is the most significant aspect of my experience as a film director – is the result of a human experience, that was analogous to the one in the story, that I experienced with my team. The journey (for Daniel Pratt and also for me as the director) led to the discovery of a final vision that I was able to conquer only during the course of the realization of the film. Also for me, this was a "search for infinity".

Aesthetics and Estimated Expenses

Most of the film presents the same deprecated problem: the lack of resources. And so it is classified as a "low cost film", a term that is often utilised in order to classify works in a precise category that appears to receive special attention from the public and the critics. Also "Moebius" was classified in this category but I think it would be appropriate to reflect a bit on the matter.

The cost of "Moebius" amounted to 250,000 dollars, and it is difficult to establish if this amount is excessive or insufficient for the realization of a film. It could be an exorbitant amount for a film with two actors, that was shot entirely in a hotel room, and it could also very well be considered insufficient in order to realize "2001: space odyssey" correctly.

In analysing the cost of "Moebius" it must be borne in mind that no member of the technical team, nor of the actors, and not even of the production or the film-direction received any compensation for the entire period of the work, so that the cost of 250,000 dollars cannot be considered the "real cost" of the film. The same occurred in the case of the publicity of "El Mariachi" for which the possibility of its realization was based on the use of an enormous quantity of materials that were lent or donated. This is not the way to evaluate the cost of a film nor its result. However, given the above, I am interested in taking care of a more important fact that is tied to this theme, that is the alchemy that a film director creates for his film.

It is well known that it is not the availability of – a lot or little – money that determines the "grace" of a film, but not many people pay attention to the manner in which the different elements – whether there are many or few – all match in the structure that the film director is building. The "magic" act, is therefore the manner in which these elements unite – whether few or many – so that no harm occurs to the verisimilitude.

With reference to the old mathematical formulations, it may be stated that a determined number of resources will correspond to some – and only those – possible choices, that the film director can realize so that the result appears to be verisimilar to the spectator.

The talent will therefore be directly proportional to the determination of the correct formula, considering whatever the screen play requires to be verisimilar for the spectator, independently of the amount of money that one may avail of. However, undoubtedly, the more cuts are made in the estimated costs, the more the quantity of possible mixes are decreased, leaving a smaller number of alternatives for the film director to use.

One always tends to think that the alchemies that remain beyond our possibilities are the ones that would have worked best. (Even this cannot be verified, and no one will ever know, because a film can be made once, there is no second chance for the same film-director to deal with the same topic, in order to make any comparisons). It is surely difficult for their number to become zero due to a decrease in the estimated expenses. Rather, the unhappy result usually derives from the attempt to continue to apply a formula in the new situation, that was designed for a higher estimated cost of the film, while a decided reformulation would be necessary.

I have no doubts that the saddest task for any film creator is that of renouncing a dream that he has already envisaged and to reformulate it to make it possible with what he is offered, and that may only be seen by the spectator ... the completed film.

For anyone who has the task of realizing a film, the best way to find a solution in any case is that of doing everything over and over again as many times as the dream can support, and to continue to discard and reconstruct, time after time, till the time of preproduction comes.

Which is the case of "Moebius". I do not believe it would have been possible for me to obtain this result with a limited amount of resources if I had not reformulated and viewed the idea of the final result of the film that I was searching, hundreds of times, while the seven members of the team laid out different ideas and dialogues, and while I also had the unpleasant task of discarding hundreds that were equally valid, because they were difficult to adapt to the general set up, without giving any logical explanations for doing so.

At this stage, even the time factor became an essential weapon. I was able to count on an team with very flexible working times, and I was able to keep more hours available for work in the galleries than what a normal production would have envisaged.

I suppose there are other elements to be highlighted, and every production of a film inevitably has its own, but I think I have summarised the ones I consider most important at the time the work was conceived.

How to Create a Film

Peter Greenaway

Some Organizing Principles – Part one

I am constantly looking for something more substantial than narrative to hold the vocabulary of cinema together.

I have constantly looked for, quoted, and invented organising principles that reflect temporal passing more successfully than narrative, that code behaviour more abstractly than narrative, and perform these tasks with some form of passionate detachment. I want to find a cinema that gives us an overview over ephemerality, acknowledges its own presence readily, is non-manipulative in any insidious and negative emotional sense, and that accepts readily and self-consciously the responsibility of the fakery of illusion. I enjoy self-conscious art activity; the sort of activity that gives you skeleton and flesh, blueprint and finished object at one and the same time, synchronously and simultaneously.

Numbers help. Numbers can mean definable structure, readily understandable around the world. And numbers essentially carry no emotional overload. Curiously cinema itself, a notoriously artificial medium, has always been familiarly couched in numbers – 8 mm, 9.5 mm, 16 mm, 35 mm, 6 by 8, 1 to 1.33 and, most famously, to quote Godard, it is a medium that is supposed to give you the truth 24 frames a second, though, for many years, it is ironic that it was known that 24 frames a second – in camera and then in projector – only gives you a blurred and soft-focus sort of truth. 24 frames a second was a figure-speed arrived at through a determination to be economical on film-stock; it was the lowest possible speed that would give you an imitation of reality that would suffice. 60 frames a second would be a better facsimile of the truth, but nobody is going to undo a convention that saves you money, if the eye and the mind is already unthinkingly prepared to accept a compromise. Besides, films shot and projected at 60 frames a second produced a nausea of reality for many experimental viewers – they left the cinema reeling with too keen a simulation of the real world.

So, first axiom, (in the present series) the cinema is currently and knowingly, a poor simulated visual presentation of the world. We know we could do better.

The first film I still allow out walking was made in Venice. Called *Intervals,* this short ten minutes film was just that – time spaces. Time is nicely appreciated by film – space equals time – 24 frames of space is regularly and dependably equivalent to one second of time.

Since this film about buildings, doorways, porches, pavements and gridded architecture, was made in and about Venice – Vivaldi was its appropriate arbiter. I had learnt that Vivaldi composed occasionally around the excitements of the number 13. This film, Intervals, consists of groups of 13 visual images arranged in 13 sections, the whole being repeated three times with an evermore increasingly sophisticated soundtrack. The images were waterless – a provocation for a film about Venice – though there was plenty of water to be heard splashing and rushing, trickling and dripping on the soundtrack. I was excited about the acerbity of the editing plan, but it became quickly apparent that elegant mathematical schemes are not visible to the viewer – not even to the more than above average discernible viewer. It is more than difficult for an audience to count the shots, and appreciate the temporal symmetries – the shots are all of different lengths, come in varying clusters, and, most importantly, their content tends to distract good and accurate counting. Which of course is as it should be.

However, an elegance in editing structure was eminently satisfying. I suppose the excitement was cabalistic. I continued the experiment. I made a film called Dear Phone – an essay about the uses and abuses of the telephone, a divertissement where the narrative content was singular, appertaining to a single idea but split up into twelve parts. A narrator reads twelve stories whilst the viewer watches the same (almost the same) twelve texts on screen.

It is here an aside, (the first aside in the present series) but for me an important pointer for the future – to emphasise this particular war (with truces) between image and text, and this peace (with outbreaks of violence) of the text as image.

In *Dear Phone* the texts start almost illegibly on the back of the equivalent of an envelope, or as they used to say in thirties novels and movies "on the back of a cigarette-packet", and progress to being manufactured in set-type in a book – a charting of an idea (with small variations) from rough draft to published text. Running in my mind was the idea that most cinema is illustrated text, so why, to avoid being merely an interpreter when I wanted to be a prime creator, not make a film of entirely unillustrated text. However, that was not completely true, because at intervals, clutches of illustrations of telephones and telephone-boxes, cropped up. It was similar to an illustrated text-book of the early 20th century where, due to the economic (again) vagaries of publication finance, text and illustrations were separated. You, the gentle reader, by flipping the pages, were obliged to join that which had been separated by the sheer financial-technological characteristics of book publication.

Though I had a pleasure in numbers again in this film, I doubt whether the audience shared so directly in that particular pleasure. I hoped, of course, that the audience's delight and fascination was alive to other devices and constructs in the film, but I was greedy. I wanted them to delight in my arithmetical delights as well. I persisted in this ambition.

In the following two years, the films grew longer and my conclusive piece to end that run of experiment was a film called *Vertical Features Remake.* The social content of this film was the encroachment of the city on the country, the urban on the rural. The visual content were simply filmed verticals – posts, poles, goal-posts, props, fence-posts, set up in fields, gardens, backyards and farmyards, on the open moor and on the open heath. The aesthetic content was a contemplation of a hinterland of town-village-country wilderness staked out by man, and filmed with an eye to weather and landscape and (somewhat melancholic) beauty. Whereas the social, visual and aesthetic content was alive and sensitive to intuition, the editing structure was rigorously not at all so. There was an exacting scheme that should not, under any circumstances, be subjected to any intuitive delight or any arbitary pleasure principle. The verticals were arranged (almost) in three mathematically exact editing schemes, first in exactly rendered multiple frame-counts, second in progressively lengthening and diminishing frames-counts, and third, in an asymmetrical combination of the two. The different edits were accompanied, and in some cases slaved to a musical soundtrack by Michael Nyman in an acerbic and decidedly metronomic mood. I was exultant, soaked on the beauty of a combination of mathematics and English pastoral traditions writ anew.

But the delightful rigours were not easily available to even an astutely trained eye. It was the old story of frustrations. To combat it on this occasion, I invented what Robert Wilson would have called Knee-plays – sections, so to speak, for the jointing of the lower limbs – sections of explanation arranged as consultation and dispute located between the three films, and arranged fictively between film archival academics arguing the very question of mathematical versus aesthetic interpretation. I posited the internal films of Vertical Feature Remake as just that – three films organised on Cagian principles of a given system taken to exhaustion without the comfort or consolation of intuition – three films remade with great care from lost originals according to structures deemed correct by different academics.

As an aside (the second in the present series), these warring academics became character fodder for a mythical generation of characters – Tulse Luper, Cissie Colpitts, Lephrenic, GangLion, the Keeper of the Amsterdam map, and many more. These characters live on still, and are about to be reborn in a grandiose new project based on the atomic number of uranium 92 – but more of that later.

I had arrived at a cross-roads of contradiction – the plausibility of, and enjoyment potential of, notions of an abstracted mathematically constructed cinema, set against the requirements of conventional cinema enjoyments. I side-stepped. If numbers and Western counting systems – local and national – were now comprehended all over the world, were there other universally understood codings? Well, the Western alphabet, give or take 26 letters, has become ubiquitous, even in non-phonetic language-countries where planes land, money is counted, books are written, and international communication is deemed essential.

I made several films based on the usages of the 26 letters of the English alphabet, bearing in mind, as the film *H is for House* did, that, in English at least, His Holiness, happiness, hysteria, headache, Hitchcock, heaven and hell can all be legitimately and absurdly closeted in intimate proximity according to their shared initial. As a film, *H is for House* was a suggested training-manual for a child intent on naming things, like Adam at the beginning of the world, ruminating on the question that does an object always demand a name, and if an object does not have a name, can it exist? Thinking of simple anatomy for example, what languages have a name for the space at the back of the knee, or that space between nose and mouth? If we have no name for such anatomical spaces, will they ever, do they ever, make a significance on our imagination or on our experience? Another film in this personally invented alphabetical genre was the film *26 Bathrooms* – a film of no especial mystery – though S is for the Samuel Becket bathroom seems a little perverse – but, there again, perhaps not as perverse as in H is for House where H stood for Bean – Haricot bean and Has-been. There are difficulties of course – even in English bathrooms where much happens outside of brushing your teeth, Q and X and even Z are rare initials to be seen naked with.

Who was responsible for all this shunning of the text and the plot and the story and the narrative in favour of numbers? Well, there were several personages to blame -and John Cage was an especial hero. The working years 1976 to 1980 for me were somewhat based on a mathematical error occasioned by John Cage. Some time in the 1950s he had published many of his short anecdotal narratives on a vinyl disc – two sides of exotic tales about his macrobiotic diet, his friendship with John Tudor, his watering of cacti, his fascination for fungi and anechoic chambers, and Confucius saying that a beautiful woman only serves to frighten the fish when she jumps in the water ... his strategy for telling these short tales was to restrict, and even constrict, each one within the confines of time – 60 seconds each; the one-line stories thus had to be stretched to fit the minute, and the two paragraph stories had to be garbled to meet the distance. Either way the stories became for the most part unintelligible, abstracted, even musicalised – certainly the attention of the listener was filled with a certain sort of anxiety that claimed time as an element in each story. I enjoyed this tyranny of time over narrative. It suited my antagonism to sloppy story-telling, facile story-telling, arbitrary story-telling. And I borrowed the structure in certain personally fashioned ways, first in a film called A Walk Through H which had as subtitle, *The Reincarnation of an Ornithologist.* This film dealt again with the excitements of the English variable landscape, about my father (an ornithologist) and his death, and about migrating birds, but primarily about the excitement of maps which would stretch from a remembrance of maps of Chinese mountains – all brown hatching and no words – to Borges's manufacture of a map on the same scale as the world – to the idea that a map is always presented in three tenses – it tells you where you are, where you have been and where you can go.

Secondly, and more importantly, I used a variation on the Cageian narration-time conceit in a long, three and a quarter hour film called *The Falls*. A *Walk Through H* contained 92 maps and *The Falls* had 92 characters. (Fig. 1)

The Falls was very ambitious. Birdlore, flying dreams, Icarus, the Wright Brothers, the lousy ending to Hitchcock's *The Birds,* disaster theory, as many ways as I could think of to make films – and that particular parlour game of the time – as many ways as you can think of, of how the world would end. The mistake I made in all this obsessive number-crunching, was a very simple one. Because of the ambiguous intervals (intervals again) between the Cage stories on his vinyl disc – I had counted his stories to make a total of 92, whereas there were, in fact, only 90. My homage to Cage had been a genuine counting mistake.

Small enough for the world, big for me. Several years later I made a documentary with and about Cage and he laughed a loud laugh when I explained the error

Fig. 1.
The Falls © P. Greenaway

I could lay indirectly at his feet. *The Falls* was shown at a Washington University Film-Club , and my mistake was comforted by the suggestion of an attentive enthusiast that 92 – being the atomic number of uranium – was singularly appropriate in a film about ways the world might end. This mistake has since become a credo. Let me quote Cage again, using numbers. He said, "If you introduce 20 per cent of novelty into a new art work you are immediately going to lose 80 per cent of your audience". He was correct to suggest that introducing bare arithmetic into narrative feature-film-making can have a direct bearing on your audience percentages. I have box-office evidence to prove his theory.

With films like *Vertical Features Remake* and *The Falls*, I was coming to the end of a much-veined mine. To assist my dilemma, I was encouraged to write a script where the characters talked to one another and not to the camera. The film was *The Draughtsman's Contract* which could be described as an account of a seventeenth century (just – it is set in 1694) draughtsman who makes 13 drawings in a pattern of 12 plus 1. (Fig. 2)

Fig. 2.
The Draughtsman's Contract © P. Greenaway

He is a man who believes in veracity. He draws country houses and gardens, and he prides himself on an exactness to nature or at least to the human eye looking at nature. He uses an optical device. Canaletto, his near contemporary, made use of one, as did da Vinci and Dürer at an earlier time. This device – for us – was a squared grid of wire set up on an easel to be viewed with a monocular eye-piece. The eye transfers the grided landscape to a piece of similarly grided paper. But if this tool is an indication of the draughtsman's ferocious desire to make truth-to-Nature drawings, his work-method is perhaps far more exacting and even more controlled. He has noticed, as certainly I had, performing a similiar task, that shadows on architecture on a bright day move very fast – so to draw accurately, one hour or two hours at most in a given viewpoint, was enough before an accuracy to depict changing shadows made no more sense. Consequently, the draughtsman sets a regime in motion to move his apparatus and himself every two hours, following the sun, aiming to return to the same spot and the same monocular eye-piece, the following day, to continue his task. Landscapes of English country-houses are full of servants, gardeners and sheep. And they move. And they move objects. The draughtsman's demand is that all should be frozen for the purposes of his pencil. When this regime has finally been made clear, he signs his draughtsman's contract – 12 (plus 1) drawings of a country-house in return for 12 (plus 1 – a bonus) sexual liasons with his commissioner, the lady of the house of Compton Anstey, a lady though we do not know it yet, who is looking for an heir. After much drawing and pea-cocking, the draughtsman finds himself the victim of a frame-up. His drawing frames grid him into an accusation of murder. Unwittingly, his prowess as a stud has been confirmed, his prowess as an exacting draughtsman dismissed. His drawings are destroyed. (Fig. 3)

Julian Calendars and atomic clocks notwithstanding, the plot, with many asides to discuss many things, held some water, and, notwithstanding, a mysterious finale. Audiences were amused and the film received plaudits from many sources. Its comedy of manners, its status-seeking ambitions and its hierarchies made it amenable to the English, the baroque flourishes and extravagances and sexual audacity made it amenable to the Italians, and the hard edge of arithmetic and the sharp Descartian method made it very acceptable to the French. (Fig. 4)

The next film *A Zed and Two Noughts* had eight Darwinian evolutionary periods as its arithmetical strategy. I was an eager neo-Darwinist at that time, after being an amateur paleontologist as a teenager and a great collector of fossils (and a collector of British insects like the young Darwin when he was thinking of becoming a priest, if only to find an occupation that gave him time to be an entomologist). I had, and have, come certainly to believe that there is no better system to answer unanswerable questions than that introduced in the *Origin of Species*. "The Beagle" has become an important password. I suspect the ordered systems of paleontology, daring to segment grand thirty-million year chunks of chaos into ordered pieces – is invigorating and encouraging, and the prospect of the ordered ranks of insect specimens pinned with dates and names in a camphorus box is a sight to set a satisfaction that Nature can be classified, tidied up, arranged, tamed. But the irony stays. Darwin and his system is as good – very good – as far as it

Fig. 3. *The Draughtsman's Contract*
© P. Greenaway

Fig. 4. *The Draughtsman's Contract*
© P. Greenaway

goes. And I am very pleased, at last, that we have a universal system that offers answers to the big questions without having at the same time to offer rewards, consolations and condolences. But the very tidy Victorian minds seeking confirmation of the observational sciences and determined to exclude God, created it out of a special need that we will eventually find to be local and replete with vested interest.

The film *Zed and Two Noughts* deals with the exploits of two animal behaviourists determined to use Darwin to explain the violent unnecessary deaths of their wives in a car crash which is associated with all manner of circumstantial clues. Twinship and the Dutchman Vermeer play significant roles in the film – but all subjects, great and small, obey the pattern of a simple narrative promulgated by Darwin, though massively simplified as though it were to be filtered through a primary school text book. The eight subject-areas of the film follow an evolutionary development of a sort, made manifest in a series of putrefactions. First we watch the time-lapse decay of an apple – Adam's apple, or should we say Eve's apple – an easy poke at Genesis but a nod towards the primary division of the animal and plant kingdoms. Then, in order, we witness the decay of shrimps (invertebrates), angelfish (water vertebrates), a crocodile (reptiles), and then in an uneasy chronology that makes no great evolutionary leaps – but implies at least a growth in size and exotica – a bird and four mammals – a swan, a dog (a Dalmatian), and a zebra (all noticeably and significantly black and white creatures) – moving to a black ape and finally to the potential expectation at least of the putrefaction of a white man – indeed two white men, the animal behaviourists themselves – making a self-sacrifice in the name of (bogus) science. There is also an alphabetical count. A small child is taught to regard the world as an alphabetical zoo. But that is a game and a small ironic parallel. Alphabetical listing may be neat and even elegant and certainly suggests order, but it hardly offers explanations by its listings. To relate an aardvark and an antelope by their initials is to make no evolutionary comment at all. Like the child in the film who placed a spider and a fly in the same specimen jar because they were both brown, thus ignoring important principles rapidly observed when one creature eats the other, all organising principles are faulty in one way or another. And little of course is achieved by the animal behaviourist brothers in the way of understanding sudden death – but the certain closure that is reached is that all systems are negotiable, all constructs are local in geography and history, all organising schemes are only as good as they are because for the present no better ones exist. But of course there will be better systems. We can take comfort that they will arise when needed. The film, *Zed and Two Noughts* concludes – as the Devil's advocate – that Darwin could well become a credulous Adam, and his stories could become no more or less than pretty myths and romantic metaphors. It was a parable for me for much to come. Systems are valuable but mutable.

There are no verities. Even 2 plus 2 can sometimes equal 5. We seek the truth but the tools of which we are so often so very proud, are notoriously weak.

The film *The Belly of An Architect* followed. It too was an "eighter" – this time a project divided into eight classical architectural periods, most pertinently to be seen in Rome, at the Augusteum, the Roman Forum behind the Capitoline Hill, the Pantheon, the Villa Adriana, St Peter's, Bernini's Piazza Navona, the Piazza del Popolo, the Vittoriano, the Foro Italico and the "square colosseum" at EUR. The film was a disposition on the pleasures of the harmony of classical architecture, seen in terms of a classical architect's imagination. Symmetrical elevations, ground-plans and the grid, with very strong verticals and horizontals, governed its camera-strategy. All the metaphors and the narrative links joined in the person of the arch-exponent and most potent theoretician of the classical platonic solids, Etienne-Louis Boullée, lover of the sphere, the cube and the pyramid, and the admirer of Newton – perpetual hero of gravity who makes all architects both free and, at the same time, disciplined, by keeping their feet on the ground so that their minds can be free in the clouds. Paving the way for future considerations, there were other organising principles at work in the film – some ordered colour coding – for example, a reduction of colour to the hues of Rome and the tints of human flesh, making green inimical to both – trees uproot buildings, and green is the colour of decay in the human corpse.

Drowning by Numbers followed. With such a title, numbering cannot be ignored. The principle was to unite narrative and a number count. Sometimes they pitch and toss together, sometimes they metaphorically unite, sometime they dramatically fall out of synchronisation. The parallel was the rubbing together of the shoulders of Fate and Freewill. (Fig. 5)

Fate is the numbering. Freewill is the narrative.

The story concerned a number-sensitive trio, three women all called Cissie Colpitts, the same woman three times over. Why do women always come in threes? Do men always come in sevens? Three is sanctified by Dante and the Catholic Church, though, on the illegitimate side, there are the Three Fates, spinning, measuring and snipping our lives into carefully measured sections, and the three witches in Macbeth, ordering death by prophecy. Seven is a much more maligned but exciting number – days of the week, ages of man, colours of the rainbow,

Fig. 5.
Drowning by Numbers
© P. Greenaway

oceans, continents and wonders of the world, the largest number perceived without counting, apparent number of the most significant wave, snore and period of menstruation. The women of *Drowning By Numbers* fit a mediaeval pattern-book – virgin, pregnant wife and witch, though, in the film, this trio of female types was overturned, the virgin is anything but a virgin, the pregnant wife is barren and the witch is a glorious and glamorous grandmother. These women count in threes at funerals to warn off importunate grief, and of course, there are three funerals. These women are wooed and courted by a coroner, a man of death, who is always unsuccessful in his romantic and sexual self-presentations. He is used by the women as a pawn to free them from the crimes of drowning their husbands. Death by drowning can so easily look like an accident. In return for hypothetical sexual attentions, he is to sign the death certificates with invented innocent causes. The coroner has a son, a twelve-year old, keen to imitate his father's investigations into death, using road corpses as his metier. He makes a death count, celebrating each small animal corpse by lighting a rocket to permit the dead creature's soul – the soul of a squashed hedgehog, a crushed fox, a broken-backed squirrel, a ploughed-over mole – to reach heaven as quickly as possible.

And over all these fanatical counting games is the film's impersonal counting system. One to a hundred. The numbers are strictly chronological but an audience has to be awake to follow the strict count. The numbers appear visually on tree-trunks, butterfly collections, the back of a honey bee, the rump of a dairy cow. They appear aurally in dialogue and song. An audience will know for certain that when the number count reaches 50, that the film has half expired.

And number-count and narrative come elegantly together when inevitably the coroner is persuaded to keep a certain suicidal appointment by water. His sinking boat has no name but it has a number, and it is, of course, one hundred. (Fig. 6)

With such overt genuflection to the excitment of sheer numbering for its own sake, I had reached another significant sign-post.

Some Organizing Principles – Part two

The film, *The Cook, the Thief, His Wife and Her Lover* has no numerical skeleton, it has a colour coding system instead – seven spaces coloured to metaphorically invoke associations, amongst others, of the blue polar regions, the chlorophyll green jungle, and a crimson, scarlet and vermilion carnivorous restaurant. As the characters move from one room to another, as if to really emphasise the organising colour insistence, their clothing colour changes to match the new environment. All seven Newtonian spectrum colours unite to make white light, orthodox symbolic colour of Heaven when and where, with heavy irony, the lovers carnally meet in the blinding whiteness of the restaurant's toilet, though the inferences are obvious that the colours of a toilet ought to be, should be, might well be, are expected to be, the colours of defecation. A smaller organising principle in the film is a day by day menu card working its way through the movie towards fish on Friday, and working its way through an ultimate imaginary meal – a Last Supper – of many courses from hors d'oeuvre to coffee with a gesture of cannibalism to be imbibed with the last liquer, before retiring to bed, or in the thief's case, before retiring to death. A still smaller conceit of numerology is the organisation of the lover's library. His shelves of books, as the camera tracks towards the librarian

Fig. 6. *Drowning by Numbers*
© P. Greenaway

lover's murder by ingestion of his own volumes, map out a chronology of sensational French history – from the French Revolution through the Years of Terror to Napoleon's Consulate and ultimately his demise at Waterloo. It is a small metaphor of a reign of violence and blood – safely (and almost fictively) digestible as text to be placed against a real reign of terror impossibly ingested in a contemporary restaurant in some European capital city of the 1980s.

This large scale melodrama of sex and digestion was a baroque prologue to further excesses in the next film, deliberately titled Prospero's Books to make sure that it was not strictly, wholly and only, a film of Shakespeare's *The Tempest.* The film was criticised by English purists as having too many Greenaway elements and not enough elements of Shakespeare. It is true that there was one invention that was extra-Shakespearean, but legitimised by his text. This concerned another library. The library of Alonso's books. Prospero, Duke of Milan, exiled by his brother, put into a leaking boat with his young daughter Miranda, is cast off to certain presumed death by drowning. However, his friend Alonso in a hopeless but well-meaning, clandestine gesture fills the boat with a little food and a few books. Those few books in this film, and perhaps unstated in the play, become 24 important volumes. They are the foundation of Prospero's power and magic. They will see him through the fifteen years until rescue is possible. There is a Book of Water, a Book of Mirrors, a Book of Mythologies, a primer of the Small Stars, an Atlas belonging to Orpheus, a Harsh Book of Geometry, a Book of Colours, a Vesalius Book of Anatomy, an Alphabetical Inventory of the Dead, A Book of Traveller's Tales, a Book of the Earth, A Book of Architecture and Other Music, The Ninety-Conceits of the Minotaur, a Book of Language, A Herbal called End-Plants, a Book of Love, a Bestiary of Past, Present and Future Animals, a Book of Utopias, a Book of Universal Cosmographies, a Book called Love of Ruins, a Pornography called the Autobiography of Pasiphae and Semiramis, a Book of Motion, a Book of Games and a Book called Thirty-Six Plays which indeed was the complete works of Shakespeare. For our interest in numbers here, the primer of the Small Stars glittered with ascensions and declensions, the Orphean Atlas was full of figures to guide you across the Styx and down into the multi-layers of an admittedly Dantean Hell, and the Harsh Book of Geometry, thanks to modern graphic technologies, flapped and flickered with moving diagrams, ameliorating its description of "harsh" – meaning non-negotiable – and introducing its attendant, the MathsBoy, whose page-white body was stencilled and tattooed with a living abacus. The Book of Architecture and Other Music made alliances between those two arts, with annotated scores, the Ninety-Two Conceits of the Minotaur made reference to the magic number of Uranium in association with destruction and the possibilities of radiation-mutating aberrations. Finally, the Book of Universal Cosmography and the Book of Motion both culled arithmetical conundrums from everything and everyone mathematical and systematic from Euclid to Robert Fludd, Kirchner to Paracelsus, Babbage to Buckminster Fuller, and Muybridge to Hawkings. There was to be no respector of 1611, the apparent year of Prospero's rescue and the apparent year of Shakespeare's play, for magic has no chronological boundaries.

These twenty-four volumes could easily be seen to contain all the world's information.

There was only so much space in a two-hour film for a dissertation on the 24 volumes and the 100 allegorical characters that Prospero created on his island of sounds that delight and hurt not. The twelve books and the one hundred characters became the subject of some considerable scrutiny in many exhibitions of the former (including the Palazzo Fortuny in Venice) and in a still unfinished novel of the latter, called *Prospero's Creatures,* an enterprise that introduced me to millenial fantasia and a plethora of projects assisted by the prop of the number one hundred, including an exhibition of 100 screens in Munich and a manifestation of 100 staircases in Geneva.

But the prime exhibit in this new ordering enthusiasm was, first an installation-exhibition, and then a travelling opera, both going under the same name of *One Hundred Objects to Represent the World.* The installation-exhibition was held in the Hofburg Palace and the Semper Depot in Vienna, and the opera was presented in Saltzburg, Paris, Palermo, Naples, Munich, Rio, San Paulo, Copenhagen, Stockholm and Corunna. There were a great many games, conundrums and conceits in this opera which pastiched the education of Adam and Eve by a mentor called Thrope – a shortening of misanthrope – assisted and hindered by God represented as being no more than a severed head – literally a figure-head – on a table, and Satan, an Otto Dix-dressed female whose wit – hopefully – matched her sense of evil. Object Number One in this work was the Sun and Object Number 100 – Ice, so a viewer could quickly see the drift from heat to cold, life to death, optimism to pessimism. Half-way item number 50 was Rubbish. Item number 24 was appropriately Cinema, celebrating Godard's dictum that cinema was truth 24 frames a second.

Why one hundred? Well, it is a number of orthodox convenience. We chop up time into centuries. We think and dream and count our money in centuries. And as the small Velasquez skipping-girl at the front of *Drowning by Numbers* suggested – all hundreds are the same – once you have counted one hundred, other hundreds are merely repeats. Graphically it is a number that is most satisfying – a bold vertical and two bold circles. Putting present millenial celebrations to one side, I am sure the number 100 would have been paramount in any other decade, and besides – what do numbers mean when Vatican theologians can now tell us that Christ was born in 6 AD, because the nearest likely star in the east was a comet flying by in the Mediterranean winter of that year? But the necessary King Herod who demands the Massacre of the Innocents is dead by 3BC. So we are either celebrating the millenium six years too early or three years too late.

The putative novel *Prospero's Creatures* itself certainly started as an examination of the creation and consequent histories of 100 creatures invented by Prospero in the fifteen years of his exile. They were a mixture of the sacred and the profane, the Judo-Christian and the Romano-Greek, with a number of eccentricities – both private and public – thrown in to largely underline my fascination with libraries. The manufacture of images soon overcame the production of publishable text, and at the invitation of the University of Strasbourg, a book was announced and consequently published, based on the same criteria. The 100 allegories were manufactured pictorially on equipment of the new technologies – 100 full-page images of collaged reference built around an allegorical proposition.

Some 200 Strasbourg citizens stripped off their clothes to pose nude to make 100 photographic images which I could then clothe with collaging made possible with computer technology. There were many minor numerical games played in the classifications. Number 54, the Chronologist, is stamped with the number possibilities of the United Kingdom's lottery, Number 84, King Midas is set before the first America Bank Note valued at six pounds. Number 76, Semiramis, is set up before a numerically strong checklist of her military lovers, Number 25, the Epicuress, is set forward with notated details of her clothing bills.

During this period, 1993 to 1998, often acting in association with the films, often acting independently, I was variously associated with a series of exhibitions of curatorial interest, taking objects and items from international collections and subjectively reorganising their contents to make comment about museumology, collating and collecting, classification and cultural prejudices. Alphabetical counts, numerical counts and colour-coding were often used as structure.

In 1994, the most explicit of these exhibitions was presented in South Wales titled *Some Organising Principles.* I was given permission and encouragement to search some thirty small city museums of South Wales to collect items of organisation – modest 17th, 18th, 19th and early 20th century machinery of commerce, agriculture and industry – weighing machines, counting machines, measuring machines, clocks, cash-registers, keyboards, educational tools, card and file indexes, printing and publishing equipment. South Wales, with its copious supply of coal and cheap labour was an experimental ground for the Industrial Revolution. There is much in this period that is shameful, not least in its haughty regimentational attitude to things, events, places, ecology, and most pertinently, people. For example, there was a large phrenological collection, satisfying middle-class prejudice about predetermined criminality. Care for the coal miners's well-being was rudimentary. Simple gas-escape equipment was crudely calibrated. Numbers, and not names, governed the miner's life, identifying his pay-tin, his clocking-in equipment, his miner's lamp, surrounding him with numbered tags, keys and punched-cards. Whilst the poorly-paid miner worked long hours in miserable conditions, his employers and their associates dreamed up artificial heroics for so-called Welsh indigenous culture, inventing runic alphabets, and romantic confabulations for the many ancient Welsh monuments often considered to be Celtic calendars. To parallel, parry and pastiche the exhibits and their meanings, we brought numerical and organisational systems of our own, in manufacturing counting and numbering devices to be seen and experienced through the elements of air, water, fire and earth that had been exploited to make that part of the world a temporary paradise for some, and a continuing hell for others. To give some equivocal perspective to this exhibition of some three thousand items, the catalogue cover was printed with Dürer's print of *Melencholia* I which contains depictions of platonic solids, Time, weighing and measuring, bells, and a most elegant Numbers Square based on 34, where any reading of the consecutive numbers 1 to 16, up and along, across and down, diagonally, sideways and in minor squares, adds up to 34. (Fig. 7)

By now, a total self-consciousness as to the efficacy of numbers in the manufacture of the films was paramount, and number-counting became almost an oblig-

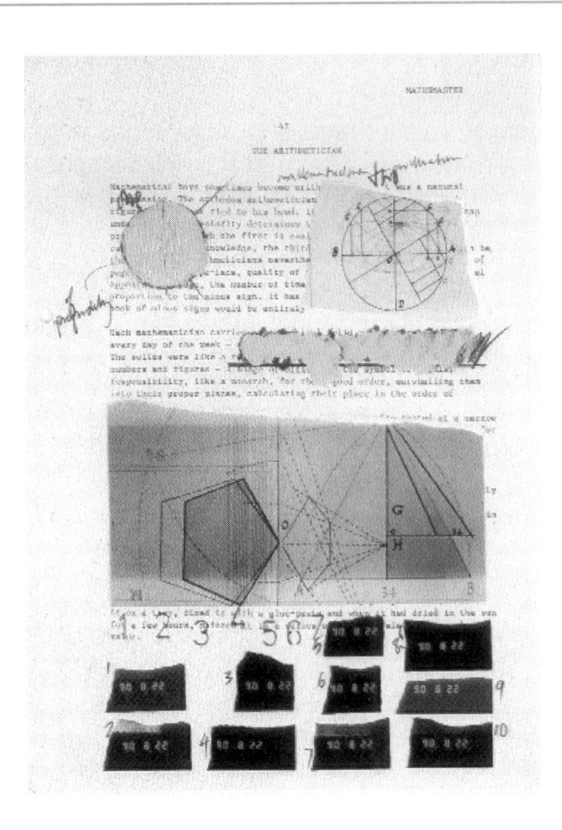

Fig. 7. © P. Greenaway

ation and certainly often self-parody. The film *The Baby of Macon* escaped this obsession, wishing to soak itself in other intense fascinations. The film was overridden with the three colours Gold, Red and Black, a trinity of the Church, Blood and Death, and by the organisational metaphor of the Petrarchian Procession – a sort of hierarchal counting system in itself. And the following film, *The Pillow-Book,* was more interested in texts, mainly calligraphic texts, than numbers, though there was a progression that measured itself in a steadily increasing number of lovers forever receding in age and foreigness. Degrees of foreigness and indeed degrees of efficacy as a lover are not easily calibrated.

The opera, *Rosa, A Horse Drama,* and more readily, the film of the opera, was obsessed with the violent deaths of ten composers starting with the accidental killing of Anton Webern in 1945, and ending with the assassination of John Lennon in 1980. In the case of these two composers and the other eight that linked them, there were always present ten identical clues, and these were constantly reiterated and listed, with variations and different sortings.

And then the film *Eight and a Half Women.* This was a homage of sorts to that most celebrated film of the early 1960s, Fellini's *Otto e Mezzo,* and most particularly to its male fantasy sequence presented via Marcello Mastroianni. In *Eight and a Half Women,* eight and a half archetypal male sexual fantasies are blatantly presented. It is also my eighth and a half feature film, and represents one eighth of the number of films I have made in 31 years.

And the next project is *The Tulse Luper Suitcase,* supported by the secondary title of *A Fictional History of Uranium.* This , in sum, is a hopeful return to early mythologies and the magic atomic number 92. I can discuss its parameters, intimate its numerical ambitions, but I resist the temptation, wary that the final computations of its structure and syntax might turn out differently from its neat, and I hope, elegant, numerical beginnings. Suffice also to say, one of its many counting devices is related to the re-writing of the celebrated *1001 Tales of Scherherazade,* one story a night for three years. At the end of three years I will report back from the numerical frontier, hopeful that the counting did not lapse, since metaphorically, I am under the same interdict as Scherherazade herself, whose penalty for default was a beheading.

All Illustrations by Courtesy of Peter Greenaway.

Filmography Related to the Article

ALL FILMS BY PETER GREENAWAY

[1] *Intervals,* subject and screenplay by Peter Greenaway, produced by Peter Greenaway, distributed by British Film Institute, Great Britain (1969), 7'

[2] *Dear Phone,* subject and screenplay by Peter Greenaway, music by Michael Nyman, produced by Peter Greenaway, distributed by British Film Institute, Great Britain (1977), 17 '

[3] *Vertical Features Remake,* subject and screenplay by Peter Greenaway, music by Michael Nyman, produced by Peter Greenaway & Arts Council of Britain, distributed by British Film Institute, Great Britain (1978), 45'

[4] *A Walk through H,* subject and screenplay by Peter Greenaway, music by Michael Nyman, produced and distributed by British Film Institute, Great Britain (1978), 41'

[5] *26 Bathrooms,* subject and screenplay by Peter Greenaway, music by Michael Nyman, produced by Sophie Balhetchet for Channel Four, Great Britain (1985), 28'

[6] *The Falls,* subject and screenplay by Peter Greenaway, music by Michael Nyman, produced and distributed by British Film Institute, Great Britain (1980), 185'

[7] *The Draghtsman's Contract,* subject and screenplay by Peter Greenaway, music by Michael Nyman, produced by British Film Institute and Channel Four, distributed by British Film Institute, Great Britain (1982), 108'

[8] *A Zed and Two Noughts,* subject and screenplay by Peter Greenaway, music by Michael Nyman, produced by British Film Institute, Film Four International, Artficial Aye Prodcutions, Great Britain- Holland (1986), 115'

[9] *The Belly of an Architect,* subject and screenplay by Peter Greenaway, music by Wim Mertens, produced by Callender Company, Mondial Lmtd, Tangram Film, Great Britain-Italy (1987), 118'

[10] *Drowing by Numbers,* subject and screenplay by Peter Greenaway, music by Michael Nyman, produced by Allarts productions, Film Four Int Elsevier Vendex Film, Great Britain (1988), 119'

[11] *The Cook, the Thief, his Wife and her Lover,* subject and screenplay by Peter Greenaway, music by Michael Nyman, produced by Allarts CXook Ldt., Erato Film, Film Inc., Great Britain-(1989), 124'

[12] *Prospero's Books,* subject and screenplay by Peter Greenaway, music by Michael Nyman, produced by Allarts productions, Great Britain-Holland-France (1991), 130'
[13] *The Baby of Macon,* subject and screenplay by Peter Greenaway, produced by Allarts, IGC, Cine Electra, Channel Four, La Sept Cinéma, Filmstiftung Nordrhein Westfalen, Great Britain-Holland-France-Germany (1993), 122'
[14] *The Pillow Book,* subject and screenplay by Peter Greenaway, produced by Kasander & Wigman Productions, Great Britain (1996), 118'

Mathematical Centers

A Museum of Mathematics: How, Why and Where

Enrico Giusti

One of the things that immediately hit visitors to any museum of science is the almost complete absence of mathematics, or at least its totally marginal role compared to other scientific disciplines.

So, visitors to a museum of science will come into contact with the most recent discoveries in the field of physics, astronomy, biology, chemistry, ecology, pharmacology and technology; on the other hand, at the end of their visit, they will know little more mathematics than they did when they entered the museum.

The same can be said for magazines or television programmes of popular science, in which each new discovery – often even when fake – causes a stir and raises comments and discussions, but in which mathematics is rarely present.

The proof of Fermat's last theorem – with the considerable quantity of related articles and books – is more the exception than the rule; it is an exception probably due to the simplicity of its statement and the exceptional length of time between its formulation and the demonstration of the theorem.

The main reason for this disinterest is neither curators' or editors' lack of willingness, nor is it mathematicians' deplored disinterest for popularisation, but rather the very nature of mathematics, a science in which – to paraphrase Bertrand Russell – one does not know what one is talking about, neither does one know whether what one is saying is true or not, or alternatively, one does not know what to talk about.

In fact, in comparison with other sciences, mathematics is characterised by the lack of objects, or at least of material objects that can be proved, described, touched and that are worth knowing about.

The objects of mathematics are immaterial objects – situated in a region half way between the real and the imaginary – that can hardly be exposed in a museum. Moreover, even when one can talk about them, who would be interested in the structure of such abstract entities that are so far removed from daily experience? One can develop a passion for the structure of the atom or for the evolution of galaxies, but even a well-disposed visitor will find it difficult to develop an enthusiasm for the axioms of group theory.

Compared with other sciences *on their turf,* mathematics is destined to succumb, or at least to be reduced to its folk aspects.

I stressed *on their turf,* because I believe that the real difference between museums of science and popular mathematics lies right here. The structure of a museum of science, but also of most popularisation literature, revolves around objects and phenomena – objects to be shown and phenomena to describe and amaze with. This is the starting point from which to lead visitors to enlarge their horizons and familiarise themselves with current scientific theories. Museums of science are mostly organised around phenomena, history of science museum around objects; these "units of science communication" are the starting points on which museums of science can count to widen their discourse and promote the knowledge of science or history of science.

In museums that are thus organised, mathematics will always find itself ill-at-ease. In fact, apart from rare exceptions, mathematics has neither objects that can interest a large audience, nor phenomena that can grab them. By forcing mathematics into a schema modelled on other sciences, one risks reducing it to its most visible, but also marginal, aspects. Even worse, it might distort it completely, thus giving a totally misleading impression, reflecting the popular image of mathematicians (and mathematics) as being alienated from the events and necessities of daily life. Just as ordinary people imagine mathematicians busily carrying out long and difficult calculations (for which they will soon be replaced by computers), so museums exhibit the first million digits of π.

Therefore, mathematics cannot follow the lead of museums of science, at least not those we are familiar with. It must not be organised like a shop window, but rather like a mine where visitors can dig into the mathematics hidden within the exhibits. These exhibits will necessarily come from the outside: from physics, technology and daily life. In any case, a museum of mathematics is not a show of objects; at the most, these objects, in the form of models or tools (for instance, surface models, measuring or calculation tools), will constitute a historical section of the museum.

One cannot stick a book onto a wall either. The purpose is not to teach mathematics, especially not proving theorems. On the contrary, long and formal explanations will have to be absolutely avoided. From this viewpoint, the programme of the museum is complementary to the school curriculum: at school there is teaching, at the museum people are brought closer to mathematics. From this fundamentally complementary relationship, the possibility of a fruitful collaboration is born between museum and school, without any confusion as to roles.

In conclusion, museum mathematics is not perceived immediately, just as it sometimes happens for other scientific disciplines; neither can it be explained with words, for this is the task of school mathematics. Therefore, the museum unit is neither an object, nor a phenomenon (neither of which speak the language of mathematics), nor is it written text that must accompany but not replace direct contact with exhibits. In order for exhibits to show their mathematical content, it is necessary for them to be presented not individually, but in structured, homogeneous progressions from which (at a higher level) the individual objects composing it fade away and the mathematical structure that unites it and brings it to life can emerge.

The itinerary then becomes the composing unit of the museum. A composite unit represented by the objects that are functional to the description of the mathematical idea underlying them, but also by objects with more marked game-like characteristics that are necessary to keep visitors interested. It is impossible to overemphasise the importance of interactivity for exhibits.

There is a conflict between the necessity for an objectification of the mathematical concepts present in the museum and the abstract character of mathematics. It is a challenge that anyone intending to organise a museum of mathematics has to face: how to make concrete an idea or mathematical theory? I believe that the answer must be rather pragmatic: one starts off from the areas of mathematics that best lend themselves to this move to the concrete, leaving to the future, even a remote future, those that are less suitable. In short, no pretence of completeness: museum mathematics is and will be only a subset of mathematics and not necessarily the most important or the most modern. As much as this is a painful choice, it is a necessary one.

Another problem in setting up a museum of mathematics is the level of attention required of visitors: this level of attention is higher than that required in other scientific museums, because of the very nature of the intellectual operation to be carried out. Thus, in a museum of science, visitors can limit themselves to reading exhibits, as they speaks with their very material nature; on the other hand, in a museum of mathematics, a second reading is required to penetrate a mathematical structure that cannot be immediately perceived. Hence the need for greater and constant attention that is difficult to maintain. The experience of the "Beyond the Compass" exhibition has taught that the maximum threshold is one hour; beyond this limit, visitors' attention becomes patchier and is limited to the more immediate aspects of the exhibition. Of course, one can alternate more intense sections requiring greater attention with more descriptive sections (such as historical, biographical or documentary), but even this method does not allow extending time much above two hours. As a consequence, a museum that is too big will be visited superficially, at least in part.

The museum project named "Archimedes's garden" keeps these considerations in mind. Firstly, the exposition structure is more interactive and this applies not just to the mathematical sections but also, although to a lesser extent, to the historical and documentary sections. It is organised in self-sufficient units, each of which deals with a subject independently. Of these, no more than four or five (of which one or two are of the more intense kind) will be exhibited simultaneously; the rest, if any, can be utilised in other shows elsewhere. The optimal size of these units is around 150–200 square metres, which gives a total of about 1,000 square metres of exhibition area.

This cellular structure of the museum allows its decentralised territorial organisation, with one or more branches proposing exhibition units in turn, but possibly also guest exhibitions conceived and realised outside the museum on mathematical themes. Of course, not all exhibitions are the same. The museum core is obviously constituted by the more mathematical exhibitions, by those exhibitions that illustrate mathematical theories and methods. It is important not to allow oneself to be conditioned by the theoretical level of the proposed mathematics: if we consider it from the point of view of exposition, elementary mathematics is no easier to explain than higher mathematics. In both cases, it is not important what the difficulty level of the tools used is, but rather the degree of communicability of the obtained results. In other words, it is not so much the mathematical theory that is of interest, as much as the possibility of its description in terms of the exhibits. This description will be all the more effective, the more familiar the objects and the phenomena that convey it.

In addition to this "mathematical" core, there may well be other more "cultural" exhibitions. These might be historical or documentary in type: what is important is that the interactive aspect is favoured. Natural complements to the exhibitions will be a multimedia section – basically mathematics video and hypertext – and a mathematics laboratory, where visitors can spend some time applying themselves to simple problems that require more inventiveness than knowledge.

Depending on the types of activity involved, different parts of the museum will need to be revised more or less often. While the video section can be set up on a permanent basis (although even here theme periods could be organised, in view of the increasing number of products on offer), the laboratory will have to keep renewing its offering to prevent it rapidly becoming obsolete. Mathematics exhibitions – from which the level of museum offer depends – will be more stable than other, more descriptive exhibitions. Moreover, it will be possible to exhibit them simultaneously in several different locations, as a distinctive sign of the museum. Finally, guest exhibitions will be shown for a shorter time, depending on the needs of their authors. In any event, the various locations will have to offer propositions that are at the same time unitary and articulated, with at most one or two sections in common – to testify to their unity of intent and planning – while the rest of the sections will vary across sites, so as to guarantee the independence of the offers made by the various exhibitions.

The territorial spread has another advantage: its greater attraction to schools. Although not exclusively aimed at schools, but open to everyone, museums see schools as a privileged interlocutor and propose themselves as possible school trip destinations. In particular, the main customer group is constituted by schools within a certain distance from a museum, as this allows easy organisation of one-day school trips to the museum. A multicentre structure will allow covering a greater part of the national territory, without overlapping.

The first of these exhibition centres was opened on September 1999 in the castle of S. Martino in Priverno (LT), Italy. The castle of S. Martino represents a decentralised but promising museum location[1], given its position halfway between Rome and Naples – Priverno is about one hour from Rome and one and a half hours from Naples when travelling by train – the pleasantness of the area and its nearby 30-hectare park with picnic areas, the proximity of the Abbey of Fossanova – one of the pearls of Italian Gothic-Cistercian architecture – and of the archeological area of the old Privernum.

Another exhibition centre is planned in Florence, in a building made available by the Province of Florence that also allocated funds for the renovation of the building. The opening of the Florence centre is expected to take place in 2004.

1 *The "Giardino di Archimede" (Archimedes's Garden) is open Tuesday to Sunday. Guided tours are also available for groups of over ten people. To book, call (Italy) 0773-904601. Further details can be obtained online at: http://www.archimede.ms*

Pop Maths All Around the World

Michel Darche

In French

Since the Bourbaki School, the French Mathematical Society ("SMF" in French) has always been present to popularise maths throughout France.

After the initial experiment of the mathematical space in the "Palais de la Découverte" in Paris, interactive exhibits were created during the eighties, and a large exposition hall for maths was built in 1986, in the new "Cité des Sciences et de l'Industrie" located at La Villette in Paris.

The exhibits "Horizons Mathématiques", created in Orléans and Bourges in 1980 (and, subsequently, "Maths 2000"), were presented in more than 50 countries and 200 towns. They were visited by over one million people over twenty years. Since the World Mathematical Year 2000 (WMY), they have been updated.

In the year 2000, the International Mathematical Union launched a programme to popularise mathematics. Now, with the support of UNESCO, IMU, ICMI and EMS[1], a new travelling exhibition has proposed to continue this idea: **"Why Mathematics ?"**

This exhibit will be the interactive translation of two of the more successful regional poster series initiatives: Mathematics in Everyday Life and Mathematics in Nature produced by Centre•Sciences and the Committee for Mathematical Year 2000 as well as some of the manipulative models from *Mathematical Art,* produced by Tokai University in Japan.

It will be presented first in Paris in April-May 2004, then at ICME in Copenhague in July and later at the 4ECM in Stockholm in August 2004.

Making Maths Popular: Why Is It Necessary?

Mathematics is a part of our everyday lives. We have to show that mathematics is:
- astonishing, interesting and useful,
- accessible to everyone,
- plays a large part in daily life and has an important role in our culture, development and progress.

1 *IMU : International Mathematical Union*
ICMI : International Committee Mathematical Instruction
EMS : European Mathematical Society

Fig. 1. One of the 12 posters of "Math in the daily life"

"People should always be learning" is the basic principle of maths popularisation. Thus, we have to harmonise, throughout our life, our basic knowledge and our general culture.

Scientific training and scientific culture must lie upon experimental proceeding, sense of observation, research-like self-questioning, care for debate and for proving, and, even more nowadays, search for series of presumptions.

In maths, we have to compensate for the superiority of operating knowledge upon cultural facts, by using innovation, imagination, and abilities to lead research projects.

Our Concept of Maths Popularisation: the Search for "Right Problems"

The knowledge that young (and elder) people acquire is the answer to the questions they are asking to themselves (or to the questions they are asked).

How to proceed? By getting closer to research topics.

What is the 1st activity of a mathematician? Or, more precisely, the one in which we are interested? We think it is the one consisting in knowing how to describe "right problems".

What is a "right problem"?

For a mathematician (or, more generally, a research scientist), it is a problem which, either solved or not, should enable to develop new knowledge domains. For example, Fermat's last conjecture, which, before having been recently solved by Faltings, led to new knowledge and research subjects. On the other hand, the "4 colours theorem", which, even solved, did not provide us with new knowledge topics (except concerning the use of computers for mathematical proofs), even thought the concept of algorithm found there a new experiment field. [Référence: "Maths à venir", Palaiseau Workshop, 1992]

What is a "right problem" for vulgarising people?

The answer was given to us by D. Hilbert, during the 1st World Meeting of Mathematics in Paris in … 1900! He said, during his presentation of "the 27 problems for the 20th century", that a problem is right when you can explain it to anybody.

Of course, he meant the description of the problem, not its solution. Many scientific questions are excellent examples of this definition. Unfortunately, not all of

Fig. 2. "Maths 2000" in Tours (oct.'99)

them, and thus the job of vulgarising must consist in enabling everybody to understand complex problems.

When transposed to scientific culture, we can say nowadays that a good (scientific) question can be explained to everybody, on a desk corner or in front of a computer.

Examples:

1. Is the volume of 1 Kg of grain coffee larger or smaller than 1 Kg of moulded coffee?
2. Is 1 m^3 of dry sand heavier or lighter than 1 m^3 of wet sand?
3. M.J. Pérec runs a 200 meters race in 21 sec $3/10^{th}$. Johnson runs it in 19 sec $3/10^{th}$. If they had to run together, what delay should be given to Johnson in order to give Pérec a chance to win?

We can also find good questions based on observation:

1. In which direction does the moon rotate around the Earth?
2. In Europe, why is the weather cold in winter and warm in summer?

More experimental questions are used in our interactive exhibits. Which tools do we develop?

I will give several other examples:

- interactive hands-on exhibits, such as "Computer Spirit", The sky of Babylone, The Sun, our good star!, play with the sand ...
- 150 to 200 m^2 of travelling exhibits, with 20 to 50 hands-experiments on various topics, from classical subjects (Pythagore, perspective, binary logics) to new research topics (fractals, chaos, algorithms, complexity).
- These exhibits are now "on the road", and are very easy to install. They are proposed to anybody aged over 10 years. When we had to work in small cities, we realised that we would have problems to find halls large enough to receive these exhibits.
- That is why we created exhibits called "12 panels – 12 experiments", which are very easy to install inside a highschool, or in a library:
  80 m^2, and 2 or 3 boxes taking place in the back of car. That is the way we built small maths exhibits:
  - African games and culture
  - Order and chaos in nature
  - Games, randomness, and strategies

Fig. 3. "Maths 2ooo" in Valpareso (Chile , dec.'99)

- Luck and random in life
- From the eye to the brain
- Logical and maths games
- When science speaks Arabic
- Maths in the Mediterranean Sea

All these exhibits are circulating throughout France and abroad, especially within the 30 CCSTI ("Centres de culture scientifique, technique et Industrielle), which exist on the whole French territory and in foreign countries (Spain, Greece, ...).

Other actions are also planned for scholars. At Centre-Sciences, CCSTI of the "Centre" region (in the center of France), we make a particular effort towards pupils and teachers. We are installed in the teacher training centre ("IUFM" of the "Centre" region). Especially, we propose to highschools, even to the smallest, to welcome a researcher, who will describe his work and his passion: This is the so-called "100 researchers in 100 classrooms" operation, each year.

We work together with the education projects and the scientific workshops, also by proposing to them to be followed by a researcher serving as advisor. We develop training actions for teachers, ranging from the production of interactive exhibits (like those about time "at the 3rd top", or about communication "@, hang on").

Another Example: The Hellenic Centers for Science

In Patras, Athens, two buildings – aimed for 12 to 22 year-old students and - designed around exhibits about maths computers and physics will be available to students and teachers for a period of three years, starting in September 2001.

The students, and their teachers, can stay there from one day to one week. They organize their daytimes around three or four activities:
- visits of the interactive exhibits
- training workshops
- consulting in the multimedia room
- workshops about computer models

There you will find several evaluation elements collected by teachers and animators of these centers.

Fig. 4. "Maths 2000" in Athens (feb.'99)

Educational Process at the Hellenic Center of Sciences (HCS)

The didactic – educational function and procedure which takes place at the HCS in no way replaces whatever is being taught and achieved at school. Moreover, in noway the HCS should be regarded as a scientific museum, where the students as visitors – sightseers only peer at the exhibits.

Each of the 12 exhibits of "Maths 2000" and 12 ones of "Computer Spirit" of the permanent Exhibition at the HCS focuses on a specific thematic area, functions with a specific educational process and creates learning results ; and all these is succeeded through the perfectly organized interaction of the exhibits with the students.

Under the stimulus of creation, the student is called to seek, discover, explain and comprehend notions and phenomena most of which have already been taught or going to be taught at school. The student is called to use his critical mind and imagination and thus, he is prepared and directed to put forward the clear statement of problem, of physical laws and functions, in such a way that just afterwards he will be able to give the correct and accurate explanation and solution.

In this way the HCS plays and should play a supplementary role with the school. Furthermore the visitors – teachers are able to notice the students direct reactions related to the exhibits, as well as the educational process taking place at the HCS. The learning procedure aims so that the students would be able to

a. Interact with exhibits
b. Analyze the phenomena
c. Find the problem's solution on a paper and the board.

Fig. 5. "Data on the move" Praga (oct.'99)

More specifically the teacher who belongs to the Educational Group of the HCS (EG) will present the exhibits at the beginning of the visit and then the students, deviled into groups will focus on the exhibits they have chosen according to their own interest and thus the educational interaction starts in an organized way.

At the first stage of the educational procedure the students will be directed to an interaction with the exhibits observe the phenomena, think deeply, discover notions and put forward scientific reasoning and surmises. The teacher of the EG instruct the students and at the end of the first stage the students should have received the correct educational stimulus and put clearly in their minds the equivalent problems.

During the second stage the students are led to a room, where, through a teaching-dialogue process and by using special scientific-educational material, (e.g. video cassette, multimedia, special software), they focus on one of the phenomena-subjects that has raised any sort of problems / question to them during the interaction with the exhibits. It is worth saying that during this stage there exist debate, analysis and explanation and the students themselves are the "leading actors". It is interesting to be mentioned here that the students do impress with the quantity and the quality of their questions as well as the methods they used to face and solve the arised problems.

During the last stage of the educational procedure the students divided into groups of 5-8 are occupied with specific problems and questions and write there though on the paper. The "exercise" areas are the following:

- Informatics applications at the Computer lab
- Maths-Informatics relation
- Physics-Informatics
- Applications of the Mathematics to Sciences
- Maths-Informatics' relation with art
- Ancient Greek technology
- Educational CD / Video
- Questions arisen during the stages A and B
- Use of library-electronic, library-Internet
- Other educational activities

During this stage the students will state correctly whatever they have understood, while, at the same time, they learn how to debate, cooperate, decide and give a solution as a unit.

Afterwards the students go to a room where in groups they evaluate the educational process and present their own conclusions. In between the three educational stages there are brakes during which the students have the opportunity to listen music, enjoy refreshments and talk with other students and the teachers. Before the departure the teachers of the EG ask the visitors, both the students and their teachers to fill a questionnaire. Based on this the members of the EG will improve the educational process taking place at the HCS and take into consideration the visitors' comments.

At the end of each semester the visitors-teachers are invited in a Seminar held by the members of the EG. Members of the scientific committee and the

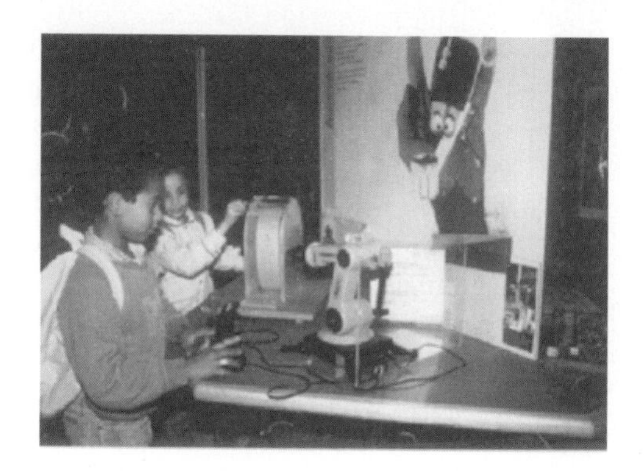

Fig. 6. "Data on the move" Dakar (1998)

Pedagogical Institute by participating at this seminar have the opportunity to express their though related to educational matters. The conclusions of the Seminar are sent to the Ministry of Education and the equivalent Departments and Institutions.

Finally, educational meetings and seminars are held by the HCS, in which greek and foreign teachers take part and debate about their teaching experiences and educational thoughts in the context of greek, european and international level.

During the first three years of its existence, more than 20,000 students have taken part in the educational visits at the HCS. (Fig. 7/Fig. 8)

The following remarks are really interesting:

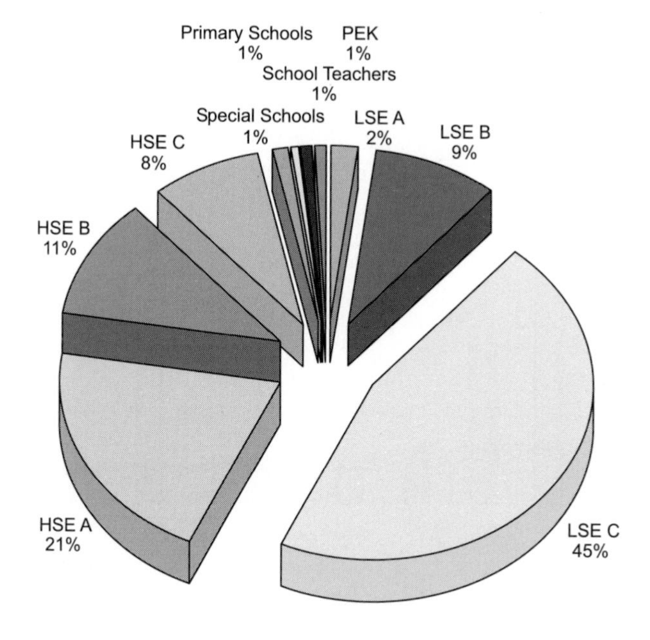

Fig. 7. The educational visits of the Primary Schools, of the Upper Secondary Education C Level and of the PEK are not included in the daily schedule at the HCS.

- A lot of students from Athens and other cities and places all over Greece have asked to visit the HCS and carry out educational activities which will last more than one day. The students of a school outside Athens, for example, after their first visit at the HCS, during which an activity had taken place focused on Symmetry, were divided into groups at school and have been working on this for the whole school year. During their second visit, the school year after, they presented their work and they did impressed us with the way they had worked and the depth of the thought and research they had reached.
- The lower percentage of the students who have visited the HCS are those who attend the C Levelof Upper Secondary Education and have to face the University Entrance Examinations. Consequently, they should follow "a standardized methodology and study", while the HCS is based on "a free thought and in quest of knowledge"
- The students who are the "bad" ones at school, were the ones who participated effectively at the HCS and gave the correct solutions relevant to the exhibits.
- Students coming from the "under priviledged areas" were more interested in the exhibits and participated more actively and positively than the ones from the "rich areas".
- Our pilot schedule concerning the primary education students was remarkably positive and encouraging.

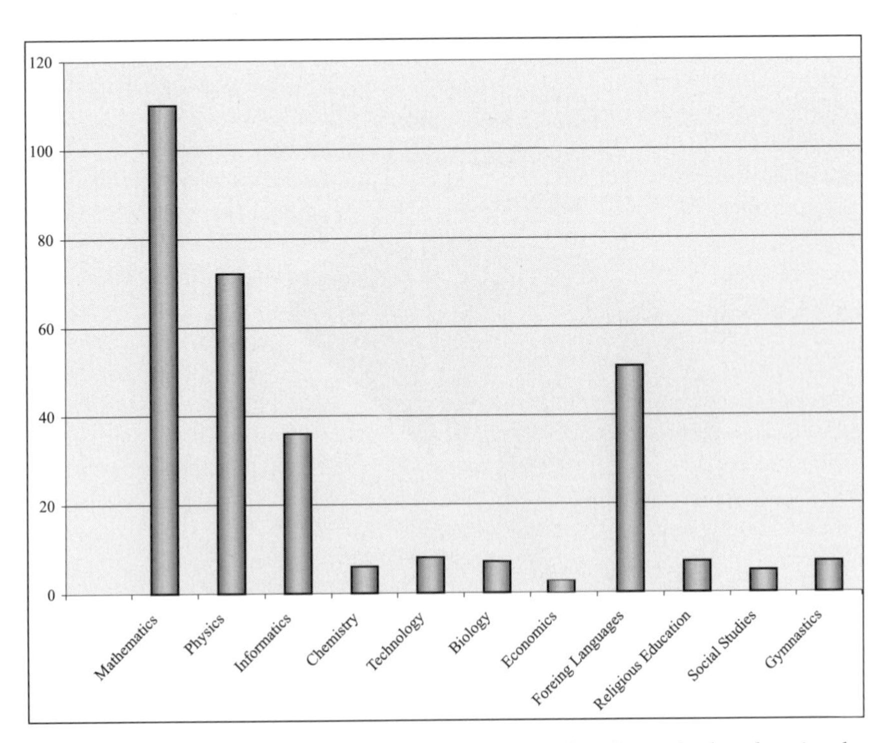

Fig. 8. Visitors-teachers who came with their students and took part in the educational process.

- The educational visits of disabled students were the most shocking ones and this made us think of finding new ways and being better organized in such a way that we will be able to accept more and more disabled students.
- Nearly all the students who visit the HCS ask from their teachers to arrange a second visit the soonest possible, so that they will be able to analyze in depth the exhibits they were interested in and were not satisfied it because of the limited available time.
- Most of the visitors – teachers have asked to visit the HCS again on their own, in order to have the opportunity to be better informed, to study and analyze the exhibits in detail and, in general, to collaborate and be in a close scientific relationship with the teachers of HCS.
- By the end of the first month of the school year nearly all the days "have been closed" and, consequently, we can not satisfy all of the schools.

Furthermore, an evaluation is presented based on questionnaires given to 1908 students at the end of their educational visit. The following diagrams show the students' opinion concerning their preferences for the exhibits as well as their interest in connection with the educational process, which takes place at the HCS. (Fig. 9)

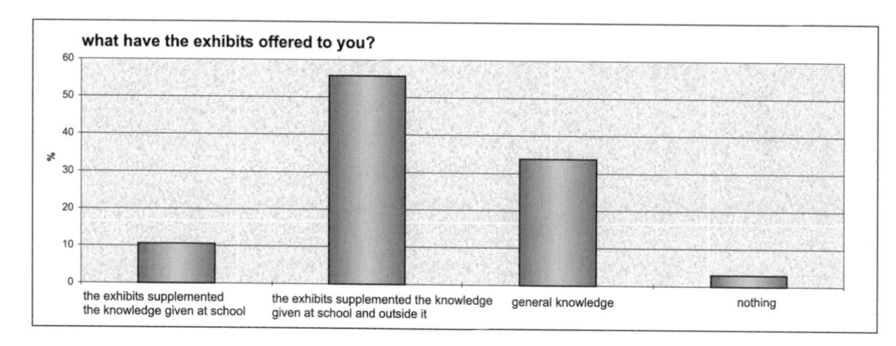

Fig. 9. What have the exhibits offered to you?

Which were the exhibits that impressed you at the Mathematics Exhibition?

1. Mathematics and Arts
2. Surfaces and Curvatures
3. Forms and Structures
4. Probability and Predictions
5. Areas and Puzzles
6. Problems and Conjectures
7. Nature and Symmetries
8. Fractals and Repetitions
9. Mathematics and Physics
10. Order and Chaos
11. Calculations and Algorithms
12. Patterns and Realities

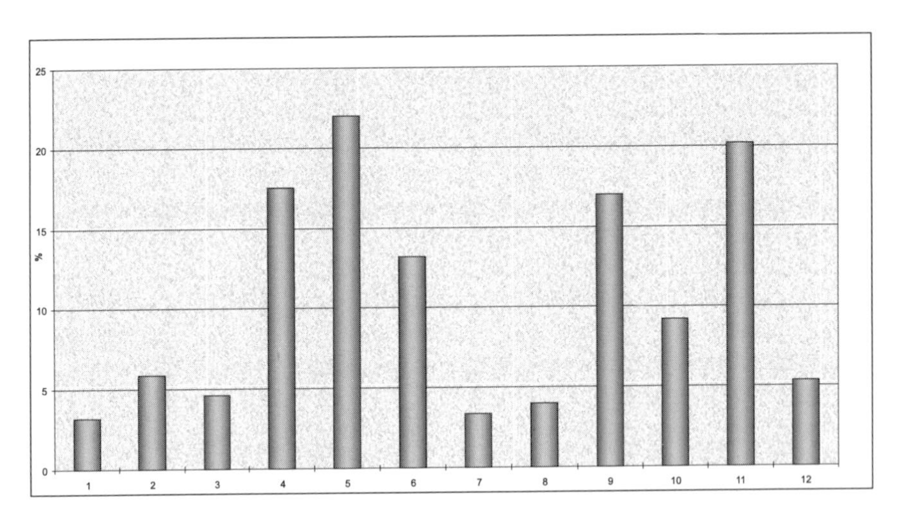

Fig. 10. Which were the exhibits that impressed you at the Exhibition

Which were the exhibits that impressed you at the Exhibition of "Computer Spirit"?

1. Everything is a matter of codification
2. The first algorithms
3. Algorithms of the year 2000
4. Pile-Unpile
5. Did you say logic?
6. Everything is on the tree
7. Are you well programmed?
8. Can everything be calculated?

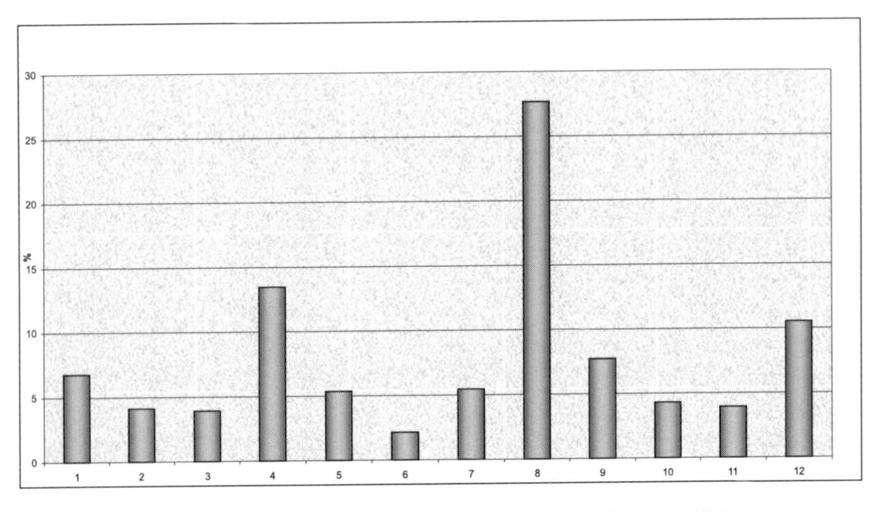

Fig. 11. Which were the exhibits that impressed you at the Exhibition of "Computer Spirit"?

9. Automata and Robots
10. Man's communication with machines
11. The computer: is it an intelligent tool?
12. A trip into the inner world of computers

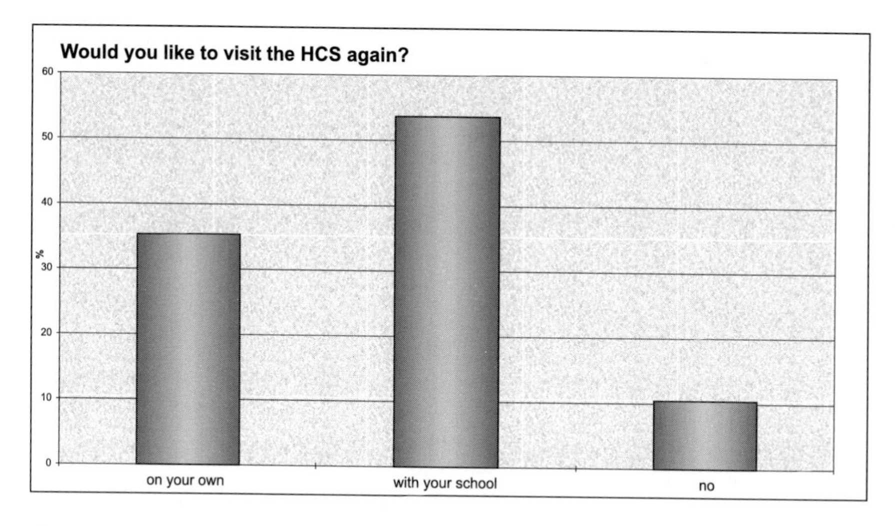

Fig. 12. Would you like to visit the HCS again? ...

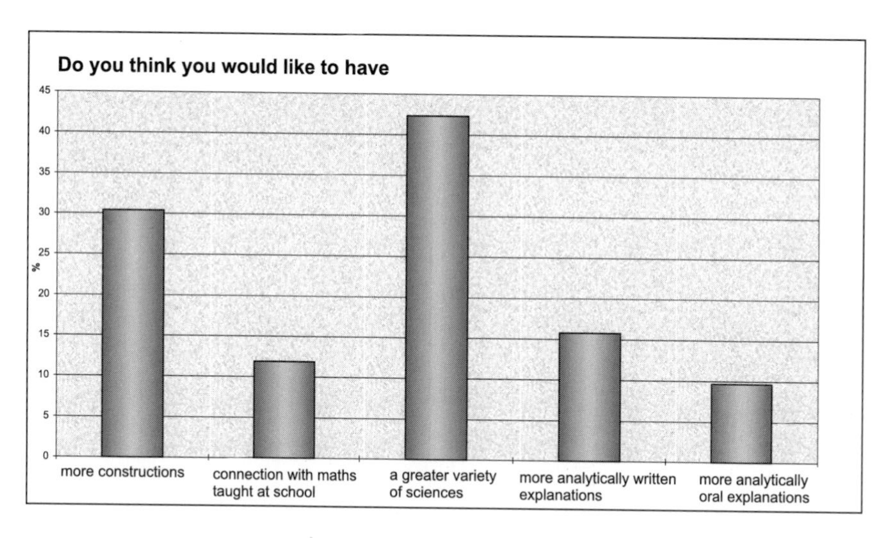

Fig. 13. Do you think you would like to have ...

Do you think you had a benefit of? **Yes: 94%** **No: 6%**

The Mathemagical Mystery Tour: Maths in the Malls

Richard Mankiewicz

The Mathemagical Mystery Tour *(MathsTour)* is a travelling exhibition promoting mathematics to the general public and designed to take place within shopping centres. We have now visited London, Edinburgh and Livingston, and hope to do a national tour during Maths Year 2000, the British government's initiative for World Mathematical Year. The first pilot exhibition was funded by the Millennium Awards Scheme, which is administered by the Royal Society and the British Association for the Advancement of Science and funded by the Millennium Commission.

I would like to describe some of the experiences of holding the exhibition and some thoughts about possible future developments in these public events. I do not intend this to be a scholarly or thorough article, but a record of my current thinking on raising public awareness of mathematics in the light of observations of the public's interaction with the *MathsTour.* The exhibition consists of a central cube with sides 2 metres in length with activities radiating out from this. It is a flexible structure and we usually require a total area of about 20-30 sqm to take account of visitor flow. We wanted to try a number of different types of activities so the *MathsTour* contains hands-on mathematical games and puzzles, a video installation, computers, a small shop (if the mall consented) and a number of leaflet distribution points.

We were heartened to see that by far the most popular activities were the games & puzzles, with over 80% of visitors interviewed saying that they were both the reason for stopping at the stand and also why they stayed for so long. The exhibition was very colourful and attractive, covered with posters of mathematical structures, and the whole feel was very much like a mathematical playground. I think there is a very serious point to be made here; that learning through play must not be underestimated and that a playful trial-and-error approach to problem solving is far closer to the reality of being a mathematician or a scientist than the presentation of arid formulae followed by repetitious exercises. One of my favourite quotes from one of the kids was that he thought the puzzles were *tough but cool.* These sentiments were echoed by other visitors who became hooked on solving a particular game and who were not disconcerted because it appeared difficult. There was a range of activities of varying difficulty as well as some open-ended games such as tilings. We were very pleased to notice that the main criterion for a successful activity was not whether it was difficult or not but whether it was boring or not. Sometimes the easy puzzles were indeed seen as boring because they were too trivial. "It's good to be able to think about things", said one woman, "especially in a place like this." Comments like these need to be looked at more closely and if widely held would be of enormous benefit in developing truly effective curricula for our children and motivate adults to look at mathematics afresh. Attempts at making the subject easier could well be having the opposite effect to that expected and making it more boring and even less appealing.

There was a great contrast at the *MathsTour* between the interaction with the hands-on activities compared with the computer interactives. The ubiquitous use of computers at home, at school and at work, and the growing use of the Internet means that merely having a computer on show is no longer enough of an attraction. Indeed only 15% of respondents cited it as their favourite activity. This is probably high enough to retain some form of computer activity but we need to think carefully what sort this should be. The costs of adapting software for an exhibition and in ensuring the system's security need to be proportional to its popularity. In the future we may look at using the power of computers in a more imaginative way which is more impressive than a mere monitor and mouse.

In the debate about hands-on versus computers I think we just need to accept that each has its strengths and weaknesses. An exhibit on cryptography or game theory would probably require a computer, as would the visualization of many subtle mathematical structures. But rotating an on-screen polyhedron is a far inferior experience, in my opinion, than the physical handling of a real object. We are physical beings and would be wise to look at the ways in which handling and playing with things enhances our experience and understanding of even abstract ideas. The freedom to adapt (or subvert) exhibits to other ends is a freedom not available within the constructed world of a program. By far and away our most successful single exhibit was a Hoberman sphere. After the initial amazement some children took to inventing games with it such as volleyball. One amusing story not related to the *MathsTour* involves a museum that shall remain nameless. An interactive exhibition was temporarily housed within a traditional museum and was so successful that children had to be repeatedly told by security guards that the rest of the museum was absolutely non-interactive!

Returning to the contents of the *MathsTour*, the video had a number of short programmes on mathematics made for television. These programmes were produced by the BBC, Channel 4 and WQED Pittsburgh, and were aimed at a range of ages. It became fairly obvious that the main role of the video was as a beacon attracting people who could see it from afar. Very few people had the time or inclination to watch the video as a whole programme and in many ways it became more part of the set design rather than functioning as a separate activity on a par with the puzzles and computers. We also had a raffle with mathematical prizes and, where possible, had a limited range of products on sale, mainly mathematical games & puzzles and software.

The helpers at the exhibition were usually local students, either from a university or school students in their final years. This has worked generally very well and we've had a good mixture of male and female helpers, often they have all been girls. This is very helpful in dispelling the myth that mathematics is exclusively for boys. It has also been a very worthwhile experience for the helpers, gaining confidence in dealing with the general public and developing their communications skills, as well as turning the tables on teachers. The local education authorities have been very keen on this aspect of the *MathsTour* and we may also use specialist teachers as helpers in the future.

The public's response to the *MathsTour* was overwhelmingly positive. The weekends were very busy and over a week we estimated about 20,000 people stopped to engage in some way with the exhibition, even if it was just to pick up a leaflet. The visitor numbers were about 10% of the average footfall passing the exhibition. About 40% of visitors spent less than 5 minutes at the stand but over 15% spent longer than 20 minutes. People are generally in the shopping mall for very different reasons so we were very pleased at the level of interest and that the content of the exhibition was able to sustain long-term interest as well as a quick visit. I feel it has also justified the whole philosophy of bringing mathematics within high density public spaces. Where else could we possibly obtain such visitor numbers? Perhaps only within other public areas such as airport lounges and railway stations. The *MathsTour's* design and ambitions were to reach out especially to those people who perhaps would not normally attend science or mathematics events. I think we succeeded in this and look forward to further evaluations on what visitors got out of their experience after visiting the exhibition.

The information leaflets we had on the stand were a crucial aspect of the whole enterprise. After attracting visitors to the stand and making the environment intriguing and enjoyable it was important people took away with them the means to continue their interest sparked by the *MathsTour*. The information was a mixture of educational resources such as full programme listings for the BBC and Channel 4, as well as catalogues of educational mathematical products and information on local resources and activities for both children and adults. Much of this information is readily available to teachers but most parents have never seen any of this. The *MathsTour* has thus developed into an important element in bridging the worlds of school education and home learning. Both teachers and parents can have access to the same resources thereby creating a more integrated and supportive learning environment for the children. We could, of course, have just had a promotional stand with a portable screen, a table and heaps of leaflets. But this would not have attracted anywhere near the numbers of people we had at the exhibition. I have observed similar stands and people's first reaction is to walk away from them fearing they are about to be sold something, and nobody wants to be trapped into a sales pitch, even in a shopping mall. People have also in general a poor image of mathematics, born from their own educational experiences either past or present, that it was important to create an attractor. I think it worked.

Criticisms levelled at the *MathsTour* are that it does not have any *real maths* and that it does not teach visitors any maths. I have to state categorically that the aim of the *MathsTour* is not to teach mathematics but to change attitudes. If we are able to change people's attitudes for the positive then I feel our job has been well done. If people are able to take the next step towards a more supportive environment for mathematics learning then I feel our job has been well done. The whole design focuses on colour and geometry and the activities are challenging and stimulating – there is not an equation in sight. The mathematics is implicit rather than explicit. What I wish to encourage is mathematical thinking, creative thinking and problem-solving – attitudes and methods which are far more useful and relevant to most people than, for example, learning a formula to solve quadratics. I think people would run away from the exhibition if they felt educated at. One of the helpers at Livingston said that "you'd never have thought such a wee puzzle would make you think about maths". This is exactly the change of attitude I feel the exhibition does best. Formal mathematics education has to be left to institutions, or even better, should be encouraged as a personal activity.

One must also take into account the dynamics of a shopping mall and the time people have available to stop and engage with an exhibition which to many was a great surprise to see. Very few people spent more than half an hour at the stand and so the activities had to have an immediate impact. I don't think we could afford to have exhibits which were very wordy or which would take a long time to understand. In fact, I believe one of the problems with computers in an exhibition is that one only gets real benefit from a program after a little navigation through the environment and an understanding of the scope and structure of the program. This is all well and good in the comfort of one's own home but in a hectic environment there need to be more short-term payoffs. This is not a matter of decreasing attention spans but of motivation and environment. Exhibits which work well in a science centre may well not be so successful in a travelling exhibition. The converse may also be true. However, I feel that something similar to the *MathsTour* would work well as an advertisement for a local science centre wishing to increase visitor numbers. Such a *taster* would need to have the same impact as an advert.

Regarding showing the utility of mathematics in everyday life, the video showed much of this, especially the Life by the Numbers programmes from the States. But I feel that promoting the applications of mathematics requires a separate exhibition and that trying to put together too many different messages into a relatively small space would have confused the issue and diluted our main aim. Mathematics is still a very hidden subject and there are few products which are explicitly mathematical. For example, for all the mathematics that goes into engineering a car it is not one of its selling points and it is nowhere to be seen by an inexpert eye. People choose a car for form and function without really caring about the science that goes into it. Even software is not generally regarded as being a branch of mathematics, and the consumer is merely interest as to whether the piece of software does what they want and not in the underlying code. All of this is to say that these issues are important ones to tackle in our drive to promote mathematics to a wider audience, they're just ones which the *MathsTour* does not explicitly deal with.

After having designed the prototype *MathsTour*, and from my experience of it out in the public, I think that the way it has worked has led to a refinement of my initially hopeful ideas of what could be achieved and the methods required to achieve them. I think the way forward is to concentrate on making mathematics visual and tactile. This would be a mixture of hands-on activities, computer controlled interactives and an awareness of the overall design of the environment. Mathematics and science have generally not been very concerned with an external audience and communication has been largely amongst professionals (or to funding agencies). If mathematics is to be promoted at a cultural level then I feel that we need to focus far more on the audience, or audiences. We need to learn the skills of marketing, advertising, design and communications. This is not to further obscure the mathematics but to develop a broader range of communication media. I feel our task is analogous to translating ideas from one language into another where often the idioms are very different in order to express the same concept. As an analogy, if we compare mathematics with music we see that both have a specialized language and practitioners in each area can communicate with each other in a language which would be pure hieroglyphs to an outsider. And yet a composer can hear a score in a way similar to a mathematician visualizing an equation. But music has a public face, a performance can be appreciated without being able to read the score, whereas mathematics so far has had little in terms of performance. I think developments in visual mathematics, video and computer graphics, together with interactive installations may herald a new and fresh approach to the presentation of mathematics to a general audience. There are now many fine books popularizing mathematics, all I'm saying is that in order to broaden the cultural impact we need to also look at exploiting all the other media.

For the future, I hope to take the *MathsTour* around the country, probably a new version redesigned and built to accentuate the most successful elements of these pilots. There are many aspects of mathematics we could promote in the future and it will probably be a few years before the novelty has worn off with the public. It would also be interesting to see whether this is exportable to other countries. I suspect that children are the same around the world and that playing will never go out of fashion.

Mathematics and Literature

Mathematics and Literature

Lucio Russo

Translated from the Italian by Emanuela Moreale

If we exclude the obvious influence of science on the minor literature genres of scientific popularisation and sci-fi, the relationship between mathematics works and literary works seem to be secondary and exceptional, even though on various occasions it has been the case that a literary work has contained interesting observations pertaining to mathematics. One such example is that of *Young Törless.*

The aim of this contribution, however, is that of mentioning two more types of interaction between mathematics and literary world that – although often ignored – probably deserve closer attention.

Rhetoric and Mathematical Proof

In his Institutio Oratoria (I,10), Quintilian writes[1]:

> But geometry and oratory are related in a yet more important way than this. In the first place, logical development is one of the necessities of geometry. And is it not equally a necessity for oratory? Geometry arrives at its conclusions from definite premises, and by arguing from what is certain proves what was previously uncertain. Is it not just what we do in speaking? Again, are not the problems of geometry almost entirely solved by the syllogistic method [...]? But even the orator will [...] use the syllogism, and he will certainly make use of the enthymeme, which is a rhetorical form of syllogism. Further, the most absolute form of proof is [..] γραμμικαὶ ἀποδείξεις (linear or geometrical demonstration). And what is the aim of oratory if not proof? [...] An orator [...] can under no circumstances dispense with a knowledge of geometry.

Quintilian's statements can certainly appear to be a rhetorical exercise, but in fact they contain the indirect remembrance of a genetic relationship between oratory and geometry. Not only those mentioned by Quintilian, but also other

[1] *Based on translation by H. E. Butler, The Loeb Classical Library, 1921.*

forms of argumentation had appeared in rhetoric before appearing in mathematical proofs.

In reality, it is not difficult to draw a continuous line connecting the forms of argumentations studied in rhetoric schools with pre-Socrates philosophical schools' *proofs,* the theory of syllogism developed by Aristotle and the demonstrative method that became common in mathematics. The link between oratory and proof (in its *syllogistic* version) is particularly clear in Aristotle's Rhetoric, in which various types of argumentation are identified and the fact is stressed that rhetoricians' enthymemes are but syllogisms. Aristotle presents rhetoric as an application of the tools he himself had developed in his logic work, but the historical order had obviously been reversed. Since treatises on the art of rhetoric preceded work on logic by roughly a century, it would seem that the theory of syllogism must have been born, at least to some extent, as a reflection on the enthymeme developed within deliberative and forensic rhetoric. This is not the place to discuss the differences between Euclid's demonstrative method and Aristotle's syllogistic demonstration, but the existence of a tight connection between the two forms of argumentation appears to me to be totally clear.

Quintilian's final statement should therefore be somewhat reversed: before Euclid, there could not possibly be a geometrician who was not aware of the art of rhetoric. Nowadays – the term rhetoric having taken on a pejorative sense and knowledge of classical culture become a rarity among mathematicians – the nearly *literary* origin of the mathematical method may appear strange to many, but it should cause reflection. The art of rhetoric consisted in the capacity to argue convincingly in assemblies with decisional powers and it is no case that it found no place in autocratic regimes. In fact, in the Egypt of pharaohs, no demonstrative method had developed and mathematical papyri of the time contain only prescriptions to be followed scrupulously on the sheer basis of the principle of authority. Instead, the art of rhetoric developed only during the Greek democracies of the V century. There is therefore an important link between the existence of some form of democracy and the development of argumentative capacities that has led to the demonstrative method.

Even in modern Europe, until some decades ago, the argumentative capacities of ruling class exponents had benefited from the study of rational geometry. The latter was considered to be part of basic secondary school education. It is sufficient to follow any television debate to verify how the abandonment of the demonstrative method in school has been accompanied by a lowering of the argumentative capacities of both politicians and audience. The media's widespread contempt towards *theorems* is one of the elements of the process under way, which is gradually eliminating the demonstrative method even from university curricula.

Literature and Scientific Misinformation

Edgar Allan Poe concludes his tale *The Mystery of Marie Roget* with the following reflections:

> [...] we must not fail to hold in view that the very Calculus of Probabilities to which I have referred, forbids all idea of the extension of the parallel: – forbids it with a positiveness strong and decided just in proportion as this parallel has already been longdrawn and exact. This is one of those anomalous propositions which, seemingly appealing to thought altogether apart from the mathematical, is yet one which only the mathematician can fully entertain. Nothing, for example, is more difficult than to convince the merely general reader that the fact of sixes having been thrown twice in succession by a player at dice, is sufficient cause for betting the largest odds that sixes will not be thrown in the third attempt. A suggestion to this effect is usually rejected by the intellect at once. It does not appear that the two throws which have been completed, and which lie now absolutely in the Past, can have influence upon the throw which exists only in the Future. The chance for throwing sixes seems to be precisely as it was at any ordinary time – that is to say, subject only to the influence of the various other throws which may be made by the dice. And this is a reflection which appears so exceedingly obvious that attempts to controvert it are received more frequently with a derisive smile than with anything like respectful attention.

This passage suggests a series of considerations. The first and obvious one is that Poe understands nothing of probability calculus. He believes that a dice can remember what faces have previously come up, tending not to show a six if this number came up in the two previous throws. I think this passage is of remarkable interest: not for these absurd statements, but – as pointed out by student Alessandro Della Corte, whom I thank – because of the acknowledgement that they appear to be absurd to the common reader, in Poe's parlance. Anyone who has studied probability theory will testify to often having had the opposite experience of unsuccessfully trying to remove Poe's absurd idea from the minds of ordinary people, as they hardly allow themselves to be guided by the common sense that Poe (inadvertently) found so abundant in his interlocutors.

The contrast between the two experiences – the one described by Poe and our own – affords two possible explanations. The first and most banal one is that not only did the American writer not understand mathematics, but he did not understand the discussions of ordinary people either, or simply made up their opinions. However, this is an explanation that is not just excessively harsh towards an undoubtedly brilliant intellectual, but hardly believable. In fact, it is extremely unlikely that correct ideas attributed to others might be produced by the person criticising the ideas without this person understanding them.

There remains another possible but disturbing explanation: non-mathematicians' intuition concerning casual phenomena has drastically worsened since

Poe's times (the tale dates back to November 1842). Several elements seem to confirm this interpretation.

First of all, Poe does not counter what he considers ignorant people's deceptive intuition with any rational argument, but only with the principle of authority. He believes that probability theory has reached some results that simply cannot be understood by non-mathematicians, in that they appear contrary to extremely obvious ideas.

It is clear that it would have been impossible to conceive a similar argument before the law of large numbers was formulated, circulated and misunderstood. It could be presumed that Poe's interlocutors had not yet been reached by the mathematical misinformation that had neutralised the writer's natural common sense. In fact, many largely common absurdities can only be explained as degenerations of scientific statements. In order to remain within the field of probabilities, it could be noted that even today people playing the lottery on *late* numbers and supporters of similar ideas do not defend their theories with arguments based on common sense, but rather claim to be applying mathematical results and invoke the principle of authority.

The habit of living with a science whose internal logic is not understood – but whose power is admired – greatly weakens ordinary people's trust in common sense. It is feasible to think that literature has played some part in this process. Poe's story itself, for instance, must have had some misleading effect on his numerous readers' understanding of probabilities. Of course, nowadays the new *media* allow a much more pervasive and powerful work of misinformation to take place.

From Galilee to Galileo

Piergiorgio Odifreddi

Translated from the Italian by Emanuela Moreale

Mathematics exerts its more immediate and direct influence over beliefs – and not just religious ones – through numerology. The most typical example is *perfect numbers,* that is those numbers that are equal to the sum of their divisors (including the number 1, but obviously excluding the number itself). The first two examples of perfect numbers are 6 and 28, respectively the sum of 1, 2 and 3 and 1, 2, 4, 7 and 14.

Already the very choice of name testifies to the reverence that the Greeks had for perfect numbers. In his *Creation of the World,* the first century Jewish philosopher Philo Judaeus even claims that God created the world in six days exactly because six is a perfect number: a gentle reminder of the perfection of the whole affair. In *City of God* (XI, 30), St. Augustine associates himself with this view by commenting:

> [T]herefore, we must not despise the science of numbers, which, in many passages of holy Scripture, is found to be of eminent service to the careful interpreter.

Starting from the Middle Ages, philosophical and mathematical questions connected with the notion of infinity would often digress into the theological field. One of the problems that afflicted scholars was, for instance, the tension between God's omnipotence – asserted by the Gospel according to Luke (I, 37) – on the one hand and Aristotle's claim about the impossibility of infinity in nature on the other.

> If God is really omnipotent, why could He not create an infinite stone? St. Thomas Aquinas gives an answer to this in his *Summa Theologiae* (VII, 2) where he states that God, although omnipotent, cannot create something non-being, because this would create a contradiction, so something absolutely infinite cannot be created.

In other words, an omnipotent being can do everything possible, but not even an omnipotent being can do the impossible, since otherwise the impossible would not be impossible anymore.

Gregorius from Rimini instead believed that Zenonian infinite regress could help even the Almighty. He demonstrated that, had He wanted, God would have been able to create an infinite stone within a mere hour: it was sufficient for Him to start with a one kilogram stone and add half a kilogram after half an hour, another kilogram after a quarter of an hour and so on.

Giovanni Buridano was not persuaded by this argument: he believed that the argument only showed that God could create stones of unlimited size in less than one hour, but not that He could complete the work.

Nicola Cusano was the first theologician who really faced the problem of infinity in theology. He attempted to base Christianity on a philosophical system inspired by mathematics: both geometry – in his *De docta ignorantia* – and arithmetic – in *De conjectures.* (From now on these works will be referred to as D and C respectively).

According to Cusano, just like other mystics, God is ineffable and can be talked about only in a negative sense (D, 87): theology is therefore necessarily "clear and brief" (C, 20) and this could (and should) be about it. However, Cusano finds a way out: if the ineffable cannot be arrived at directly, it can be arrived at metaphorically through the infinite – since this is in fact *non*-finite – and therefore through mathematics (D, 33). This provides him with a whole series of images that should serve to show what cannot be talked about. For instance, the fact that a straight line is made of as many points as the number of its segments, shows that God can simultaneously be wholly in each of His creatures and contain them all (D, 51).

Similarly, the fact that a straight line is an infinite triangle shows that God can indeed be one and a Trinity at the same time (D, 56). The way that Cusano arrives at the odd link between straight line and infinite triangle is through the following reasoning: in a triangle, one side is smaller than the sum of the other two sides, so if one side is infinite, then the other two are infinite too. A similar argument applied to angles. However, since everything coincides in infinity, an infinite triangle has only one side and one angle and they coincide (D, 37).

A comparison that Cusano uses obsessively is the circle. Cusano uses it to describe the flow descending from God to the intellect, reason and senses, before flowing back in the opposite direction (C, 64). Given the impossibility of approximating it by using polygons, he found a parable of the vain attempt of the intellect to comprehend infinity, as well as a description of the relationship between man and God (D,10 and 206). In an infinite circle – whose centre, diameter and circumference coincide – he saw an image of God, who is simultaneously inside everything, penetrates it and surrounds it (D, 64).

Similar considerations apply to an infinite circle, with centre anywhere and circumference nowhere (D, 162), that constitutes one of the most commonly used images in theology, both before and after Cusano. Its thrilling story, written by Jorge Luis Borges in *The Sphere of Pascal* – part of the *Other Inquisitions* collection – involves master Eckhart, Charles de Bovelles, Giordano Bruno and Blaise Pascal. As to the geometric shape, in Pascal – another essay belonging to the same collection – Borges notes that the story records spherical gods, but only conic, cubic or pyramidal idols.

Cusano was aware that the preceding images were destined to being unsuccessful, just like the whole of theology that, as usual, is also circular (D, 66). He was however the first who attempted to use the concept of infinity in a positive way, by applying it not only to God, but also partially to the universe: the latter is in fact finite from God's external viewpoint, but appears to be infinite from our internal point of view (D, 97).

In 1584, his disciple Giordano Bruno went even further in his *De l'infinito Universo et Mondi* where he stated:

> I claim that the universe is infinite, because it has no borders, limits or surface. I claim that the universe is not totally infinite, because each part of it we might select is indeed finite and – of all the innumerable worlds it contains – each is finite.
>
> I claim that God is totally infinite, because He excludes any term and each of His attributes is one and infinite. I call God totally infinite, because all of Him is in all the world and in each part infinitely and totally.

According to Bruno, the universe is therefore infinite, but less infinite than God: thus he anticipates the existence of varying degrees of infinity that will be discovered by Georg Cantor in 1874. But Bruno is guilty of the worst crime: saying the right thing for the wrong reason – as Thomas Eliot would put it – or – as Sartre would put it – of being wrong in his way of being right. In fact, he sees the difference between the universe and God in the fact that each limited part of the former is finite, while each limited part of the latter is infinite. But the example of integers and rational numbers shows that the two properties do not necessarily imply the existence of different orders of infinity.

On the other hand, Charles de Bovelles was still bogged down in the finiteness of universe and the uniqueness of infinity. The first publisher of Cusano's works, in the *Book of Nothingness*, he followed the route of mathematical theology shown by his master. From the (incorrect) hypothesis that two infinities can only coexist if they are equal, he derived that God coincides with nothingness. From the (correct) assumption that a point has no size – and that therefore removing it or adding it to a metric quantity does not modify it – he deduced that creation does not diminish either God or nothingness. In particular, God did not have to recede to make room for the world and, had He wanted to, He would have had enough room to create an infinite number of other worlds.

In *Thoughts* (162 and 164) – published posthumously in 1670 – Pascal trusts his superior mathematical ability to search for more adequate metaphors. A first metaphor is aimed at showing the possibility that God is, at the same time, *infinite and without parts;* he did this by comparing God to a point moving everywhere with an infinite velocity. God is therefore one in all locations and whole in each location. Up to this point, Pascal does not surpass the level of the worst Cusano.

A second metaphor aims to use mathematical analogy to prove that *it is possible to know that God exists, without knowing what God is.* The idea is that there exist numbers whose existence is known, but whose properties are not known: for instance, since there is an infinite amount of integers, it is known that there exist an infinite number determining their quantity. However, since adding one to infinity does not change it, we do not know whether that number is odd or even.

Today we would say that the problem is not that we do not know if an infinite number is odd or even, but rather that not all properties of finite numbers apply to infinite numbers. The mathematical metaphor should therefore be interpreted theologically by saying that *our properties do not apply to God*. One of such properties might well be existence; it goes without saying that this is probably even more the case with clearly anthropomorphic properties, such as goodness, justice or knowledge.

The passing of time taught mathematicians to adjust to the idea of infinity and to manipulate it in a (sometimes too) casual manner. For instance, thanks mostly to Newton's approach to infinitesimal calculus, the notion of infinite sum stopped being considered paradoxical and the idea that it can correspond to a finite number was accepted. The problem was, however, *what finite number*. In this respect, the alternate series:

$$1 - 1 + 1 - 1 + \dots$$

was perhaps the most talked about series in the seventeen hundreds. By rearranging the brackets, it gives rise to the following irritating paradox:

$$\begin{aligned} 0 &= (1 - 1) + (1 - 1) + \dots \\ &= 1 + (-1 + 1) + (-1 + 1) + \dots \end{aligned}$$

Not only did Guido Grandi happily accept the result, but – in his work *Quadratura Circuli et Hyperbolae* – he claimed that this was an explanation of the way in which God had created the world out of nothing. This statement should have sounded blasphemous, as it reduced creation to a question of brackets!

This incredible argument should obviously be assessed in context: paradoxes later allowed the discovery of a precise definition of the sum of a series as a limit of partial sums and also the comprehension of the fact that the ambiguity above derives from assigning a definite sum to a series whose partial sums oscillate between 0 and 1.

It was Niels Henrik Abel who wrote the "final word" in the history of the alternate series in 1828. Going back to theology, from which Grandi had started off, he declared that series were in fact devil's creations and should be treated as such.

Among those that dealt with the issue of the value of alternate series was also Leibniz, who Solomonically decided on ½, this being the average of the partial sums and therefore the most likely value. This is remindful of the story of the chicken that statistics would considered eaten half each by two people, even though one of them ate the whole thing and the other fasted. Leibniz admitted that his reasoning was more metaphysical than mathematical, but claimed that mathematics was effectively more metaphysics than is normally conceded.

In fact, in 1734, he published *Mathematical Proof of the Creation and Ordering of the World*. This time, the mathematics was correct and was based on the discovery of binary notation that allowed writing any integer using only 0s and 1s: for instance, 2 becomes 10, 3 becomes 11, 4 becomes 100 and so on. Leibniz had the idea while meditating on the hexagrams of classic Taoist *I Ching*, whom he had learnt about through some acquaintances of his who were missionaries in China.

The binary system is the foundation of computer logic and thus nowadays it needs no further justifications. In Leibniz's times, however, it seemed only a curiosity, which he rushed to embellish with his usual metaphysics, summarised by the motto on the cover reading *omnibus ex nihilo ducendis sufficit unum*, "to generate everything from nothing, the number one is sufficient". In other words, the possibility to reduce the representation of any number to 0 and 1 became an image of the creation of everything starting from nothingness and God.

Another great mathematician who took such an interest in theology that he became Doctor in Divinity in Oxford was John Wallis. In his 1690 work entitled *Doctrine of the Blessed Trinity Briefly Explained*, he proposed the following image: in the same way as having three distinct dimensions (width, depth and height) does not prevent a cube from being just one cube, so having three distinct persons does not stop the Trinity from being one God.

The superficiality of Wallis's metaphor is uncovered by adding one more dimension: it would be sufficient to replace a three-dimensional cube with a four-dimensional hypercube (for instance, to accommodate Bernardo di Chiaravalle's *Considerations*, according to which "God is length, width, depth and height") to obtain a God that is one and four at the same time, which sounds like a bar joke.

It would be wrong to believe that all mathematicians living through the Seventeen Hundreds were willing to use their art to dabble in theology, even if in odd ways. On the contrary, many had to be unwilling to deal with theological concepts and arguments, but well disposed towards mathematical ones, because in 1734 bishop Berkeley wrote a treatise just to criticise them. The complete title of his work is meaningful: *The ANALYST: or a DISCOURSE Addressed to an Infidel Mathematician, wherein it is examined whether the Object, Principles and Inferences of the modern Analysis are more distinctly conceived, or more evidently deduced, than Religious Mysterie and Points of Faith... "First cast out the beam out of thine own Eye; and then shall thou see clearly to cast out the mote of thy brother's eye".*

Berkeley's attack was aimed at the new discipline of mathematical analysis and in particular at infinities and infinitesimals, that the bishop claimed did not exist because they could not be directly perceived: his motto was in fact *esse est percipi* "to be is to be perceived". He was attacked in turn and then realised that he was holding the same mathematical positions he had criticised in his theological adversaries; therefore he felt it was not inconsistent of him to respond with *A Defense of Free-Thinking in Mathematics.*

If even infinitesimal calculus has shown us religious aspects, set theory – being the arrival point of a centuries-old argument on infinity – could not help doing the same. The theological involvement of this theory began in 1874, when Cantor proved his most famous theorem: that the power of an infinite set has more elements than the starting set. The introduction of an infinite therefore generates an infinite number of greater and greater borderless infinities.

This worried the Roman Curia that became suspicious: not only had mathematics taken possession of the infinite, that could have been reserved to God, but it was even talking of several infinities, with an obvious risk of polytheism. Therefore a group of Dominicans was tasked to study the matter.

In 1886, Cantor – who had been baptised a Christian – sent his defence to cardinal Franzelin in the Vatican and explained that the infinities being talked about in mathematics were in fact the various omegas – the transfinites – beyond the finite. What worried the cardinal was the only Omega, the *absolute infinite* found beyond the transfinite.

The difference was essential: far from diminishing the essence of God with their proliferation, transfinites extolled it and tried to approximate it, without however being able to reach it: the paradoxes of infinity were in fact a sign of this inaccessibility. The same justification of transfinites was derived directly from God, as they "exist at the highest degree of reality as eternal ideas in the *Intellectus Divinus*". The cardinal was convinced and decreed: "The theory of the *Transfinitum* does not endanger religious truth".

What was endangered instead were Cantor's mental faculties: Cantor openly conceded that he was nothing more than a scribe at the service of God and that it was God who supplied him with inspiration and left him only the choice of the style to be used in writing out the revealed truths. Part of his sacred mission was to help the Church to correctly conceive the concept of infinity, by correcting the errors made by scholasticism:

> If I am right in asserting the truth or the possibility of the *Transfinitum*, then (undoubtedly) there would be a sure risk of religious error in remaining of the opposite opinion.

Unsurprisingly, Cantor ended up going in and out of mental hospitals. More surprisingly, after one of such visits, he wrote *Ex Oriente Lux*, a pamphlet that was meant to prove that Christ was the biological son of Joseph of Arimathea. In this respect, he emulated Newton, who – while studying for a chair at Trinity College (which, at the time, meant taking Orders) – had lost faith in the Trinity and converted to Aryanism, thus beginning a new career as a religious writer.

Newton's theological works, totalling one million words, represented a true scandal. They were universally and repeatedly refuted by the Royal Society, British Museum and many universities, including Harvard and Princeton. Eventually, in 1936, they were auctioned and now lay in Cambridge and Jerusalem, still mostly unpublished.

Although they do not explicitly use mathematics, they still represented an integral part of Newton's thought. In fact, he believed that the universe and the Scriptures, respectively God's work and word, were to be investigated and interpreted using the same method – demonstrations – in order to reach the same aim: the Truth. The latter was respectively: the true map of the Jerusalem Temple in the *Chronology of Ancient Reigns;* betrayals of the evangelic tradition in *Two Remarkable Corruptions of the Scriptures*; the identification of the Church with the Beast and the Pope with the Antichrist in the *Treatise on the Apocalypse*.

Nowadays, logic-mathematical theology is out of fashion. Gone are its times of glory and delirium: it is now mostly the object of ironic and sarcastic remarks that – as usual – contain a certain degree of truth. Let's consider the example of *Argumentum Ornithologicum*, a 1960 provocation by Jorge Luis Borges:

> I close my eyes and see a flock of birds. The vision lasts a second, or perhaps less; I am not sure how many birds I saw. Was the number of birds definite or indefinite? The problem involves the existence of God. If God exists, the number is definite, because God knows how many birds I saw. If God does not exist, the number is indefinite, because no one can have counted. In this case I saw fewer than ten birds (let us say) and more than one, but did not see nine, eight, seven, six, five, four, three, or two birds. I saw a number between ten and one, which was not nine, eight, seven, six, five, etc. That integer – not-nine, not-eight, not-seven, not-six, not-five, etc. – is inconceivable. *Ergo,* God exists.

It is obviously not a good idea to split hairs on this topic, since its title explicitly declares the semi-serious character of this work. It is however worth noting that the existence of God is deduced from the impossibility of the existence of "indefinite" numbers between 1 and 10, but different from 2, 3, 4, 5, 6, 7, 8 and 9.

Now, in 1934, Thoralf Skolem discovered that there are strange arithmetic universes – called *non-standard* – in which there exist such "indefinite" integers, different from all definite integers 0, 1, 2, 3, and so on, but having exactly the same properties that can be described in the language of arithmetic.

Borges's statement that, if God exists, then all numbers are definite, turns against him, as it is equivalent to the statement that, if a certain number is indefinite – as is the case in non-standard universes – then God does not exist.

Few years earlier, in 1920, Skolem had found an apparent paradox. In fact, there also exist some strange set universes, with a number of elements equal to the smallest possible infinity. On the one hand, within them no set can be bigger than the whole universe and this latter has only a "small" number of elements. On the other hand, from Cantor's theorem we know that there must exist in these universes infinite sets of every dimension and, in particular, also sets with "large" number of elements.

However, the contradiction is only apparent. In fact, an infinite set has – as number of elements – a bigger infinity than that of integers, if it cannot be put in biunivocal relationship with the set of these numbers. Now, there are two ways this can happen: one involves God and the other man. If God says that there is no biunivocal relationship, this is an absolute fact regarding reality. If man says the same thing, this is a relative fact that has to do with human knowledge.

Infinite sets in Skolem's infinity are all "small" from God's point of view, as God sees, from the outside, a way to put each of these sets in a biunivocal relationship with integers. But – from the point of view of humanity – these sets are "large", as none of these biunivocal relationships is inside the universe and is therefore at our disposal. Infinity, that has for some time seemed a thought at the limit, becomes again what it was for the Greeks: a limit to thought.

This subtle fact has strong theological implications, since it shows that, from inside the universe, things can appear more complex than from the outside and therefore more complex than they actually are. In particular, by paraphrasing the words of the last man on the moon: what can be a small infinity to God, can appear as a gigantic infinity to humanity. As Cusano used to say of the universe, what appears infinite to us may well actually be finite: which is almost to say that perhaps God is an optical illusion and that God's transcendent and necessary appearance is only the fruit of our immanent and contingent nature.

Through mathematics too, therefore, one arrives at the same limitations that philosophy and physics have highlighted: we are "thrown into the world" as observers taking part in the same reality we are observing and often many of the complications of life and universe appear to us as such only because we are too involved and limited.

"Geometrician" Dante had already perfectly realised this in his *Paradise* when he wrote (XXXIII, 137–139):

I wished to see how the image to the circle
Conformed itself, and how it there finds place;
But my own wings were not enough for this.

Bibliography

Odifreddi P (1999) *Il Vangelo Secondo la Scienza*, Einaudi, Turin

Mathematics and Technology

Mathematical Visualization and Online Experiments

Konrad Polthier

Abstract

The future of mathematical communication is strongly related with the internet. On a number of examples, the present paper gives a futuristic outlook how mathematical visualization imbedded in the internet will provide new insight into complex phenomena, influence the international cooperation of researchers, and allow to create online hyperbooks combining interactive experiments and mathematical texts.

Using the software JavaView we discuss practical aspects of online publications and give technical details on the ease of implementations. The online version of this paper is a sample interactive document with visualization examples and numerical experiments.

Introduction

Visualization has a major impact on the understanding and exploration of complex mathematical phenomena. As in other sciences, images are used in mathematics as helpful illustrations accompanying textual descriptions, as a communication medium to exchange ideas with other cooperating researchers, and to explain deep mathematical results in a comprehensive way to non-experts, just to mention only a few applications.

The process of visualization is a synonym for a broader process than the production of images or video animations. Prior the generation of images, abstract mathematical concepts must be translated into discrete descriptions which are numerical data structures to hold the mathematical information. For example, a smooth surface must be discretized in a set of triangles, and algorithms must be translated from their smooth counterparts acting on smooth geometries to discrete geometries.

In general these conversions are delicate tasks. For example a Riemann surface may have different smooth descriptions requiring non-obvious discrete structures such as triangulated piecewise linear surfaces with specific discrete properties. Continuous Riemann surfaces are identified under conformal maps, and so one would like to perform a similar identification on the corresponding discrete objects. It is of central importance in mathematical visualization to have

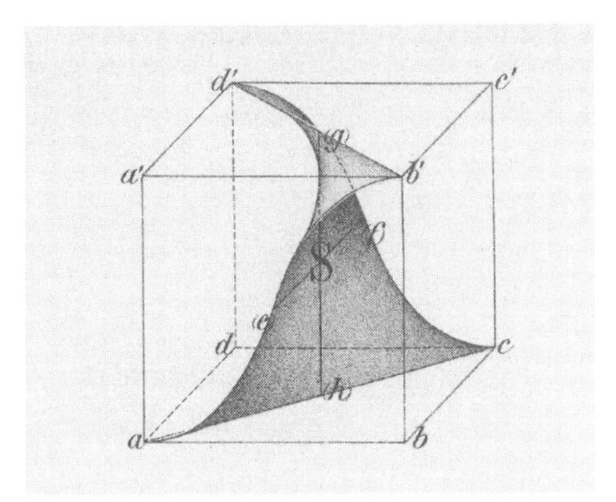

Fig. 1. Copper plate engraving of the Gergonne surface by Hermann Amandus Schwarz. Image taken from [1].

discrete equivalents of smooth concepts, and discrete algorithms on discrete surfaces which perform the same operation as their counterparts on smooth concepts.

The process of defining useful discrete data structures, finding good discretization algorithms, and deriving methods operating on discrete geometries belongs to the same category of mathematical tasks which have been so important for smooth geometries. Good visualization and numerics rely on perfect discrete definitions which are the basis to assign, measure and transform mathematical properties of discrete geometries. Numerical mathematics has gone a long way in discretizing function spaces but, for example, the understanding of the intrinsic mathematical properties of discrete geometric shapes still leads to challenging questions.

It is important to distinguish between the process of visualizing precomputed numerical data and, in other sciences, of experimentally measured data, and the process of doing numerical experiments and simulations. The first process is an analytic process trying to understand large sets of numbers by the means of finding good visual representations. This includes hiding and emphasizing of data as well as the feature extraction of detail information. Many visualization tools operate in a post-processing step by evaluating large data sets and generating new smaller sets with feature information. For example, one of the current major problems is the visualization of turbulent 3d flows. Because of its complexity the flow might be inaccessible to a direct visual representation but the path of moving vortices may be computed and easier to display. Here the correct definition of vortices in a discrete flow is an essential prerequisite before starting the development of a tracing algorithm. The second process of doing numerical experiments and simulations is a constructive and repetitive process where an experiment is analyzed during runtime and information obtained to steer parameters and interact with the numerical process. Here visualization is used to obtain insight into the behavior of a running algorithm rather than into geometric dataset.

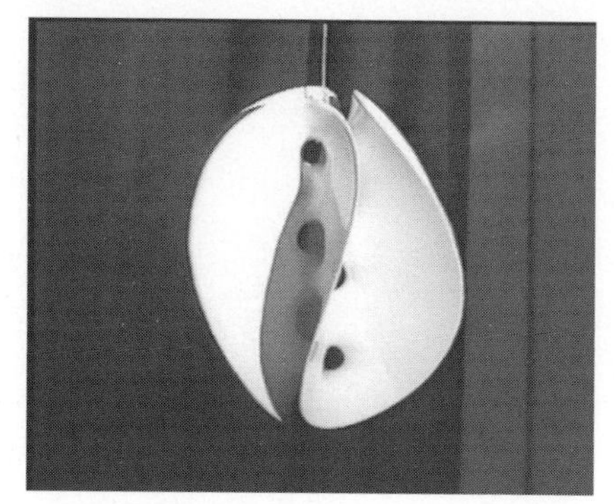

Fig. 2. Plaster model of the Kuen surface. (courtesy Gerd Fischer) (left). Stereo lithography model of Chen-Gackstatter-Karcher-Thayer surface. (right).

Over the past years, visualization has proven to be a successful tool for mathematicians in the investigation of difficult mathematical problems which seem to be inaccessible by standard mathematical tools. It has proven its potential by substantially contributing to the solution of hard mathematical problems.

Mathematicians often have a concrete imagination of abstracts shapes even in higher dimensional spaces, and especially geometers have a deeply visually related thinking. Images are very helpful during their research, they give insight into unknown phenomena and suggest directions for further investigations. Nevertheless there is a remarkable contrast between the visual thinking, and the occurrence of pictures in research publications and inner-mathematical talks. Often images are extremely rare, for example, books on differential geometry, a mathematical subject with one of the largest source of easily visualizable complex geometric shapes, contain only a handful figures with most simple shapes. The lectures on differential geometry of Luigi Bianchi is a comprehensive and very influential book on surface theory, whose German translation was published in 1910, and the classic book on Riemannian Geometry by Gromoll, Klingenberg, and Meyer from 1969, which was the book for a generation of geometers, both books contain no single image of a surface or a geometric shape. Similar examples up to modern days could be listed.

In some sense images have the character of an experiment: they suggest a truth or result but usually do not have the power of a proof. An image can illustrate a mathematical proof but an experiment or image implicitly requests a formal proof of its suggested result. Nevertheless, important experimental results exist which have not been rigorously proven, but they direct theoretical investigations and often give final hints. Experiments have been performed in the whole history of mathematics but only since 1991 there exists a publication media, the journal of Experimental Mathematics.

In contrast to a wide-spread opinion, mathematical history contains a rich set of visual examples. Archimedes was drawing figures into the sand when being bothered by the Romans during the capturing of Syracuse. Euclidean geometry,

Fig. 3. Deformable thread model of a hyperbolic paraboloid (left). Apollo Belvedere with parabolic curves (right). (Courtesy Gerd Fischer)

nowadays referred to as 'elementary geometry', is one of prominent examples where all kinds of drawings always played a central role in the communication and publication of results. Famous non-trivial examples are the copper plate engravings of Hermann Amandus Schwarz shown in figure 1 [13]. He discovered new minimal surfaces which solved long-standing questions in geometry and analysis on the existence of solutions to elliptic boundary value problems. Although his proof was of theoretical nature he found it worthwhile to invest a lot of energy to include visual images of the new surfaces into his research publications. This is remarkable if compared to the much less effort one needs for the generation of computer images today.

One of the most thorough approaches in mathematics to use physical models and experimental instruments in education and research is the famous collection of mathematical models in Göttingen. This model collection already had a long history when Hermann Amandus Schwarz and Felix Klein overtook the direction of the collection. Especially under the direction of Klein the collection was systematically modernized and completed for the education in geometry and geodesy. This collection was considered so important that Klein exhibited the models on the occasion of the World's Columbian Exposition 1893 in Chicago [2]. The models were produced among others by the publisher Martin Schilling in Halle a.S., see his catalog of mathematical models [12]. The price of approximately \$250 per model was relatively, and therefore, the large size of the collection of more than 500 plaster models is even more impressive. The collection can still be seen in the mathematical department in Göttingen, and a description including photos of many models is given in Fischer [4]. An example is the plaster model of the Kuen surface in figure 2 which is shown together with a modern model of the Chen-Gackstatter-Karcher-Thayer minimal surface produced in stereo lithography technique from digital data. The production of models slowed down and finally stopped in the beginning 30s. Following Fischer, the reasons were not purely economical nature but also the appearance of more general and abstract view points in mathematics. It was the time when the books of van der Waerden with only simple illustrations and of Nicolas Bourbaki with a complete ignorance of images appeared.

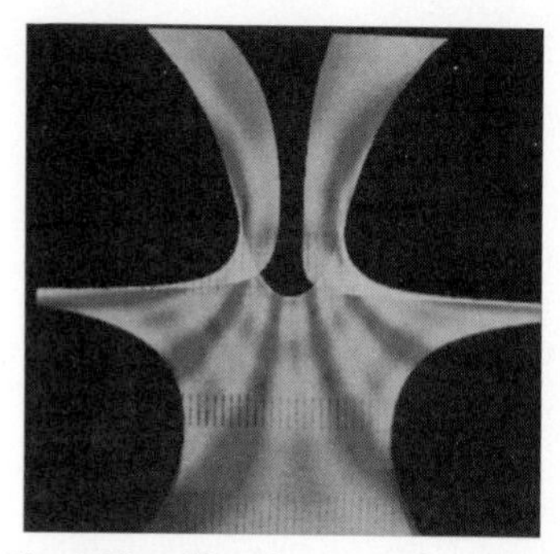

Fig. 4. Unveiling the Costa formulas: first image of the surface over not appropriate parameter domain and frame buffer artifacts (courtesy Jim and David Hoffman) (left). Years later, raytraced image of the Costa-Hoffman-Meeks minimal surface. (right).

It seems that minimal surfaces have been among those geometric shapes which often urged mathematicians and physicists to produce images. In the sequel to Schwarz and the experiments of Joseph Antoine Ferdinand Plateau, it was Richard Courant, Johannes C.C. Nitsche, and Alan H. Schoen who experimented with physical soap films and even produced real models for permanent display and for easier communication. Among the first breakthroughs of mathematical visualization was the proof of embeddedness of another minimal surface. Celsoe Costa discovered the mathematical formulae of a genus 1 minimal surface which was a candidate to solve a 200-year long question: whether there exists a third embedded and complete minimal surface with finite total curvature beside the trivial examples, the flat plane and the rotational symmetric catenoid. David Hoffman and William Meeks developed computer programs to visualize Costa's surfaces and watch surface properties which they could later successfully *prove* after having enough insight into the complex shape of the surface, see figure 4 and [7].

Experimental Online Geometry

It has been pointed out by David Epstein and Silvio Levy [3] that "the English word prove – as its Old French and Latin ancestors – has two basic meanings: to try or test, and to establish beyond doubt. The first meaning is largely archaic, though it survives in technical expressions (printers proofs) and adages (the exception proves the rule, the proof of the pudding). That these two meanings could have coexisted for so long may seem strange to us mathematicians today, accustomed as we are to thinking of proof as an unambiguous term. But it is in fact quite natural, because the most common way to establish something in everyday life is to examine it, test it, probe it, experiment with it."

Computers are nowadays a powerful machinery to perform mathematical experiments, and even do automatic proving as soon as it is possible to formulate axioms and rules in a formal computer language. A large part of applied

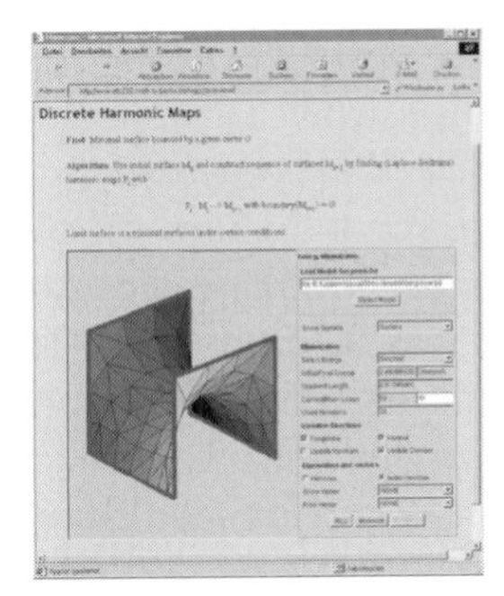

Fig. 5. Soap film machine from video *Touching Soap Films* [1] (left). Computing minimal surfaces in an interactive applet on the JavaView web site [10] (right).

mathematics is devoted to the simulation of practical physical phenomena which often are still inaccessible to formal proves of, say, convergence.

Geometry on the Internet

For a number of years the internet has been a technical infrastructure connecting computers to a global network, but in recent years it has emerged as the world-wide-web, a global information network similar to a gigantic hypertext book. Hypertext books have been around for quite some years, for example, digital help systems of software applications have been among the first to use software links between different sets of information. Mainly the success of hypertext books made written manuals obsolete so that nowadays software is solely accompanied by digital manuals. Classical hypertext books have a similar linked structure as the web but the information is usually stored locally. The new dimension of the web allows to imagine the human's knowledge as a worldwide digital encyclopedia which is directly accessible by all people rather than by those living in the vicinity of the local library.

The success of the world-wide web was not planned. It is one of the achievements that happen in the modern computer world where suddenly a far-reaching answer to the problems of a large number of people appears. Among the major reasons of the success of the world-wide-web are the global standards which have been established for the document format HTML, the network protocol HTTP, and the software browsers. While the network protocol is based on the internet protocol TCP/IP which has been used in the academic domains for a number of years, it was the simple new document format HTML which helped to attract so many people. This format is easy to read and to write on nearly all computer systems, and it allows to include multimedia elements like images, sound, video, and other data. From the beginning of world-wide-web, browsers had a simple user interface which allowed immediate access to all kinds of information and thereby succeeded to replace complicated software tools previously used in the academic internet.

The origin of the world-wide-web is a good example of another aspect of the web beside the publication and presentation of information, namely, communication. In fact, the HTTP protocol, which is the underlying technical protocol for internet connections, was defined at CERN in Switzerland to support communication and exchange among large groups of scientists involved in physical experiments. Often these groups involve more than hundreds of scientists located at different places in the world who must exchange experimental data and communicate research results in efficient ways. This means that the origin of the web is not related with presentation, which made the web popular, but with communication among people. The communicative aspect of the web is becoming again its original importance and is the main attraction of the web nowadays.

When considering the web as a global digital library, is there still a place for maintained libraries and encyclopedias? In fact, such managed collections are still important since they ensure a certain quality, a managed database, and well-defined access. The global web contains a huge amount of information but it is often hard, and becomes even harder, to find the good information among the non-relevant. Information must be preselected to be really useful, and current search engines are just starting with methods for ranking and sorting information by quality.

Geometry is the content in this setting which we are now going to exhibit in greater detail. The special characteristic of geometry is its rich fundus of shapes, images, and dynamic applications which make it an ideal candidate for profiting from the internet.

One of the most attractive new components of the web is the ability to present and communicate complex interactive experiments. Performing experiments has always been the domain of experts, more concrete, of those experts who wrote the simulation software. One reason is the user interface design which has been difficult to create for UNIX computers since user interface builders are not so common in this area. Another reason is the lack of incentive for a programmer to create a well-thought user interface since scientific programs usually require specialized hardware and will seldom reach a wide-spread audience. Since the arrival of the new programming language Java in 1995 the situation changed dramatically. For the first time it is possible to create software which runs on any computer platform and operating system without the delicate process of rewriting and adjusting source code for each new system.

Java is usually installed automatically on a computer at the same time when a browser is installed. In contrast to other high-level programming languages like FORTRAN, C or C++, the language Java comes also with a full-featured set of graphical user interface structures. These structures are not supplied by other parties but incorporated into the definition of the language itself. The design of a user interface is one of the major tasks when implementing a reusable experiment, and in previous times it was the one of the reasons why software was restricted to special computer platforms. Platform dependent interface code restricts software to distinguished platforms. Since Java includes the graphics user interface directly in its language specifications, Java programs run per default on any computer with installed Java.

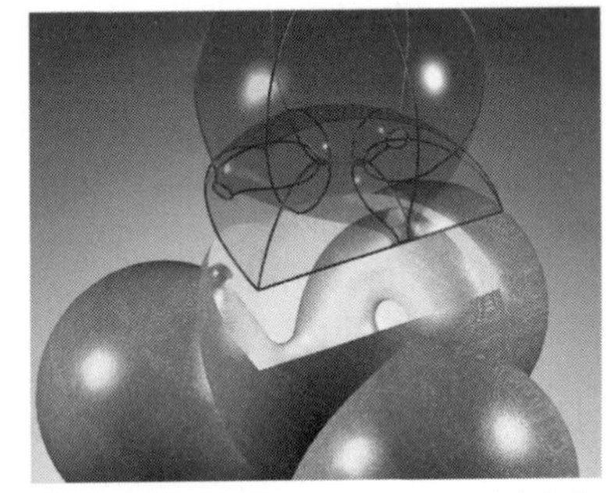

Fig. 6. Compact soap bubble with genus four (left) and tetrahedral symmetry (right).

Visualization and Online Experiments

Mathematical visualization has proven to be an efficient tool for analyzing complex mathematical phenomena, and it has given decisive hints leading to rigorous mathematical proofs of long-standing problems. Visualization is not only a tool to visualize complex objects but in combination with modern numerical methods allows to perform mathematical experiments and simulations in an artificially clean environment. For example, the unveiling of the Costa-Hoffman-Meeks surface [6] (figure 4), or the first numerical examples of compact constant mean curvature surfaces with genus greater than two [5] (figure 6) are among the most prominent results of the fruitful interaction of mathematics with the new toolkit mathematical visualization.

Mathematical experiments require the following goals to be accepted in the mathematical community similar as experiments in physics and chemistry:

1. Validation of experimental data by independent groups
2. Publication and storage of experimental results and data sets
3. Cooperation of researchers on the same experiment while sitting at different places

Up to now visualization has required high-end workstations combined with mainframe computers for numerical computations. Each research lab had developed own software dedicated to specific graphics hardware. The specialization of the software for a specific visualization task or numerical problem as well as the dependence on a specific hardware platform had major drawbacks for the scientific communication. The major reason for the current non-fulfillment of the goals 1. and 2. is a missing world-wide standard and interface for the different software packages, and for goal 3. the missing standard to exchange data sets between experimental software packages and research publications. Currently research publications are paper-based where there is no way back from written publications to digital data sets. Even worse, in order to fit into a restricted page layout and fulfilling the allowed number of pages, publications cannot include all data but restrict to the so-called 'most relevant data'. The presentation of experimental results in publications is incomplete and often moved to the appendix so that it is usually not possible to validate or reproduce them.

In order to allow validation of experiments and of numerical data sets it is essential to allow direct public access to the data in an electronic format in the same way as public access to research publications is given through libraries. This joint publication of experimental and numerical data requires to insist on the similar principles known from the publication of scientific research results:

1. Unambiguous, self-contained digital representation of the data, i.e. no dependence on existing software packages.
2. Reviewing of data sets, e.g. to ensure technical correctness and scientific relevance.
3. Indexing data sets to allow unique references, e.g. when data is used in other experiments.

In the following sections we discuss how online visualization might look like and about the possible difficulties to encounter.

Workstation versus Online Visualization

For a long time scientific visualization was beyond the budget of many mathematical departments. Large research institutes, military organizations, and commercial companies were among the first who could afford specialized graphics hardware. In science, specially funded research groups where able to afford high-end graphics workstations including the necessary staff to manage the machines and simultaneously do the scientific experiment. In the meantime, the computational power of personal computers with relatively cheap graphics card suffices to perform most of the scientific visualization tasks found in research. Nevertheless, one still encounters the following drawbacks of the current software running on specialized workstations and mainframes:

1. Specialized and expensive graphics hardware.
2. Large program size since operating system just supports basic functionality.
3. Usually only the programmer is able to run the experiments.
4. Installation at other sites requires experts, and does not allow regular update.
5. Advantage: extremely fast execution speed.

These drawbacks are in strong contrast to the situation we have encountered during the development and usage of the software JavaView. JavaView is a scientific visualization software completely written in the programming language Java. Java is an object-oriented programming language similar to the language C and C++ but different in the sense that Java is designed to run on any computer. Further, Java programs may run inside web browsers. Both properties are the reason that Java has become the major programming language for interactive web applications since its first presentation in 1995. A program written in Java has the following advantages:

1. Runs on Standard PC and Workstation.
2. Tiny program size because Java base classes are already installed.
3. Each application has a user interface per default since it runs in a browser.

4. No installation beside a browser with Java since browser performs the data transfer.
5. Speed: depends on the type of application.

These advantages have the following reasons:

1. Java is automatically installed on a computer if a web browser is installed. Therefore, the popularity of web browsers helped to install Java on nearly any computer world-wide.
2. The size of Java programs is usually very small compared to classical stand-alone application software since the Java base classes, which are comparable to software libraries, are already installed. Therefore, an application must only deliver its additional functionality, and not system routines.
3. An application inside a web page must have a well-designed graphical user interface since it is by default used by some other people than the programmer. This is in contrast to classical experimentation software, and leads to a great benefit in the design of better products.
4. The installation of classical software systems has often been a pain. The customer often needed to compile the package again on his machine, or make special adjustments depending on his specialized hardware. The author was in an even worse situation. He needed to offer and maintain different versions for different platforms. When using Java then there exists only one version of the software independent of the hardware platform and operating system. This is possible since the Java virtual machine must cope with system differences, so the responsibility is transferred from the author of applications to the supplier of the Java virtual machine. Therefore, the installation process of a Java application such as JavaView is reduced to downloading an archive, i.e. one or more library files, which is done automatically through a web browser. This allows the author to concentrate on the development of the software without keeping to much care on the destination platform, and it frees him from providing installation mechanisms. The user is freed from any installation task, he just starts his browser and selects a Java enhance web page.
5. The speed of Java applications not only depends on the hardware but to a large extend on the quality of the installed Java virtual machine (JVM). A Java application consists of machine independent byte code which is interpreted by a JVM and executed on a local computer. JVMs differ largely in quality, for example, when loading a Java application some JVMs compile the byte code into machine dependent code, which leads to a drastic increase in execution speed.

Java View

JavaView [11][10] is a software for sophisticated experiments and visualization of 2- and 3-dimensional geometric objects on a local computer as well as online in a web browser. Students, teachers and researcher can use JavaView as a tool for general scientific visualization, in a distant learning environment, for online exchange of scientific results among researchers, and for electronic publication of mathematical experiments.

JavaView is a numerical software library with a 3D geometry viewer written in Java. It allows to add interactive 3D geometries to any HTML document, and to present numerical experiments online. The future of mathematical communication is strongly related to the internet, and JavaView enhances classical textual descriptions not only with images and videos but additionally with interactive geometries and online experiments.

JavaView has been developed to solve the following technical tasks:

1. Visualization of mathematical data sets inside web pages.
2. Interactive experiments and simulations inside web pages.
3. Inclusion of mathematical experiments and simulation data in electronic publications.

The first version of JavaView fulfilling these tasks was released in November 1999 after development versions had been used in geometric research projects at the Technische Universität Berlin for over a year. JavaView is now used at different places world-wide. There exist a multitude of mathematical demonstrations to give an outlook of the range of new applications possible with web-based experimental software.

The first two tasks provide the technical basis for the third task. As a proof of usability, JavaView has been selected by the project "Dissertation Online" of the German science foundation DFG to produce a reference online dissertation in

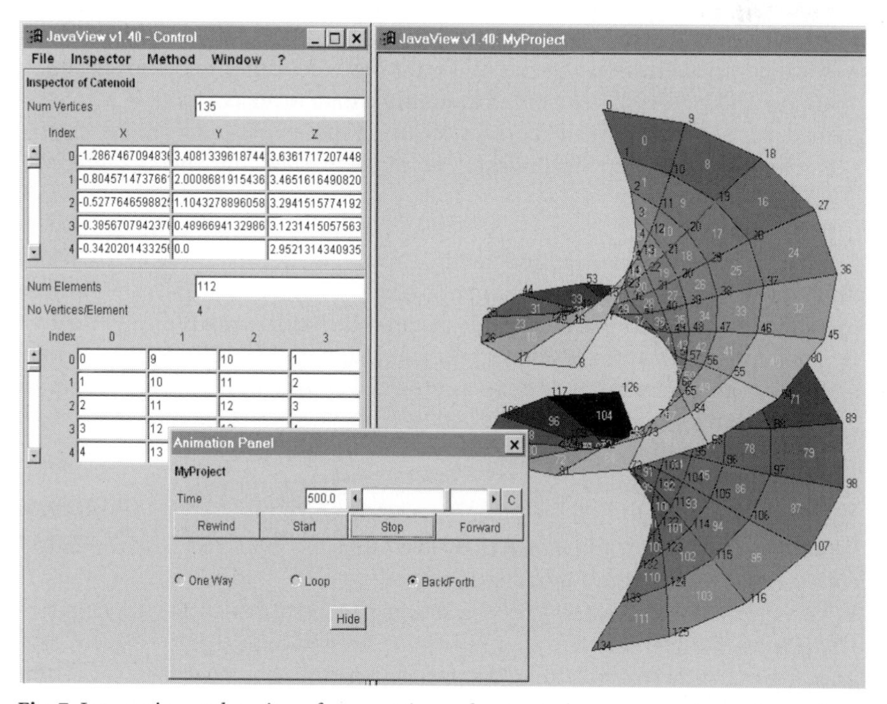

Fig. 7. Interactive exploration of geometries and animated sequences of transformations.

mathematics. There are many other issues to solve for electronic research publications, and JavaView is only one component that is connected with the inclusion of experiments and interactive visualization. For example, mathematical journals currently require a paper based version of an article even if they include the article in an electronic version of the journal too. An author must create two versions of an article, usually a TeX version for paper-based printing and an online version including, say, JavaView experiments. But even electronic versions of journals are currently not well-suited for web-based usage since they are just PostScript or PDF documents. Such documents do not fit with other internet technologies since, for example, they are hardly searchable and do not allow inclusion of Java applets or video elements.

In the following subsections we discuss some properties of JavaView and give a number of sample experiments. Since this book is paper-based we refer to an online version of this article at the web site http://www.javaview.de/ which contains interactive versions of the presented applets.

Properties

The viewer of JavaView supports most interaction features of an advanced 3D viewer. For example,

- Rotation, translation, zoom, camera control, picking
- Inspection of geometries and material properties
- Selective display of vertices, edges, faces, vector fields
- Animations, keyframes, autorotation
- Subdivision and simplification of meshes, adaptive and hierarchical triangulations
- Advanced visualization algorithms, vector field rendering, textured surfaces
- Import and export of geometries in multiple data formats
- PostScript and image file export for inclusion into paper publications
- Frontend for other applications, for example, to view *Mathematica* graphics [14] and Maple plots [9].

Images and geometries exported with JavaView may easily be included in TeX publications as well as online documents. Additionally, JavaView is a class library for advanced numerics in differential geometry including tools for finite element numerics. A standalone version of JavaView runs in a Unix or Windows shell from the command prompt, and can be attached as 3D viewer to other programs like *Mathematica* [14] and Maple [9].

Online Example Applets

This paper-based publication must restrict to a few sample applications and verbally describe the possible interaction. The following examples describe different aspects of the usage of JavaView:

1. Visualization and evaluation of precomputed models which are stored somewhere on the internet.
2. Interactive tutorials explaining simpler numeric or geometric facts to accompany classical lectures or online learnshops.

3. Sophisticated numerical research projects with intimate combination of numerics and advanced visualization.
4. Joint research of authors at different universities doing numerical research experiments embedded into web pages.

Electronic Geometric Models Online Nowadays it is even hard to obtain datasets of simple examples. One reason is the absence of published numerical data sets stored at some publishing house, or a data set which accompanies a research article published in a scientific journal. This is a serious drawback in the scientific validation of numerical experiments but also leads to the more elementary drawback that models are not available, for example, as initial data sets to perform own experiments. Often one would just like to have a digital model with certain properties to test one's own numerical algorithm or implementation. Compare applet 7 for an online animation which may be interactively investigated.
Another aspect of an online model collection is the educational benefit. Geometry books usually contain only a few images, and students still have a hard time to find good visual material. The software JavaView provides a simple way for every scientist to have mathematical models at his fingertips.
Since the publication of the Italian version of this paper, the Electronic Geometry Models server was installed at http://www.eg-models.de/. EG-Models is a new peer-reviewed journal for the publication of electronic geometry models [8].

Interactive Tutorial: Root Finder. Interactive tutorials may be easily designed to explain simpler facts from numerics or geometry. They may be made available to students online as the supplemental material of classical university and high school courses. Additional to other course notes the students may regularly look at web pages accompanying the current course. Distant learning projects must have well prepared course materials since the direct contact with students is less intensive. These projects will be among the first to include interactive experiments, and maybe even the driving forces for the development of whole packages of interactive online experiments.
The applet 8 demonstrates the numerical algorithm of finding the zero-crossings of functions. The method subdivides the original interval and uses Brent's method to find the roots on each subinterval. Here, the user is assumed to read a description of the algorithm and simultaneously study the online example.

Numerical Experiments and Visualization. Java has the same properties than other programming languages used in numerical mathematics and visualization. It has some structural speed limitations since it is an interpreted language but modern just-in-time compiler are able to provide a compilation process on the fly while loading the program. Java is currently not a language of choice for high-performance computing but maybe used in a large section of nearly all numerical areas. The great benefit of Java is its machine and operating system independence, and since most of the time used in numerical research projects is spent in program

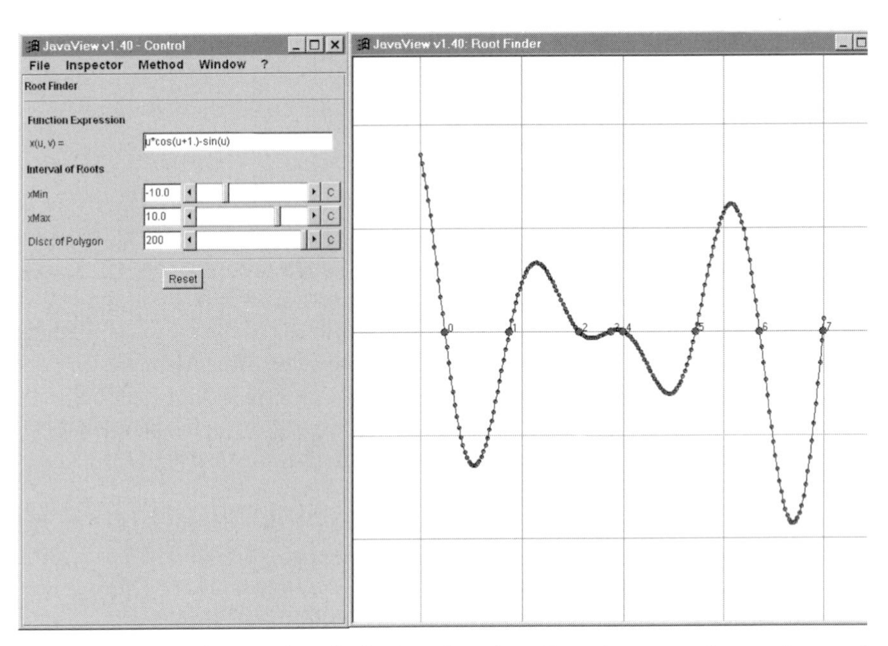

Fig. 8. Applet to find roots of explicitly given functions. Function expression maybe typed and the roots immediately searched.

development and code maintenance it is a great tool even in these domains. [tbh] Research applications like the eigenvalue computations in figure 9 demonstrate that numerical computations and scientific visualization are possible with JavaView. Further, the coherent interface of Java applets immediately make these experiments accessible by other people, and their inclusion in digital publications.

Case Studies: Research Cooperation Online. The cooperation of researchers who are located at different places usually lacks a certain amount of communication. For example, joint numerical experiments require the exchange of newly developed software and their local installation. Here we encounter one of the major benefits of Java: first, the platform and operating system independence, and second, the invisible installation of Java software via automatic web-based mechanisms.

Index of Minimal Surfaces. The index computations of the second variation of area of unstable minimal and constant mean curvature surfaces was a joint project with Wayne Rossman in Koebe in Japan. The cooperation started during his visit in Berlin, and was continued via putting our experimental applets online to be accessible world-wide. The software development was continued in Berlin and the software archive on our web site was regularly updated with the newest version. Each time Rossman did an experiment using the web based software on our site he was sure to use the latest software without ever installing any local version.

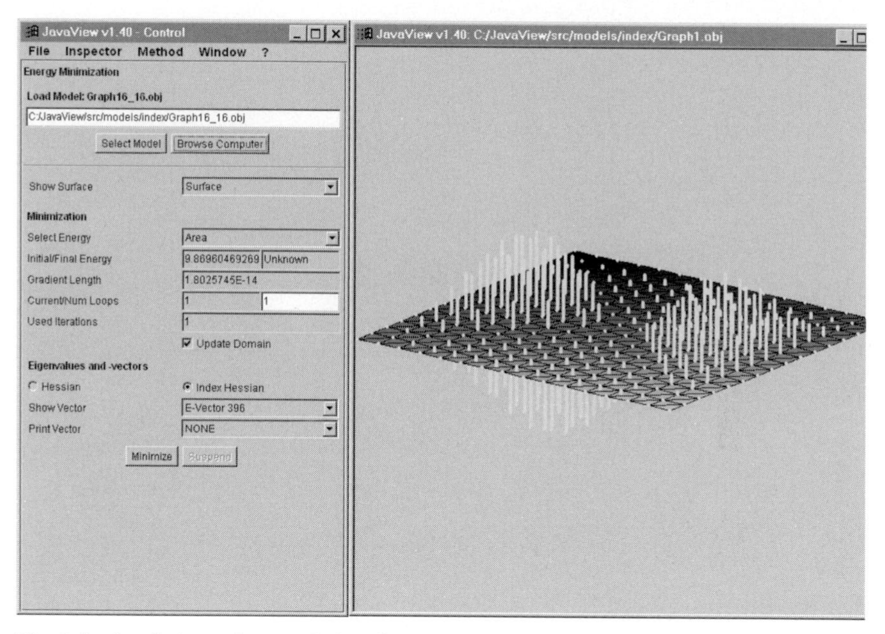

Fig. 9. Study of eigenvalues and eigenfunctions of the Laplace-Beltrami operator on surfaces. Here the known cos and sin functions are reproduced on a planar square. Other applets show the eigenfunctions of the second variation of area.

Social Behavior of Honey Bees. In the meantime, JavaView is used in other projects too. For example, biologists in South Africa use one of our applets to compute shortest curves in a beehive to get information on the distance of the queen to selected other bees. Here the biologists load a model of the beehive into the geodesic applet, interactively select the positions of two bees, and invoke our geodesics algorithm to obtain the shortest curve connecting both positions. This example http://www-sfb288.math.tu-berlin.de/vgp/bees/ stresses the online usage of JavaView as well as the automatic installation process.

A Hands-On Example

This hands-on example describes the necessary three steps to create a JavaView enhanced web page which displays a geometry model in a small window allowing interactive modifications.

1. Insert a Java applet tag into a web document *myPage.html* referring to a geometry model *brezel.obj*.
2. Get the archive *javaview.jar* from the JavaView homepage.
3. Upload all three files to a web server.

The sample applet tag somewhere inside the document *myPage.html* looks as follows:

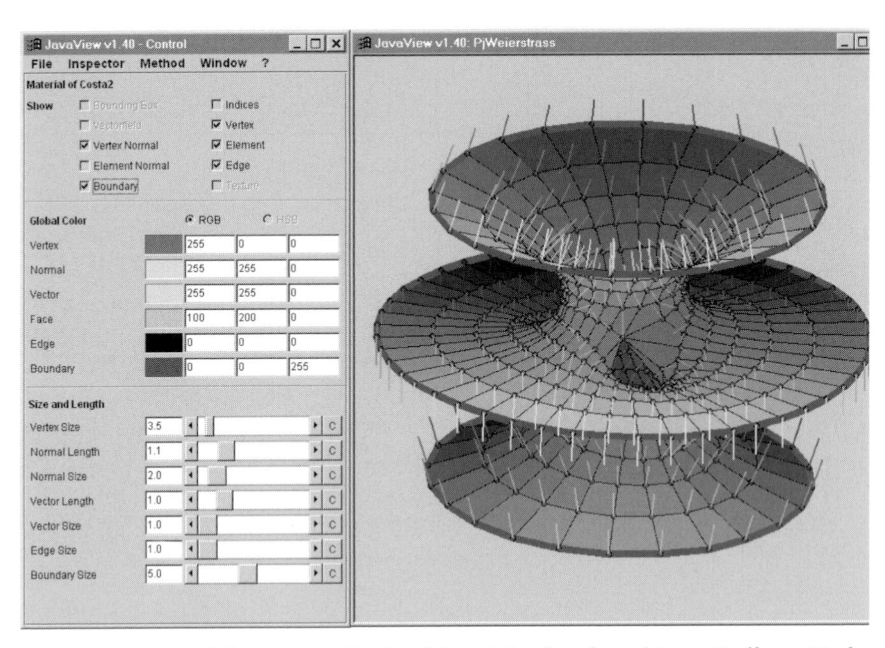

Fig. 10. Animation of the associate family of the minimal surface of Costa-Hoffman-Meeks showing the isometry of the transformation and parallelism of normal vector.


```
<applet
        code=javaview.class
        archive='' javaview.jar''
        width=200 height=200>
        <param name=model value='' brezel.obj''>
</applet>
```

This applet visualizes the geometry model inside a small window of 200*200 pixels on the web page. The model need not be a geometry file on a local computer but the model parameter may be any internet address referring to a model on an arbitrary web server.

This example stresses the fact that the installation of the JavaView software is no longer an issue compared to the installation process of other software. The browser keeps care to download the required Java archive when it encounters the archive parameter inside the applet tag. The browser also ensures that the archive is downloaded only during first usage, and later reuses the version it has stored in the browser cache.

The easy download mechanism is especially useful for library servers offering Java enhanced electronic publications. The digital article and the JavaView archive are both stored, for example, in the same directory on the library server. The files must be uploaded by the author as described above, and are automatically downloaded by a browser when a user accesses the web page. Therefore, the librarian has no additional duties related with software installation. The library must only offer the usual upload mechanism for documents which it has already installed.

Summary and Outlook

The internet will dramatically change the classical way of communicating and publishing mathematics. We have given some ideas on possible changes to expect, and the benefits which mathematics may gain from these new developments. The interactive, exploratory component of mathematics, which has been removed from mathematical publications for a too long time, is now available in the form of Java enabled software. We have given several examples of multimedia enhanced experiments which allow to imagine the possibilities waiting at the horizon.

For more information and interactive versions of the experiments described in this paper we refer to the JavaView home page [10]. These pages also include tutorial material how to include interactive geometries into own web pages.

References

[1] A. Arnez, K. Polthier, M. Steffens, and C. Teitzel. *Touching Soap Films*. Springer VideoMATH Series, 1999. German Version "Palast der Seifenhäute", Bild der Wissenschaft (1995).

[2] Chicago Historical Society. The World's Columbian Exposition. http://www.chicagohs.org/history/expo.html.

[3] D. Epstein and S. Levy. Experimentation and proof in mathematics. *Notices of the AMS*, June 1995.

[4] G. Fischer. *Mathematische Modelle/Mathematical Models*. Vieweg Verlag, Braunschweig, 1986.

[5] K. Große-Brauckmann and K. Polthier. Compact constant mean curvature surfaces with low genus. *Experimental Mathematics*, 6(1):13–32, 1997.

[6] D. Hoffman. Computer-aided discovery of new embedded minimal surfaces. *Mathematical Intelligencer*, 9(3), 1987.

[7] D. Hoffman and W. Meeks. Complete embedded minimal surfaces of finite total curvature. *Bull. AMS*, pages 134–136, 1985.

[8] M. Joswig and K. Polthier. EG-Models – a new journal for digital geometry models. In J. Borwein, M. H. Morales, K. Polthier, and J. F. Rodrigues, editors, *Multimedia Tools for Communicating Mathematics*, pages 165–190. Springer Verlag, 2002. http://www.eg-models.de.

[9] Maple Waterloo. Homepage. http://www.maplesoft.com.

[10] K. Polthier. JavaView homepage, 1998–2002. http://www.javaview.de/.

[11] K. Polthier, S. Khadem-Al-Charieh, E. Preuß, and U. Reitebuch. Publication of interactive visualizations with JavaView. In J. Borwein, M. H. Morales, K. Polthier, and J. F. Rodrigues, editors, *Multimedia Tools for Communicating Mathematics*, pages 241–264. Springer Verlag, 2002. http://www.javaview.de.

[12] M. Schilling. Catalog *Mathematischer Modelle Für Den Höheren Mathematischen Unterricht*. Schilling Verlag, Leipzig, 1911.

[13] H. A. Schwarz. *Gesammelte Mathematische Abhandlungen*. Springer, Berlin, 1890.

[14] Wolfram Research. Homepage. http://www.wolfram.com

Prime Numbers and Cryptography

ALESSANDRO LANGUASCO, ALBERTO PERELLI

Translated from the Italian by Emanuela Moreale

On the one hand, the study of numbers – and especially of prime numbers – has fascinated mathematicians since ancient times; on the other hand, humans have always felt the need for security in the transmission of information. In the last twenty years, thanks to the discovery of new mathematical methods and the remarkable progress in computers, a strict relationship has gradually developed between the two disciplines. At present, the most secure methods for the transmission of information that have recently been boosted by the development of e-commerce, are based on algorithms that depend on remarkable properties of prime numbers. In this article, we will briefly outline the development of the theory of prime numbers; then we will describe an application to the problem of security of data transmission, that is cryptography.

Prime Numbers

First of all, let us remember that a natural number $n > 1$ is said to be a prime number if it is divisible only by 1 and by itself: for instance, the numbers 2, 3, 5, 7, 11, 13, 17 and 19 are prime numbers.

Already the ancient Greeks had taken an interest in the determination of prime numbers. A technique developed in that period is the well-known Sieve of Erastothenes: this method allows the calculation of all prime numbers between 2 and x, where x is any fixed real number. After writing down all natural numbers between 2 and x, one should take the number 2 and cross off every multiple of that number; the same should be done for the next non-crossed out number on the list (3) and so on. One proceeds this way (the next step involves considering the number 5 and crossing out all its multiples) up to the biggest natural number smaller than $\sqrt{x}$. Since, by the end of this procedure, we have crossed out all natural numbers with proper divisors smaller than $\sqrt{x}$. the remaining natural numbers are all prime numbers in the interval [2,x].

Another problem that interested the Greeks is whether there are infinitely many prime numbers. The answer is affirmative and several proofs of this are known. Here we will present Euclid's arithmetic proof:

Theorem (Euclid) *There exist infinitely many prime numbers.*

Proof. By contradiction, let us suppose that there exist a finite number of prime numbers $p_1 < p_2 < \ldots < p_k$. Let us now consider the number

$$N = p_1 p_2 \cdots p_k + 1 .$$

N clearly cannot be a prime number since it is greater than p_k. On the other hand, N is not divisible by any p_j and therefore N is a prime number, which contradicts what above. The theorem is thereby proved. ■

At this point, one might wonder: *why are prime numbers interesting?* The most immediate reply to this question is that prime numbers are, in some way, the building blocks all integers are built. Formally, this statement is expressed by the following well-known theorem:

Fundamental Theorem of Arithmetic: *Every positive integer can be written uniquely as a product of prime numbers.*

Remarks: (1) The proof of the Fundamental Theorem of Arithmetic consists of two parts:

a) existence of factorisation (direct consequence of the definition of prime number),

b) uniqueness of factorisation (simple but not wholly trivial).

As to b), we recall that Hilbert's example shows how it is possible to build simple "numerical systems" in which the *uniqueness of factorisation does not hold.* Let us consider the integers of the form $4k + 1$, $k = 0, 1, \ldots$, that constitute a closed system with respect to multiplication. It can be easily verified that

$$693 = 9 \cdot 77 = 21 \cdot 33$$

provides two distinct factorisations of 693 as product of "primes" in this system: in fact, 9, 77, 21 and 33 do not allow a non-trivial factorisation as product of natural numbers of the form $4k + 1$.

(2) From the Fundamental Theorem of Arithmetic, it follows that

Corollary: $\sqrt{2}$ *is irrational.*

Proof. Let us suppose that $\sqrt{2} = \frac{m}{n}$. Then $n\sqrt{2} = m$, and therefore

$$2n^2 = m^2.$$

Let us observe now that the factor 2 on the left side of the last equation has an odd exponent, whereas it has an even exponent on the right hand side, which contradicts the Fundamental Theorem of Arithmetic. ■

Roughly, we can group problems relating to prime numbers into two major categories:

- *algebraic problems:* concerning mainly the behaviour of prime numbers in algebraic extensions of rational numbers;

– *analytic problems:* concerning mainly distribution of primes among natural numbers.

In this article, we only deal with analytic problems.

It is natural to wonder: *how many prime numbers are there?* We already know that there are infinitely many primes, but what we are asking here is what is the order of magnitude of the quantity

$$\pi(x) = \text{number of primes between 1 and } x.$$

The first attempt to solve such problem was made by Gauss towards the end of the XVIII century. Using tables of primes he himself had calculated, Gauss conjectured that the number of primes not exceeding x is asymptotic to $x/\log x$:

$$\frac{\pi(x)}{x/\log x} \to 1 \quad \text{for} \quad x \to \infty .$$

As we will see later, Gauss's conjecture turned out to be correct and is nowadays know as the *Prime Numbers Theorem* (PNT).

The first steps towards proving Gauss's conjecture were made by Chebyshev towards the midnineteenth century.

Theorem (Chebyshev) *There exist two constants $0<c<1<C$ such that, for a sufficiently large x:*

$$c\frac{x}{\log x} \le \pi(x) \le C\frac{x}{\log x}.$$

The proof of Chebyshev's Theorem is based on an elementary but clever technique involving some properties of binomial coefficients.

The decisive step towards proving the PNT was taken by Riemann just a few years after Chebyshev. The fundamental novelty of Riemann's method was that of studying the function $\pi(x)$ using *complex analysis* (hence the "analytical" adjective used for such type of research).

Riemann introduced the function of the *complex* variable s

$$\zeta(s) = \sum_{n=1}^{\infty} \frac{1}{n^s} \quad \text{for} \quad \mathrm{Re}(s) > 1$$

that is nowadays known as the *Riemann zeta function.* The Riemann zeta function is connected to prime numbers by means of *Euler's identity:*

$$\zeta(s) = \prod_{p} \left(1 - \frac{1}{p^s}\right)^{-1} \quad \text{for} \quad \mathrm{Re}(s) > 1$$

where the product is extended to all prime numbers. Euler's identity is a simple consequence of the Fundamental Theorem of Arithmetic and is actually considered to be the *analytical equivalent* of the unique factorisation of integers.

The crucial point about Euler's identity is that prime numbers explicitly appear on the right hand side, while the left hand side is defined independently of them. Riemann's method therefore paves the way for the possibility to obtain information about prime numbers through the study of the analytical properties of the function $\zeta(s)$. For instance, by taking advantage of the fact that

$$\lim_{s \to 1^+} \zeta(s) = +\infty ,$$

it is possible to easily obtain an analytical proof of Euclid's Theorem on the infinitude of prime numbers.

Such analytical proof is due to Euler, who consider $\zeta(s)$ as a function of the *real* variable s. Riemann showed that the function $\zeta(s)$ is continuable to the whole of the complex plane except for a simple pole at s=1 and that the distribution of prime numbers is strictly connected to the *distribution of zeros* in the function $\zeta(s)$. This connection has given rise to some of the most profound problems in mathematics.

The PNT was proven independently by Hadamard and de la Vallée Poussin in 1896. This demonstration, based on Riemann's method, represents the apex of a strand of research on the function theory of one complex variable, carried out mainly by Hadamard. The crucial point of this proof consisted in showing that $\zeta(1+it) \neq 0$ for any real number t; nowadays, it is known that the non-cancelling out of the Riemann zeta function on the straight line Re(s)=1 is in fact equivalent to the PNT.

Chiefly due to the influence of English mathematicians Hardy and Littlewood – who made substantial contributions to analytic number theory – for most of the first half of the twentieth century it was believed that it was impossible to obtain a proof of the PNT without making use of complex analysis techniques. Such belief turned out to be incorrect when, towards 1950, Selberg and Erdös gave an *elementary proof* of the PNT using what are essentially arithmetic techniques. It should, however, be stressed that "elementary" does not at all mean "easy": in fact, Selberg and Erdös's proof is conceptually more complex than the corresponding analytical proof.

Once the PNT was known, the next step was understanding "how good" the approximation of $\pi(x)$ was through the $x/\log x$ function or, more precisely, through the integral logarithm function

$$li(x) = \int_2^x \frac{dt}{\log t}$$

(where $x/\log x$ is the first asymptotic term). Presently, it is not possible to give a definitive answer to this problem, although the famous *Riemann Hypothesis* plays a fundamental role here:

$$\zeta(s) \neq 0 \text{ for } \mathrm{Re}(s) > \frac{1}{2} .$$

It can be proved that the Riemann Hypothesis is equivalent to the approximation

$$\pi(x) = li(x) + O(\sqrt{x} \log x)$$

that is, basically, the absolute value of the error incurred in approximating $\pi(x)$ with $li(x)$ is smaller than $\sqrt{x}\log x$. It should also be noted that such approximation, if true, is optimal; that several heuristic arguments supporting the Riemann Hypothesis are known and that largescale computations support its validity.

At this moment in time, many results have been proven, but many more *remain open problems* for the research on the distribution of primes. Apart from the Riemann Hypothesis, there are also several *classical problems:*

1) *(primes represented by polynomials)* Is there an infinite number of natural numbers n for which n^2+1 is a prime? (or, more generally, "Is $P(n)$ a prime for infinite natural numbers n?", where $P(x)$ is an irreducible polynomial with no fixed divisors)
2) *(distance between two consecutive prime numbers)* Does there always exist a prime between two consecutive perfect squares?
3) *(twin primes)* Do there exist infinite prime numbers p such that $p+2$ is still a prime?
4) *(Goldbach's conjecture)* Can each even natural number greater than 2 be written as a sum of two prime numbers?

We conclude this section with the observation that various difficulties are encountered in solving the problems above: for instance, the main difficulty in problems 3) and 4) lies in the fact that prime numbers are defined through multiplicative properties, while the problems in question involve additive properties.

Cryptography

Cryptography is the study of the methods that allow the secure transmission of information. Two main types of cryptography exist:

a) *secret key:* the classical method, used since ancient Rome. It is useful only when the number of users is small, since its correct working requires each user to agree on – and exchange secret key with – every other user prior to use;
b) *public key:* the modern method. It allows secure communication even when the number of users is high, since it does not require a prior exchange of secret keys. It was first proposed by Diffie and Hellman in 1976.

At first sight, public key cryptography seems impossible. In order to persuade you of the opposite, we propose the classical example of the *double lock*. Suppose that there are two users A and B and that A wants to send a secret message to B;

1) A puts the message in a box, locks it with his lock L_A (only A has a key to this lock) and then sends it to B;
2) B receives the box locked with lock L_A and adds his own lock L_B (only B has a key to this lock) and sends everything back to A;
3) A, receives the box with double lock, removes lock L_A and resends the box to B;
4) At this point, having received the box, B can remove the lock L_B and read A's message.

The security of this method lies in the fact that the keys to open the two locks are known only to the respective owners (who have not agreed on and exchanged keys prior to the transaction).

One of the "mathematical versions" of this idea is *R.S.A. public key cryptography,* proposed by Rivest, Shamir and Adleman in 1978. Let us briefly examine how A can send a secret message to B using the R.S.A. method:

B randomly chooses:

- two large primes p, q (consisting of 200–300 digits in base 10) and *calculates* $N = pq$ and $\varphi(N) = (p-1)(q-1)$
- a natural number e that is *coprime* with $\varphi(N)$ such that $e < \varphi(N)$, and *calculates* the natural number $d < \varphi(N)$ such that $de \equiv 1 (\mathrm{mod}\ \varphi(N))$

then *B* makes public numbers N and e.

In order to send a message to *B*, *A* carries out the following operations:

1) encodes the message in the standard way using numbers $\leq N$;
2) sends to *B* each number M resulting from such encoding under the form of

$$M^e \ (\mathrm{mod}\ N).$$

In order to decode the message, *B* simply calculates

$$(M^e)^d \ (\mathrm{mod}\ N);$$

What *B* obtains is exactly M, thanks to the Fermat-Euler Theorem stating that, in this situation, $(M^e)^d \equiv M \ (\mathrm{mod}\ N)$.

The main point is now: where does the security of the system lie? From what we have seen so far, in order to decode the message, it is necessary to know d. Knowing e, in order to calculate d, it is necessary to know $\varphi(N)$; but, knowing N, *calculating* $\varphi(N)$ *is computationally equivalent to factorising* N.

Therefore, all in all, the security of the *R.S.A.* method depends on the following facts:

- in order to encode the message, it is necessary to build large primes. This operation is computationally fast. It can be shown that the computational complexity of suitable primality tests[1] – used to establish if a number n is a prime – is of the form:

$$(\log n)^{c \log\log\log n},$$

that is, it is "almost-polynomial" in $\log n$ ($\log n$ is essentially the number of digits of n);

- in order to *break* the system, it is necessary to be able to factorise large natural numbers obtained as product of two primes. Such operation is computationally "slow" and its computational complexity is conjectured to be of the form:

$$e^{c\sqrt[3]{\log n(\log\log n)^2}}$$

that is "sub-exponential" in $\log n$.

[1] Recently (August 2002) M. Agrawal, N. Kayal and N. Saxena proved that there exists an unconditional deterministic primality algorithm with complexity $O((\log n)^{12+\varepsilon})$.

It is exactly such a marked difference in the speed of execution of operations – to determine large primes on the one hand and to factorise large numbers on the other – that guarantees the security of the method, at least for a sufficiently long period of time.

For instance, at the current state of technology, a natural number of 140 digits in base 10 can be produced through multiplication of two random primes in a few seconds on a typical computer available on the market. Yet, the factorisation operation of such 140-digit natural number would require about a month when employing several supercomputers working in parallel! Increasing the number of digits further increases the security of the system: it is currently recommended that numbers of at least 220 digits in base 10 be utilised.

Bibliography

We recommend the classical works by Ingham [1] and Davenport [2] for a clear explanation of the fundamental results on the distribution of prime numbers. We also signal the excellent introduction to the elementary theory of numbers by Davenport [3] that includes a chapter on cryptography.

For further details on the history of the development of cryptography, we recommend the book by Kahn [4], while the works by Koblitz [5] [6] provide a thorough presentation of a more rigorous mathematical modelling of public key cryptography than the one presented in this paper. For a good treatment of factorisation algorithms and primality tests, see the books by Koblitz [6], Cohen [7] and Riesel [8].

[1] Ingham A E (1932) *The Distribution of Prime Numbers,* Cambridge University Press, Cambridge
[2] Davenport H (1981) *Multiplicative Number Theory,* Springer-Verlag, Berlin Heidelberg New York
[3] Davenport H (1999) *The Higher Arithmetic: An Introduction to the Theory of Numbers,* Cambridge University Press
[4] Kahn D (1967) *The Codebreakers, the Story of Secret Writing,* Macmillan, London
[5] Koblitz N (1987) *A Course in Number Theory and Cryptography,* Springer-Verlag, Berlin Heidelberg New York
[6] Koblitz N (1998) *Algebraic Aspects of Cryptography,* Springer-Verlag, Berlin Heidelberg New York
[7] Cohen H (1994) *A Course in Computational Algebraic Number Theory,* Springer-Verlag, Berlin Heidelberg New York
[8] Riesel H (1994) *Prime Numbers and Computer Method for Factorization,* Birkhäuser, Basel

Homage to Venice

Essentiality of Mathematics for Bayesian Environmental Forecasting

CAMILLO DEJAK, ROBERTO PASTRES

Translated from the Italian by Emanuela Moreale

A homage to Venice! I find it very fair and appropriate: an environmental chemist-physicist will be able to illustrate the lagoon as a frame made by nature, while a mathematician like Michele Emmer will deal with the man-made city. He will certainly describe its beauty in terms of symmetry and asymmetry. Even for us chemical physicists, symmetry and asymmetry are important properties, starting from the simplest molecules in nature, such as diazirine and isodiazirine with their five atoms. Using rotations of 180° and reflections between two orthogonal planes, we can not only differentiate the former from the latter on the basis of experimental spectroscopic data, but can also substantially reduce the complex quanto-mechanics calculations necessary to simulate their chemical and physical properties.

This is not the mathematical tool with which our discipline can easily illustrate valuable properties of what Nature gave the Venetian environment, although there is some similarity between what affects the aesthetics of nature and of works of art. In his contribution, artist Achille Perilli describes the need to cause a feeling of "uneasiness" in the observer through broken regularities: he tells of two figures seemingly square, but not actually so, turning out to be rectangular, as he – an expert in the geometrical shapes of modern art – had immediately suspected and later verified through measurement. In this sense, he talks of the "eye" of the artist, capable of provoking sensations that impose themselves onto observers, stimulating their interest and active participation. Capi Corrales Rodriganez rightly stressed that such "eye" is not only the capacity of an artist, but of each and every man-made creative cultural form.

Thus, scientific researchers with their expert "eye" must see in nature those small imperfections that arouse "uneasiness". They give nature a chance to involve laypeople by proposing not only nearly reliable interpretations, but also some choices that can influence derived forecasts and allow highlighting which simple hypothesis is at the root of each complex alternative.

It is a good idea to explain this further with some simple examples: let us carry out three successive measurements of the same quantity, keeping experimental conditions unchanged as far as possible. Let the three values obtained be: nine, six and six. In practice, old analytical chemists proceeded as follows: a first measurement (9) and its confirmation (6); if the two measurements turned out to be substantially different, as is the case in this example, one proceeded to carry out a third measurement (6). If this third measurement was the same as the second (6), then the first measurement (9) was discarded; otherwise it was the second to be thrown away.

On the other hand, statistics has a totally different approach to the problem: the three measurements are averaged: $(9 + 6 + 6) / 3 = 7$, their variance is calculated $[2^2 + (-1)^2 + (-1)^2]/2 = 3$ and its square root is extracted, thereby arriving at the measurement of $7 \pm \sqrt{3}$. This would mean a value not equal to 6, as in the preceding procedure, but rather between 8,732 and 5,268 in 68% of cases, within a normal distribution - with 7 being the most likely value.

But is the former approach old-fashioned and is the latter correct? If – instead of carrying out measurements – one decides to execute numerous small calculations, continuously copying across numbers from a certain table, it is very likely that, in few (rare) cases, a copying error may be made. This is why checks and counterchecks are very useful: thus, the best procedure in these cases is the former. During the choice of approach, it is therefore important to reflect on basic issues, such as why measurements of the same quantity, carried out under nearly identical conditions, can give rise to such different results. This paradox causes "uneasiness" and deserves a simple answer, capable of involving the receiver of such explanation in this measurement itself. But, before answering this question, it is natural to ask what is expected of these measurements, i.e. their aim.

In this particular case, it is rather obvious that the intention is to make a forecast as to the result that a later measurement – carried out under the very same conditions – might give, or as to any different effect that might derive, depending on the value of such new measurement.

Having said this, it is necessary to make a series of hypotheses, i.e. "a priori" statements contributing to the result. On the one hand, the premise is made that "a priori causes" that determine an event are not modified in time or with variables dependent on time; on the other hand, a hypothesis is formulated as to why the measurements can give different results.

Having accepted the first premise of the temporal invariance of causes, it is possible to consider different causes in the case of this latter hypothesis. In the case leading to the exclusion of the dissimilar value (9), it is supposed that the experimenter – overburdened with the task of carrying out so many repetitive simple measurements – may in some cases make a "mistake", while in all other cases the result appears univocally defined. If the hypothesis is accepted, the first procedure appears to be correct.

Another, much more important, interpretation – always based on temporal invariance – takes it for granted that the non-perfect reproducibility of the result obtained in a measurement is inevitable, no matter how accurate the measurement is. What is usually done to increase the precision of the operation is look for all the major and most frequent causes of deviation of the measurement and correct them with experimental automatisms or theoretical corrections. Both of these approaches are based on integrative measure: however, these will not need a similar extreme precision to work, but will simply require being independent of each other. The more one pushes forwards this procedure to get greater precision, the more corrections are added, which however increases their number and therefore also the number of (independent) causes. Nevertheless, these corrections will also be increasingly smaller and thus each of them will determine a nearly insignificant deviation of the "real" value of the measurement, which will become a "precision" measurement. The statistical model is the throwing of dozens, hundreds, let us say N coins at random, for which the percentage of "heads" and "tails" should give a "true" value of one half, that is equal for each of the two frequencies. In fact, after many successive throws, the distribution of these percentages will approximate the value $\mu = 50\%$, representing not only its average, but also its mode – that is the value that comes up the most often. However, its distribution is irregular, the dispersion depending on the parameter $\sigma = 1/\sqrt{4N}$: outside the interval of percentage $\mu \pm \sigma$, there will be a probability of 32 over 100, that is 16 above and 16 below; outside $\mu \pm 2\sigma$, the probability will be of 2 over one hundred (1+1) and finally, outside $\mu \pm 3\sigma$, it will be a tiny probability of just 0.1 (0.05 + 0.05). It is well known that it is quite easy to forecast not only the exact value of the measured quantity or its expected future value, but also what will be the limits outside which these rare exceptions will probably not fall and with what probability they can be excluded.

In this respect, such model is perfectly appropriate for "precision" measurements, as long as the number of divisions on the reading axis is sufficiently high, so as to assimilate it to an infinite number of throws of an infinite number of coins, that is to a "Gaussian" or normal distributions of measurements (in the case of coins, throw percentages). For numbers that are not that large, the distribution law becomes less strict around the "mode – true value" and is called *student's t distribution.*

Going back to the aim of the forecast of any future measurements – when carried out in the conditions described above and only with respect to them – the expected value will be within the limits marked by two horizontal straight lines at height $\mu \pm \sigma$ with the probability of exceptional spread even for multiples of σ.

However, in the case at hand, there can also be a third interpretation that leaves aside temporal invariance μ ("true" value) but not σ, the size of fluctuations. In this case, one will look for the most likely interpolation of experimental data that is equidistant in time. With only three measurements – if at least a minimum of statistical credibility is wanted of the forecast – only a linear trend in time can be estimated: here, it would seem to be a downward trend. In this case, though, even keeping σ constant, constant, the limit within which the "expected" value can be found with a given probability will vary when moving away from the interval that the effected measurements fall in. As a consequence, there will be a marked difference between interpolation (within this very interval) and extrapolation (outside it), with an ever-diminishing likelihood as the distance from such interval increases.

Moreover, it would seem that, by increasing the number of measurements, one gets closer to the normal Gaussian curve with an ever-diminishing σ and that, as a consequence, the forecast can always be improved on. In this case, the determining factor is the patience of the experimenter in carrying out his work an ever-increasing number of times, always in the same conditions.

We have already seen how – even in these three basic examples – this is not the case. If the aim is that of foreseeing the effects of a future event from a series of past measurements taken in nearly identical conditions – once the events have taken place – it will be necessary to derive from them their causes and then, from the latter, the forecast. All this would seem to be a futile complication in simple cases; however, in nature, not all events depend on a single cause, but are more likely to depend on a series of causes. Perhaps such a complex logic route can be too complex for man-made simple, mono-functional objects; however, when an event is not derived from human logic, but from billions of years of natural evolution, it is imperative to take into account such considerations. Therefore, between the past (series of measurements) and the future (the expectation of a correct forecast), there is a complex present, made of statistics and logic (Tab. 1, top). It is no longer a comparison of different possible events, consequences of the same cause (with their respective direct probabilities), but – for an already occurred event – it is the comparison of different causes that may have determined it (with their respective inverse probabilities or, rather, likelihoods).

Thus, the relationship will be one between (non-normalised) probabilities relating to one cause and the sum of these over all the possible causes, aimed at producing the same event. The single term, however, will be given by a composite probability, that is by the "a priori" probability that the cause can occur in any case, multiplied by the probability that such cause can determine the event at hand (that is the direct probability mentioned above). The ratio between this product and the sum of all similar products can then be multiplied (nominator and denominator) by an equal arbitrary factor; that is it can be multiplied by the "a priori" probabilities that, in this case, will take on the name of "statistical weights" if they are natural numbers (Tab. 1, middle).

The difficulty will not be in the execution of the operations constituting "Bayes' Theorem", but rather in the evaluation of the "a priori" probabilities of the causes. In fact, they must necessarily be derived from a subjective evaluation, based on multiple experiences – often only of a qualitative nature – on intuitions and elaborations of our mind, which is what we could call a "theory" in a generic sense. Therefore, the meeting of experience and theory creates not only a Bayesian evaluation, but – in a more general sense – a "model": only from this can a forecast be derived and therefore an estimate of future effects obtained. The present, being situated between the past and the future, is determined by this inevitable moment of modelling (Tab.1, bottom). In less complex cases of pure human engineering, such moment can also be replaced by a "project", provided that this is not contaminated by unforeseen causes – which is a rare event. We are reminded of the dam of Vajont, a remarkable feat of hydraulic engineering that even today towers over the village of Longarone in the Italian region of Veneto. However, due to a geological underestimation of an ancient fault, it gave rise to the worst disaster relating to electric energy production ever occurred in Italy by causing over two thousand deaths. While man usually plans in mono-disciplinary fashion (with some marginal help from other disciplines), nature only uses transdisciplinary evaluations: these, however, are too complex for us, in that they require the simultaneous presence of all necessary disciplines. They are therefore replaced by interdisciplinarity, that is by step-by-step continuous interaction between the same disciplines, present throughout with their components open to the under-

Table 1.

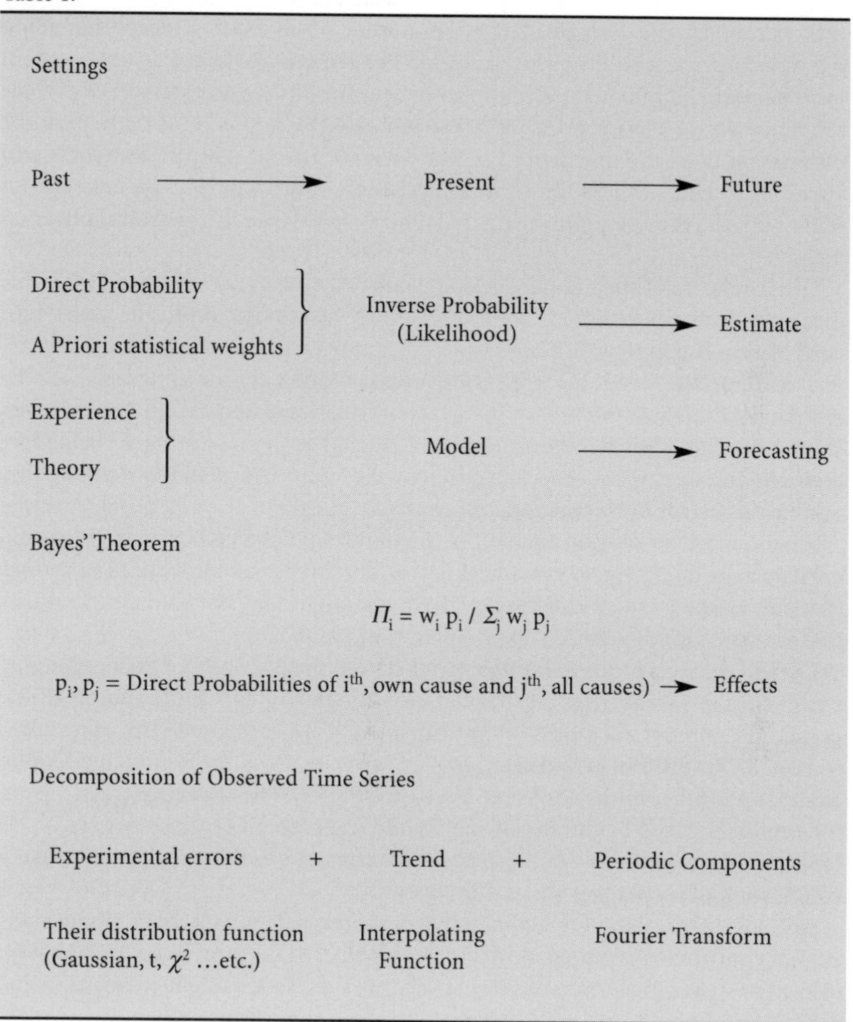

Settings

Past ⟶ Present ⟶ Future

Direct Probability } Inverse Probability (Likelihood) ⟶ Estimate
A Priori statistical weights }

Experience } Model ⟶ Forecasting
Theory }

Bayes' Theorem

$$\Pi_i = w_i\, p_i \,/\, \Sigma_j\, w_j\, p_j$$

p_i, p_j = Direct Probabilities of ith, own cause and j^{th}, all causes) ⟶ Effects

Decomposition of Observed Time Series

Experimental errors + Trend + Periodic Components

Their distribution function (Gaussian, t, χ^2 ...etc.) | Interpolating Function | Fourier Transform

standing of the knowledge of others; they are not replaced by multidisciplinarity that involves different professions separately, later assembling their respective contributions.

"Therefore, when dealing with environmental or environment-related problems in Bayesian logic, this interdisciplinarity also creates the problem of a joint evaluation across the various disciplines – of the "a priori" probability – to be able to take advantage of all the relevant experiences and intuitions. However, this is not always easy: the contribution of the various concomitant trends that can be attributed to different "disciplinary causes" should be separated out in the historical series of measurements itself. So far, in our oversimplifications above, we have examined only the components of accuracy, precision and trend.

It is exactly thanks to this apparent regular planetary and astral periodicity that astronomy can be considered the mother of all exact sciences, including mathematics. In fact, those who navigated in open seas (or in the desert) without tools helping them to maintain or find their way only had the starry sky as their reference point. However, they needed to know the annual and daily periodic movements of astronomical objects. The regularity of the daily movement is well apparent in Fig. 1, where the shutter has been left open during a whole austral night with the camera pointing towards the sky above an astronomical observatory.

The regularity of the annual cycle appears rather clearly also from Fig. 2: this illustrates the carbon dioxide concentration in the atmosphere at the observatory of Mauna Loa in Hawaii. The seasonal variations are impressive and clearly visible without the need for any calculations, thanks to the difference in the exchange of this substance with the winter vegetation compared to summer vegetation through photosynthesis. However, stochastic fluctuations are visible too: they are due to climate differences across the years and perhaps also to some other casual error or fluctuation.

Yet, the most interesting element is the multi-annual trend: at first, this trend appears to be one of linear growth. However, a trained eye soon notices an *uneasy* element: as soon as this is compared with a straight line, the data clearly shows that the trend is following a concave path towards the top.

The fact that this is due to human activity, from the discovery of the heat engine onwards, is apparent from Fig. 3, where two single points – representing all the points in the previous graph (empty squares) – are compared with concentrations of air trapped in cylindrical samples extracted from the South Pole (empty circles) and at the Antarctic base of Siple (full circles). The change in trend – from the centuries at the beginning of the second millennium to those ending it – is clearly evident, as the last two centuries are characterised by an ever-increasing combustion of carbon, oil and natural gas.

So far, we have dealt with a qualitative description, in which the periodic component contributes to giving more visual emphasis to the growth *trend*. However, in order to make a forecast and quantify also the less evident increase in the width of oscillation, one must separate the three contributions with appropriate mathematical methods. Yet, this will not be sufficient either, in that the combustion "cause" is tied to humanity's behaviour, for which there are several alternatives. Moreover, the quantity of global photosynthetic activity too is influenced by humanity's behaviours, such as deforestation. All these behaviours can change across continents, but also in time. Even if the logic is Bayesian, it must be seen within a more complex model that is open to various geographical and forecasting alternatives. It is here that mathematics – as an indispensable foundation and not only of existence and uniqueness – and numerical calculus in large computers together represent a complex forecasting tool which one can graft a predictive logic onto, if and only if this tool is suitably trustworthy and flexible.

Moreover, one should remember that every chemical reaction is given by the collision of two or more molecules and therefore its speed – proportional to the composite probability of such encounter – is the product of the respective vari-

Fig. 1. Regularity of the movements of astronomic objects above the observatory of Siding Spring in Australia. The shutter was left open for the whole, long Austral night.

able concentrations in the medium. Differential equations, on which a comprehensive model of such phenomena will be based, will then contain non-linear terms in these state variables, with possibility of bifurcation and deterministic chaos. This will have even greater incidence on enzymatic reactions in biology, where the non-linearity becomes even more pronounced. If we consider that the models required to describe complex phenomena will be mostly numerical and will therefore hide analytical logic, it becomes clear how necessary it is to proceed with mathematical rigour when dealing with these problems. All this was invented in the last few decades of the second millennium, thanks to the help of large

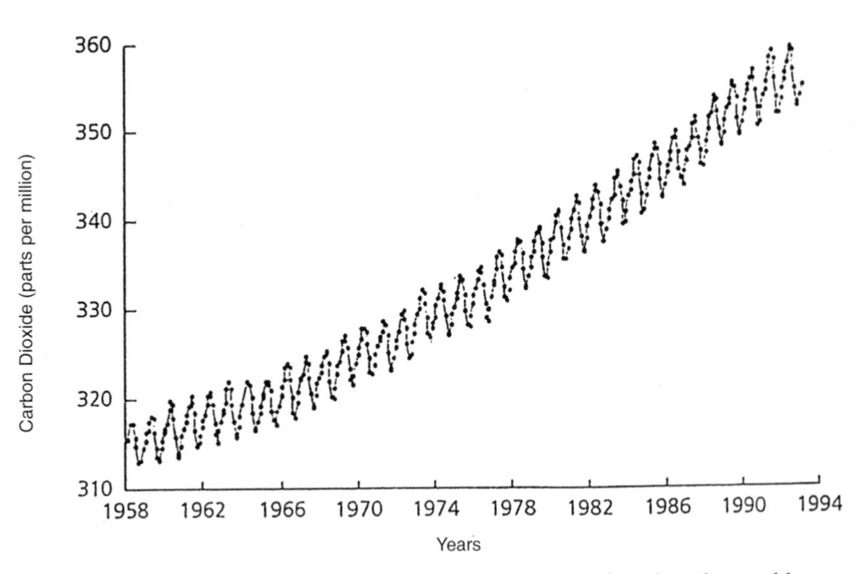

Fig. 2. Carbon dioxide/parts per million/year. Atmospheric carbon dioxide monthly concentration average at the observatory of Mauna Loa, Hawaii. The concentration is shown in parts per million of volume (ppmv) of dry air. The line joining each point with the next in this figure is not a model. The data shows two distinct variations: a seasonal trend, presumably given by the exchanges between vegetation and atmosphere, and a constant increase in yearly averages that is mainly thought to be due to human activities.

and very fast computers; it should therefore be studied further in all its implications, which are at times really unexpected. Such task requires much more work in the field of mathematical logic.

But let us go back to the analysis of series of experimental measurements: this – together with the evolution of nature – is a premise to any modelling and forecasting activity. In this extremely important sector, there is much less investment, partly because each scientific discipline believes it has different needs, at least for the different quantitative incidence of stochastic fluctuations. But even within the same discipline, there is often a need for different methods: reaction kinetics in aqueous solutions of alkaline and alkaline-earth nitrates and chlorides [1,2] shows limited fluctuations, but a different result is obtained if tetraalkylammonium salts are dissolved in it [3]. However, those who were used to the high precision levels of physics, find a more complex situation already in chemistry and particularly in kinetic chemistry. So much so, that they may have to reconsider their statistical approach to the biology of systems measured "in situ", that is on location and not in the laboratory.

Dealing with experimental data – particularly for those who need this data for modelling purposes – is a difficult job in which the eye of the researcher – trained to spot any "uneasiness" – is essential. Another essential element is their capacity to translate this uneasiness into precise mathematical formulations. On the one hand, data series that are too affected by stochastic variations do not allow –

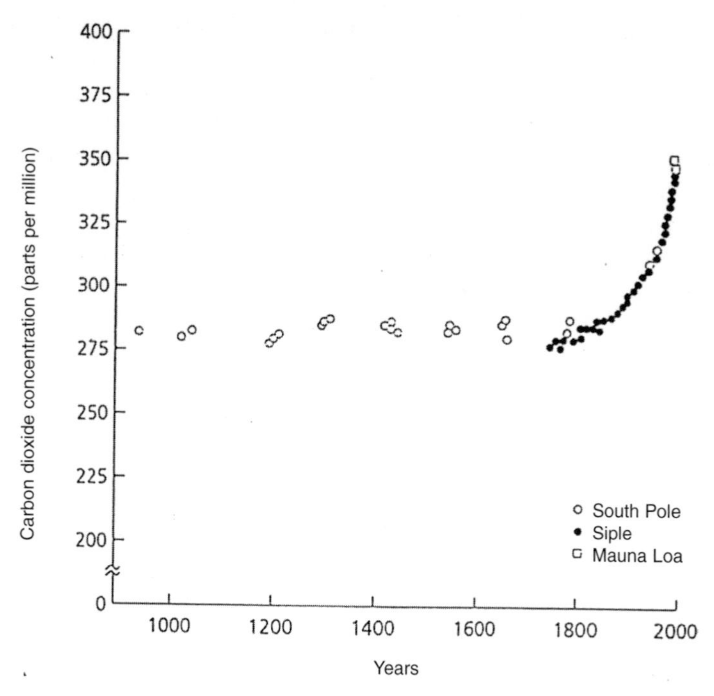

Fig. 3. Concentration of carbon dioxide/parts per million/years at the South Pole. A millenary record of carbon dioxide concentration, from cylindrical samples extracted from South Pole ice (empty circles) and from the Antarctic base of Siple (full circles), excluding the last two points that were measured in free atmosphere at Mauna Loa, Hawaii (see Fig. 2).

before a "smoothing" – a sensitivity analysis of the model with respect to semi-empirical parameters, neither do they allow a calibration of these parameters and other mathematical artifices necessary for any forecast. In general, in modelling, the laws regulating phenomena are known, while in the environmental field at least, the respective formulations contain parameters characteristic of the particular situations. They can be compared to Bayesian "statistical weights" to be determined initially "a priori", later studying the incidence on the simulation of the experimental results themselves and gathering from them optimal values through methods of mathematical calibration.

On the other hand, the first "categorical imperative" of the treatment of experimental data series lies in avoiding any subjective influence: even a subconscious choice of statistical procedures that might slightly distort the data, bringing it closer to one theoretical explanation rather than another would imply the untrust-worthiness of the whole statistical approach. The fact that, unfortunately, this artifice is often used in politics – in a way that is hardly perceived by laypeople and never admitted to by manipulators – probably represents the main reason why people are diffident towards statistics and science in general!

Going back to the separation of the three components – stochastic fluctuations, periodic increases and trends – this is not a difficult operation when the first are

Gaussian, the second sinusoidal with known period and the third linear. The non-Gaussian distribution of stochastic fluctuations is caused by reasons intrinsic to the very measurement and mathematics can do little in this case. However, analysing the distribution functions one can, at times, arrive at functional corrections. For instance, an analytical chemist insists on measuring concentrations of solutes in solvents, but also on using them as statistical variables when, under thermodynamic conditions of equilibrium (or stationary states), they are not state variables. The state variables are the chemical potentials that – for diluted solutions and non-extreme thermal variations – are proportional, within error margins, to the logarithms of the concentrations themselves. The lognormal distribution of their data can thus be easily corrected through a logarithmic transformation, so that all the usual statistical parameters can then be calculated on such chemical potentials. But, in general, distributions different from normal distributions are non-normal for much more complex reasons. It is often the case that there is heterogeneity in the component data belonging to distinct subgroups, for which a specific "a priori cause" must be identified. Alternatively, anomalous measurements are present and must be removed by resorting to notes that should appear in the original records of the researchers who took the measurements. Finally, there might be even more complex reasons, that must however be identified not mathematically, but rather with the help of those specific disciplines that can identify "a priori causes".

Separating the components of periodic variations and trends will require mostly mathematical tools, since only the latter will give integrals or series of integrals that cancel out over appropriately selected intervals. There is therefore the need to use orthogonal functions, among which Fourier's sinusoidal functions are the most commonly used (although, at times, under special conditions, also polynomials of the Chebyshev, Legendre, Laguerre, Hermite, Jacobi or ultraspherical varieties are better choices). The reasons for this lie in the intrinsic composition of the data and further details can be found in specialist textbooks in each discipline. Here we will limit ourselves to dealing with Fourier series, in that they are commonly used in chemical spectroscopy and crystal-chemistry and because there are very trust-worthy and fast programs (Fast Fourier) allowing specification of many different options.

In dealing with the last component, trend, one can extend linear curves so as to make them parabolic, cubic, and so on while, on the other hand, moving averages can be used to eliminate from periodic trends those trends that are too irregular. However, the more complex the argument, the more one should worry about its absolute objectivity; the more parameters needing to be introduced, the more the degrees of freedom characteristic of the data will be reduced, the more it will be necessary – in the end – to find a residue of Gaussian stochastic fluctuation as a confirmation of the regularity of the approach.

We will only give two examples here to illustrate the complexity of applications to environmental problems, that is to problems with high incidence of "noise" (term that we borrow from telecommunications engineering, together with its adjective "white" when its characteristics are only random, or more precisely, Gaussian). The first example refers to our research on time series of intensive measures – temperatures, solute concentrations, plankton density and so on – in water ecosystems. The second example deals with recent research on exceptionally high waters in the Venice lagoon, as already mentioned in the first example, but under the assumption that there exists just one periodic component with a cycle of eleven years. The relative standardised computer programs were applied in several technical reports [4].

Thus, the first example concerns an informatics "approach" to modelling temporal series of environmental data, through estimates of negentropy [5].

Preliminary procedures are carried out, such as obtaining equidistant data through linear interpolation (parabolic or superior interpolation usually give less trustworthy results, except for cases that are clear, due to the presence of maxima or minima with respect to time). Then, the logarithmic transformation described above is introduced because of the multiplicativity, not normal additivity, of the three contributions at hand.

The next step involves separating the three contributions and deciding which one to start from: in spite of the remarkable diversity of the four applications, it is quite clear that the first component to be eliminated is trend. Fig. 4 (top) reports the surface temperatures of water in the Venice lagoon: it is clear that there is no regularity in this trend, in that warmer and colder seasons follow each other for no apparent reason. The same is apparent from Fig. 5 (top) through the chlorophyll concentration (or, rather, logarithm of the concentration, as explained earlier), but is less evident in the concentration of ammonium nitrate (Fig. 6, top), which shows a sharp decrease. This is linked to the fact that the legislation on pollution has been adjusting more and more to the better technology now available, which translates in more and more rigid acceptability limits for the lagoon, independently of any cost-benefit analysis. Finally, in the last application – which will be dealt with further later, thanks to a decade of new measures and the uncertainties on the period arising from it – what emerged from the preceding analyses (and particularly in Fig. 6) is confirmed. In fact, there does not seem to be any regularity in the trend, even though, at times, there seems to be a growing or decreasing curve determined by "causes" that elude any rational – and therefore mathematical – treatment. In this case, the moving average is the only acceptable methodology left, particularly if the period seems to be known "a priori" from astronomical evaluations. This, however, does imply losing a semi-period at the beginning and one at the end of the record, unless some form of rather questionable extrapolation is applied. For instance, in spite of the application of strong statistical weights polarising towards the middle, Fig. 4 shows a quite questionable extrapolation, in that it was not corrected to safeguard the uniformity and therefore the objectivity of the approach, even if it could cause a deviation from experiment to simulation. This could translate into a marked unbalance when formulating forecasts and perhaps it would be more appropriate, in this case, to further reduce the already insufficient number of periods.

Having removed the trend from experimental data, the periodic component with predetermined period needs to be separated from stochastic fluctuations, which are presupposed to be "white noise". The traditional procedure (ANOVA, *analysis of variance*, ([5] p.207 and p.215) would involve mediating the single

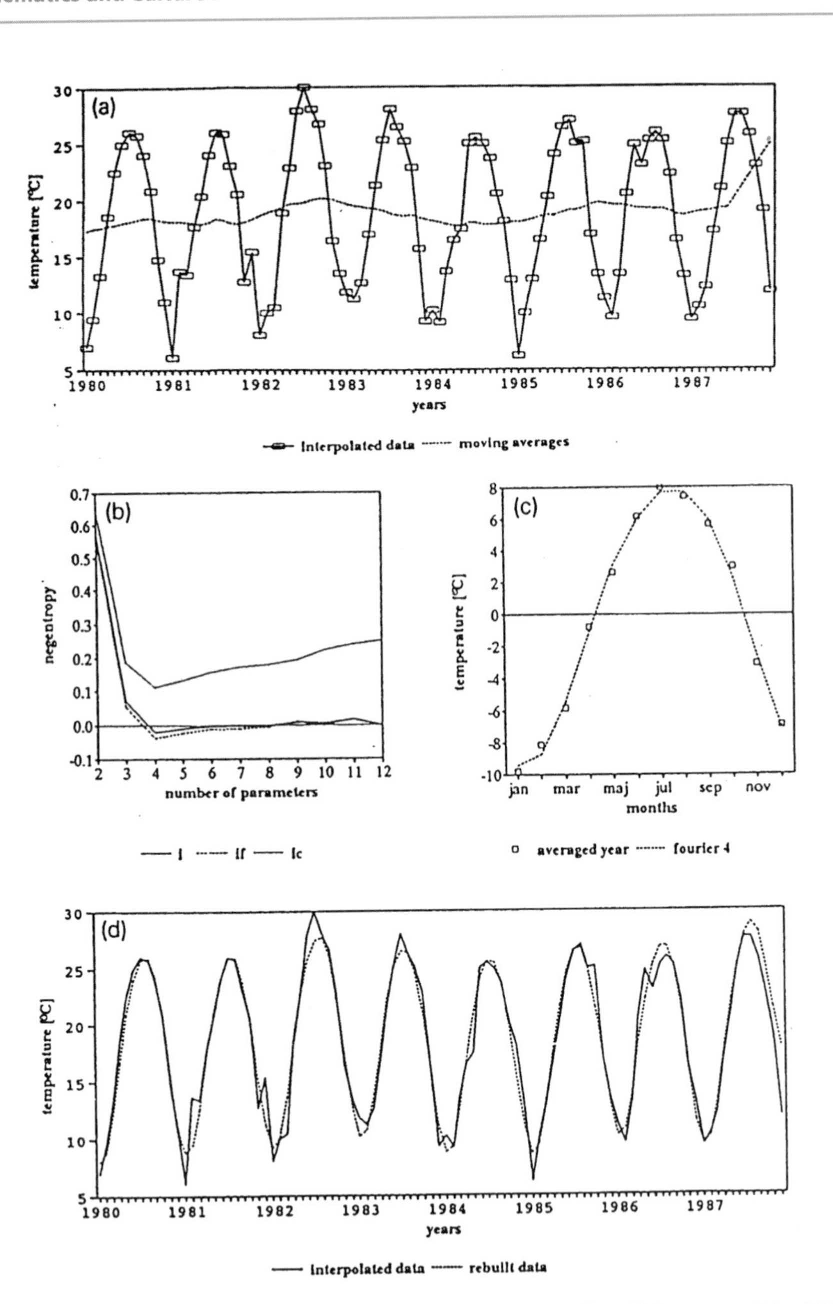

Fig. 4. Time series of Venice Lagoon surface temperatures (in °C) in years 1980– 1988, with interpolated equidistant data and yearly moving averages; **Middle:** representation of normalised negentropy (three cases described in text) in function of the number of parameters to be calibrated and inter-annual average temperature trends compared to truncated Fourier series with 4 parameters (the number that minimises negentropy in the graph opposite); **Bottom:** time series thus reconstructed are compared to original data.

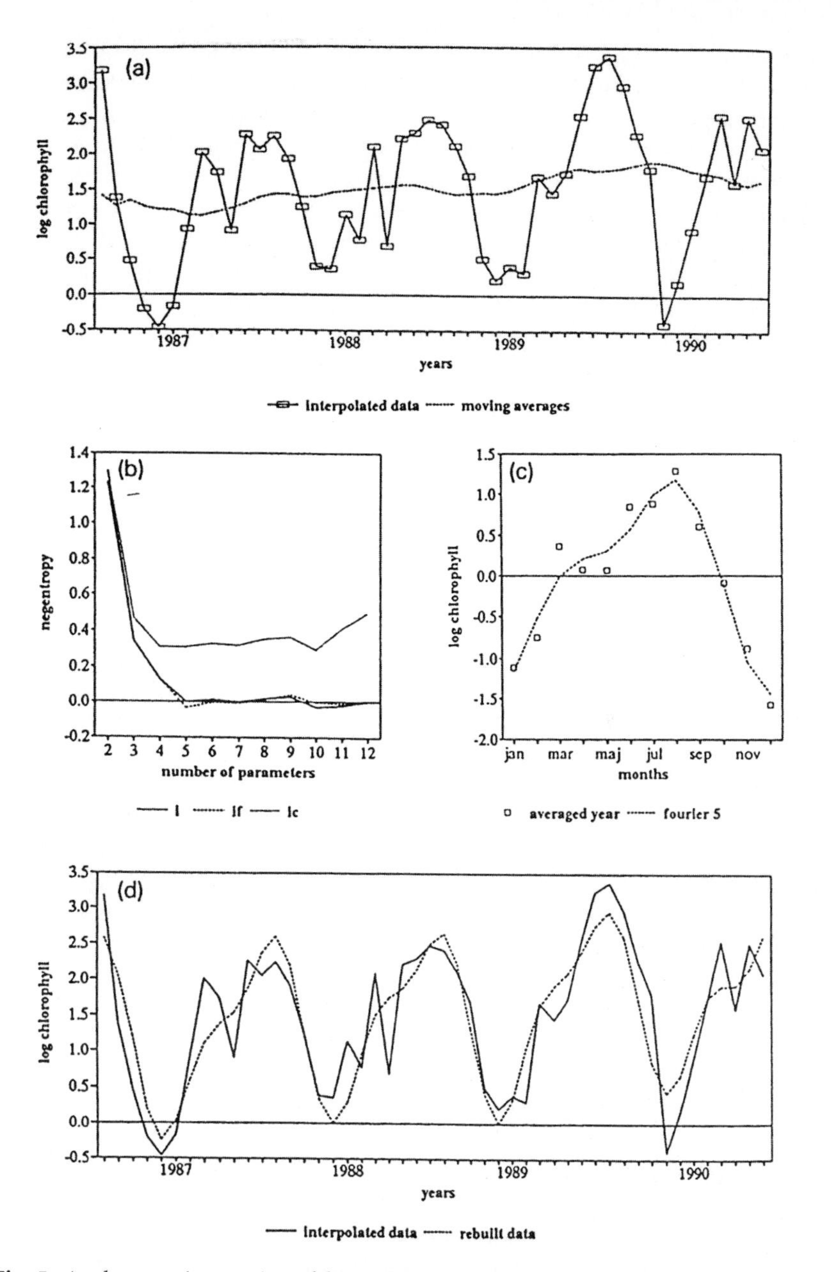

Fig. 5. Analogous time series of logarithms of chlorophyll concentration: density of phytoplankton for the years 1986–1990. Same arrangement as Fig. 4: even if the optimal truncation calculated at the second harmonic (5 parameters) proved to be effective, it is not able to show (more clearly than with a small dip) the spring blossoming (average March data) in the reconstructed trend.

periods and applying a Fourier transform to the data thus obtained. In this case, however, there is the need to truncate the resulting Fourier series, even if the predetermined period excludes all the non-harmonics. In spite of this, one cannot foresee any regularity of the coefficients of superior harmonics with those of fundamental harmonics, as is the case with mathematical and non-empirical series. The unknown quantities, Fourier parameters, would be too numerous and would therefore limit the degrees of freedom, removing from a dozen of mediated data the contribution of the first three coefficients (sine and cosine of the fundamental and the cosine of zero), as well as a couple of coefficients for each further harmonic (necessary to render the form of trends that are never perfectly sinusoidal in practice). Increasing the difference between the quantity of available data and that of such coefficients turns out to be very important when stochastic fluctuations are very strong and therefore the maximum possible number of degrees of freedom is required to overcome their disinformation effect.

It is in this case that the so called "overfitting" emerges. We mentioned overfitting at the beginning while talking of economic cycles identified in lotto results: these are stochastic fluctuations erroneously interpreted by the procedure as superior components of causal Fourier terms. There is also a further need to use a part of the periods as a "fit set" to find the coefficients of the truncated Fourier series and the (so far unused) remainder as "check set" to validate the conclusions arrived at. Given the high incidence of stochastic fluctuations, this is possible only by dividing periods in half. However, since – luckily – no regularity emerges in the series of periods, they can be assembled in all their possible combinations: if there are 2N data, the number of possible combinations will be very high: $\binom{2N}{N} = (2N)!/(N!\ N!)$, (if $2N = 8$, they will be 70, if $2N = 4$ only 6, but in the case most affected by "noise", with $2N = 12$, there will be 924!). It will thus be possible to effectively absorb the loss of the degrees of freedom, by utilising any possible piece of information contained in the time series and not only in the traditionally-used methods ([7], [8]).

It is now a question of obtaining an indicator for an optimal truncation, i.e. to determine on a case-by-case basis the appropriate number of terms in the Fourier series – such that the addition of more terms – will no longer reproduce any further "real" information. In this case, we have utilised Kullback-Leibler information measure [6]: this is identified in the case at hand with the opposite of information entropy, that is "negentropy", $-\Sigma_i p_i \ln p_i = -\int_0^1 \ln p dp$. By calculating and normalising such values, one obtains – as a function of the number of parameters of the truncated series – the values shown in Figs. 4, 5, 6 and 7 (middle left). It is thus possible to compare the reconstruction of the series itself with the averages of the experimental values and the whole data set shown in the graph at the bottom of the same figures. The first three of these graphs display negentropy cases: the values marked with an "i", by extending the treatment of ANOVA as a borderline case of the general case; the "if" case (dashed line in the figure), by using all the data as "fit set" and the "ic" case with the comparison between "check set" and "fit set". The last case presents ever-increasing values as the number of parameters introduced increases, thus highlighting the increase in uncertainty linked to the use of truncated series, since they contain an ever-increasing num-

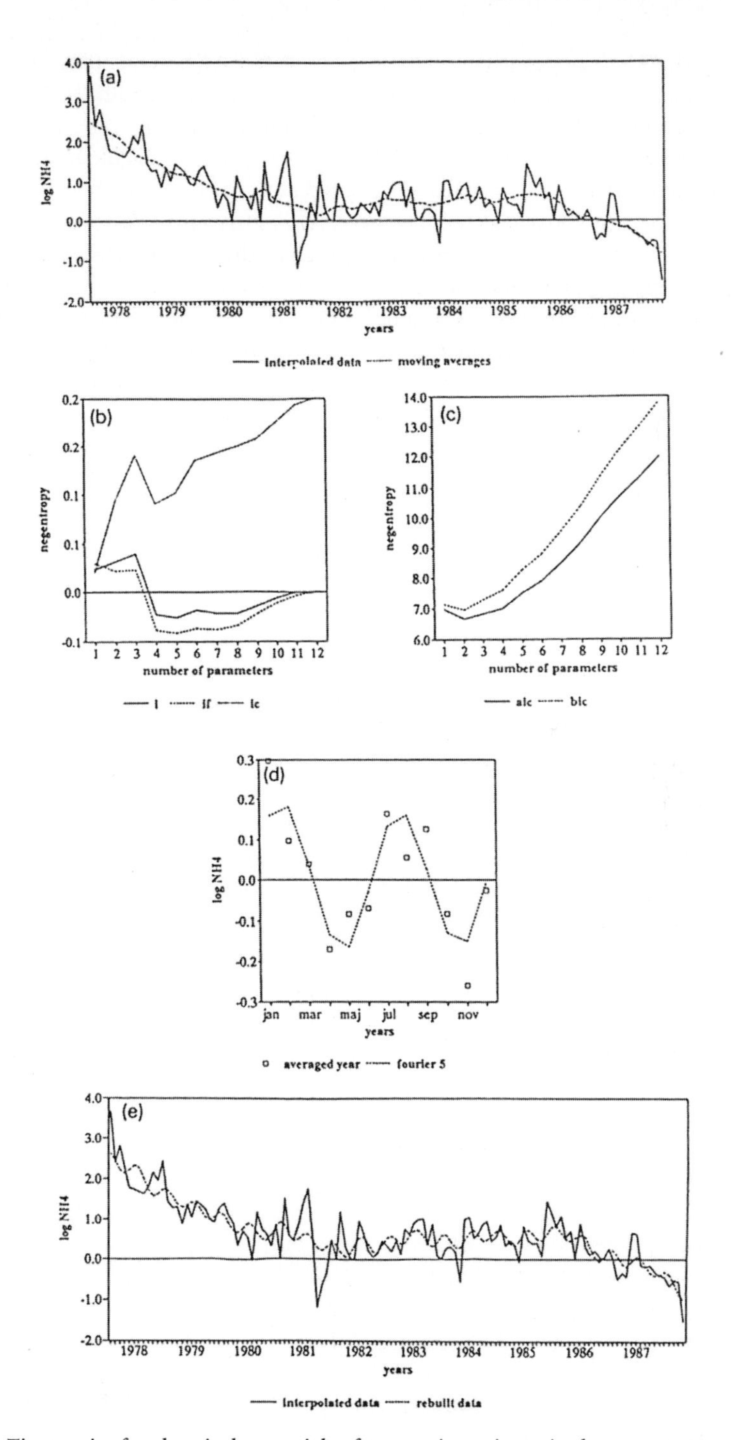

Fig. 6. Time series for chemical potentials of ammonium nitrate in the years 1977 to 1988. The additional graph compared to the previous figures shows that aic refers to H.Akaike's data [7] and bic refers to Schwarz's data [8], traditional methods.

ber of elements. The negative values of the normalised negentropies represent a significance index, but – in applications – it is the first minimum that is chosen and confirmed as the most effective indicator for the identification of the number of parameters to be calculated. In fact, the validation is visible in all four figures: in the middle right for the averages and at the bottom for the whole series. In Fig. 6 – the case most affected by stochastic fluctuations – there is also a comparison with commonly-used procedures such as those proposed in the bibliography by Akaike [7] and Schwarz [8].

It is evident that a minimum is found not with five parameters, but in the case having only two: that is, the series is truncated, maintaining the first (non-translatable) harmonic, instead of the necessary two complete harmonics, as the following figure shows.

After this procedure – which in complex real cases has been proved to render well the form factor of the periodic trend that is essential for the theoretical interpretation – it is still necessary to verify the Gaussian character of the remaining differences between interpolated and experimental data. In all four cases, the trend was verified and found adequate.

Up to this point, we have dealt with an example with predetermined period: this is normally the case in environmental problems, but is less frequent in other applications, particularly those of an econometric type. However, in the last application of the previous example, the eleven years' cycle – which is quite well-known in astronomy – does not possess the same certainty as annual cycles. In fact, it is possible that the presence of some periodic components with a longer cycle might produce phenomena of interference that are not easily highlighted with time series shorter than a century as well as remarkable influence of stochastic fluctuations. This will be dealt with in the second example, which is yet to be published.

This again deals with the number of extraordinary events per year (high water – exceeding the 1897 average by one meter and ten centimetres – in Venice). The trend for the various decades in the last century of the second millennium seems to be one of more than linear growth (the last data item must be incremented by at least a few units for the year 1999), even if the first two decades – for which it is difficult to find any certainties – are not taken into consideration (Fig. 8). The yearly trend is far less easily identifiable, because of even stronger stochastic fluctuations (Fig. 9). The data shown in Fig. 7 is limited to only forty-four years (1944–1987) and its logarithmic transformation is not extensible to the years containing zeros (years with no extraordinary events). Moreover, in the absence of predetermined periodicity, even using simple moving averages is not straight forward, at least not without influencing the possible multi-periodicity inherent in the phenomenon that, if it exists, should be highlighted.

An alternative to the logarithmic transformation is necessary here, since in this example the additive character of the contributions is more plausible, but the incidence of "white noise" is excessive, the type of accumulation is uncertain and periodic components perhaps have amplitudes that increase in time but have otherwise constant frequency. Therefore, not a functional transformation, but rather an operational one – an integration transformation to be precise – can be advan-

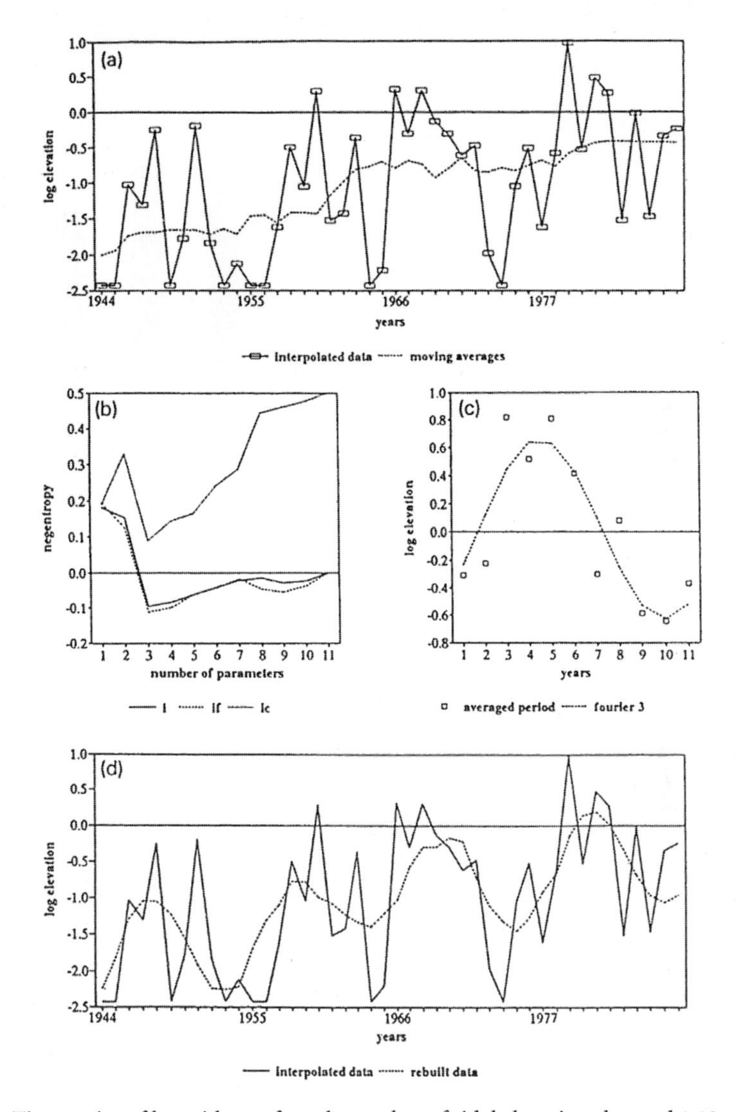

Fig. 7. Time series of logarithms of yearly number of tidal elevations beyond 1.10 m over the average 1897 sea level. The series is limited to the years 1944–1987 and therefore to a very restricted interval.

tageous here: an integral increases the polynomial trend by one degree (from constant, linear, parabolic, etc to linear, parabolic, cubic and so on) and highlights it, maintaining the sinusoidal frequency even if dampened. This is why, instead of dealing with the histogram in Fig. 9, we will examine its integral, or rather, cumulative curve in Fig. 10.

It is immediately apparent that the trend here is not linear, but must be at least parabolic and therefore the phenomenon of high water is not constant in time, but rather increases with it, since the parabola is concave towards the top. Since

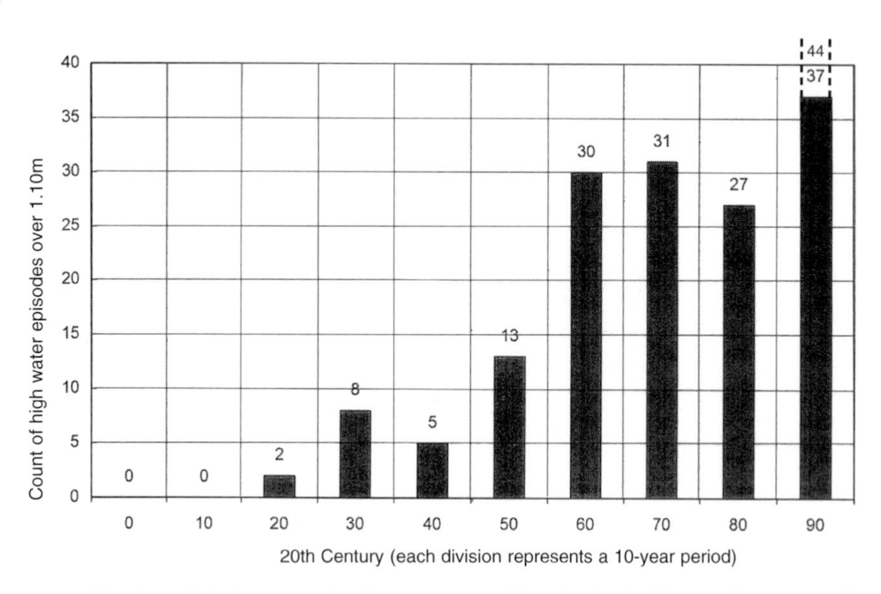

Fig. 8. Number of high water episodes over 1.10 m/decades in the Twentieth century. The number of single high water episodes over 1.10 m occurred in each decade of the Twentieth century is reported with the exception of those that happened in 1999. [1]

Fig. 8 also pointed to a more than linear trend, the integral of a parabola – a cubic trend – should also be considered. Had there also been any anomalous non-periodic curvature (at least in the span of under one century), it might have been worth trying interpolating with a fourth degree curvature. Having carried out these calculations, one notices that the residual variance – inclusive of harmonic components and stochastic fluctuations – in spite of its increase due to the reduction in the degrees of freedom, is reduced from parabolic to cubic and even to a quartic equation: therefore, this latter appears to be the optimal trend!

It was now a question of distinguishing between the original data and the three curves thus obtained: the results were more regular oscillations than those in Fig. 9 – as expected, the integration had cumulated the trend, but not the "white noise" – but their amplitude was increasing with time.

By taking the absolute values of those residual oscillations – still quite dispersed by stochastic fluctuations – a linear (not parabolic) regression was found, a sign that the oscillations increased in time, but without acceleration! Now, by dividing the original oscillating curve by the regression line, a trend with non-variable amplitude was finally obtained.

The latter was then fed to standard "Fast Fourier" programs to obtain oscillation periods: in all three cases, the main 11-year period was confirmed, thus val-

[1] *Note: in the whole of 1999, there were further 2 + 5 high water episodes over 1.10 m, raising the total from 37 to 44, as illustrated by the dashed lines in Fig. 8. The first few months of the year 2000 saw another such event at an unusual time of the year.*

idating the astronomic intuition. However, by subtracting a merely parabolic trend, three further periods are obtained; in the case of a cubic one only two more periods were obtained and finally, with a quartic one, just one period of 16 years, apart from the eleven-year period. The explanation is quite simple: the number of years in which high water has been recorded and measured is too limited to reliably highlight very long periods. The latter could be interpreted either as "one-off" curvature anomalies through cubic and quartic equations or else as periodic phenomena at such low frequency that they occurred a very low number of times but, in spite of this, were revealed through the use of the "Fast Fourier"!

What was strongly confirmed was the eleven-year period, which the most attentive among the Venetians had already spotted over the experience of many decades, independently of any astronomical explanations. What was new – instead – was a slightly longer second period: this can be considered quite likely, even if the former is repeated seven times during the interval (five times during the postwar period as in Fig. 4). In sixteen years, the repetitions become five (three in the most significant postwar interval) and therefore less certain, although still possible. Longer periods cannot therefore be highlighted in a reliable manner from such limited time series, for it is the phenomena themselves that determine this limitation: they begin following human activities after the first world war, while during the second world war there was a certain truce with nature. It is as if, when men fight among themselves, they manage to damage nature less, a phenomenon that was already noted in other environmental time series!

Overall, the analysis is yet incomplete, because there were difficulties in its synthesis and therefore new historical series are being examined concerning the dif-

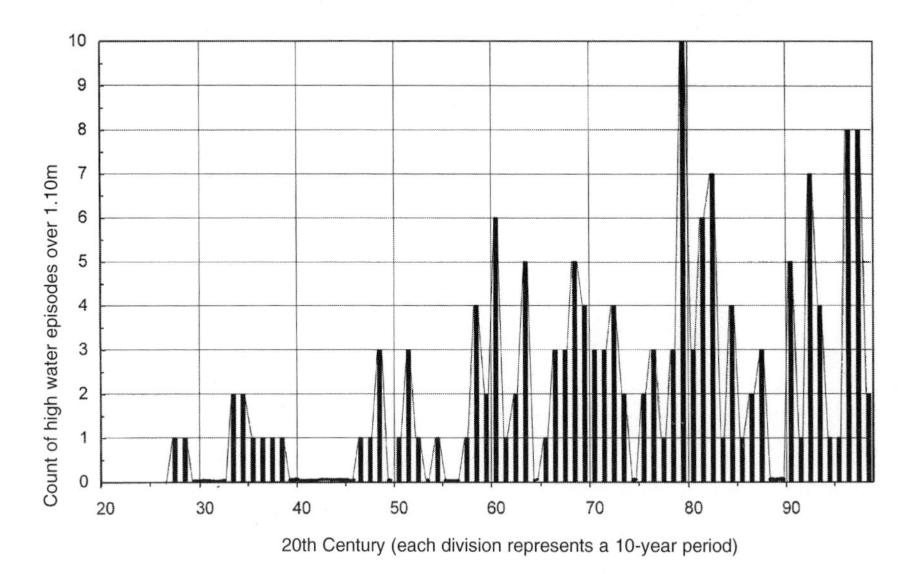

Fig. 9. Count of high water episodes as in Fig. 8, but year by year. A certain periodicity is evident, with many zeros (1929–1933, 1939–1946, 1949, 1953, 1955–56, 1964, 1974 and 1988–89) and a very strong influence of stochastic fluctuations (see also note to Fig. 8.).

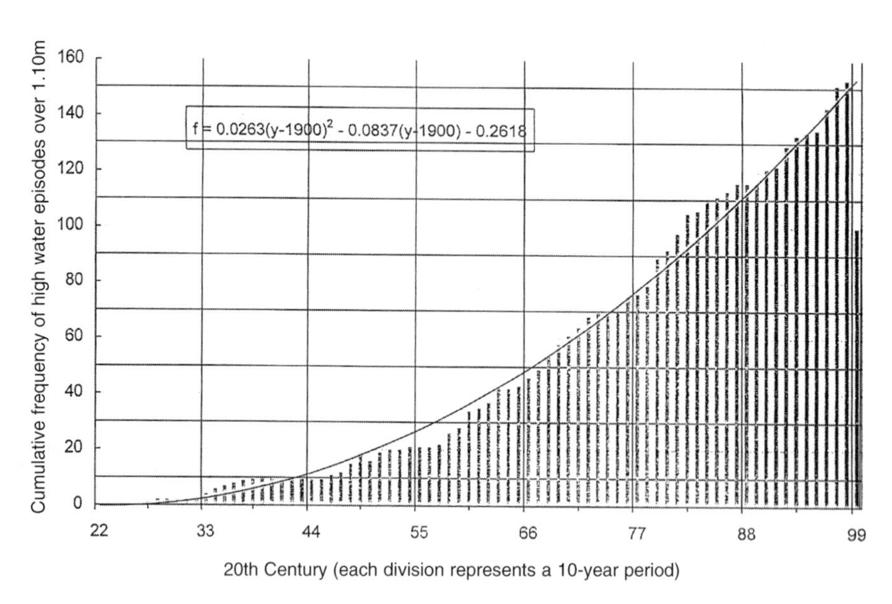

Fig. 10. Cumulative frequency of high water episodes over 1.10 m occurred before the year indicated on the x axis, compared to the regression parabola optimised between these points (for the last two years, see note to Fig. 8).

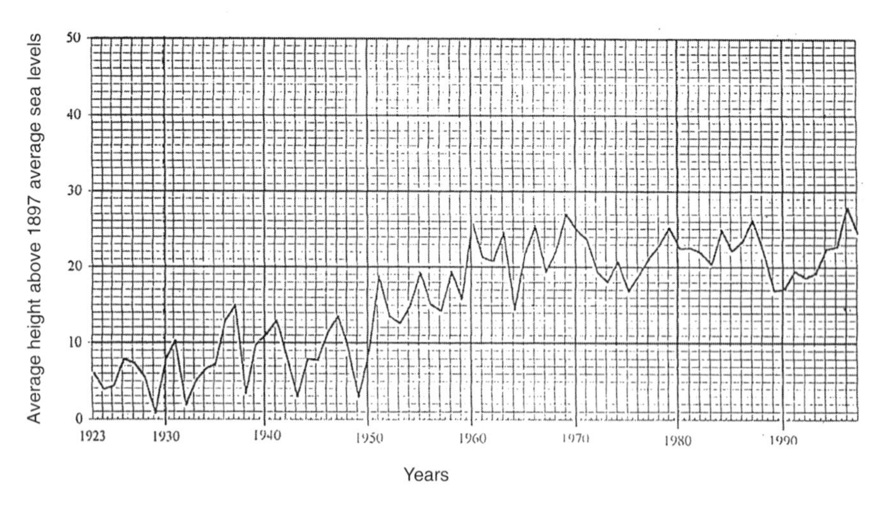

Fig. 11. Punta della Salute sea station: trend of average of tide extremes (maxima and minima) between 1923 and 1997.

ferences between high water and average sea level, which has risen as much as twenty-three centimetres (Fig. 11): new data is expected to be available for the first half of this century. Yet, from this analysis it seems to be clear that high water is rising, and more than linearly so, in a way that is independent of a certain stasis in the average sea level. Multiple periods, however, make this phenomenon less predictable and, therefore, increase the chances of superficial interpretations, particularly of the type expressed with fervour and out of political interest!

Most scientists now recognise that oscillatory phenomena increase more than average phenomena because of the greater energy produced in the atmosphere by the greenhouse effect: it is sufficient to think of hurricanes and their ascertained increased frequency! It is no longer possible for delicate environmental problems to be tackled with the superficiality of politicians or pseudo-scientists, without using mathematics to the utmost of its abilities!

In this sense, we must therefore conclude that the modern world – not in itself, in its purely human constructions, but – when it gets to the point of endangering natural "equilibria", has only one tool available to remedy the situation: more mathematics. Yet, it cannot continue to proceed only through analytical solutions, without considering the more and more powerful computers as an indispensable tool for investigation and modelling as well as for the forecasting and identification of remedies.

Mathematical innovations in the last few decades – from bifurcation in differential problems with non-linear algebraic terms, fractals, deterministic chaos and so on to the discovery of new universal constants such as Feigenbaum's d = 4.6692... – would not have been possible without the intensive use of large computers. However, it is exactly these innovations in mathematics that bring it closer to nature and its more common forms – such as a fern leaf – that have so far been inexplicable through traditional mathematics but can now be reproduced in a very surprising way! The way to go is therefore that of finding – for each instance of "uneasiness" that nature causes in man – a mathematical methodology created by the encounter of creative mind and processing power.

Bibliography

[1] Rolla M, Carassiti V (1949) Ricerche sull'Effetto Cinetico Primario di Sale: Nota I, *Bol Sc. Fac Chim Ind Bologna* 7:1–11

[2] Carassiti V, Dejak C (1957) Ricerche sull'Effetto Cinetico Primario di Sale: Nota III, *Bol Sci Fac Chim Ind Bologna* 15:63–69

[3] Dejak C, Mazzei I, Cocco G (1971) Research on the Primary Kinetic Salt Effect: XII; The Influence of Tetraalkylamonium Salts on the Iodide-Persulphate Reaction, *Gazzetta Chim It* 101:606–624

[4] Pastres R, Solidoro C, Franco D (1997) *Trattamento ed Analisi dei Dati Anche Mediante un Modello di Qualità dell'Acqua: Sintesi dei Risultati.* Consorzio "Venezia Nuova", Rapporto Tecnico sulla Sintesi dei Risultati Raggiunti

[5] Dejak C, Franco D, Pastres R, Pecenik G, Solidoro C (1993) *Ecological Modelling* 67:199–220

[6] Kullback S (1959) *Information Theory and Statistics,* J Wiley & Sons, New York USA

[7] Akaike H (1974) *IEEE Trans Automatic Control* 19:716–723

[8] Schwarz G (1978) *Ann Statist* 6(2), pp 461–464

Venice-Maths

Michele Emmer

Translated from the Italian by Emanuela Moreale

It is not forbidden [...] to speak of familiar things, and
I hold that for the true Venice-lover Venice is always in order.
There is nothing new to be said about her certainly,
but the old is better than any novelty.
Henry James [1]

Everyone believes that they have a privileged and exclusive relationship with Venice. Everyone believes that some bridge, *calle* (street) or city corner is there only for themselves and that they are the only ones aware of its existence. Each one of us has a special memory relating to the city on the water.

Of course, the inhabitants of Venice have a different view of the city. I taught at the Università Ca' Foscari in Venice for a few years and we have a house there. I am therefore not a total *foresto* (stranger, foreigner), as the Venetians put it. Just like all casual visitors, I too believe that there exist a few places and images of the city that are my own.

Between 1976 and 1990, I shot eighteen films in the *Art and Mathematics* series. Some of these were shot partly in Venice. Because they were films, it is obvious that the starting points had to be visual and that therefore I needed to ask myself the fundamental question about Venice: are there here any objects, places, works of art that have a mathematical interest? Venice is a theatre city par excellence and it is sufficient to walk around the city a while to discover remarkable geometric-mathematical shapes in architectonic structures, palaces, streets and fields. This is obviously true of nearly any Italian town.

Yet, in Venice there are specific elements that are of particular interest to the history of mathematics. Some of these works of scientific interest were realised by great Renaissance artists and this makes it sensible to think of Venice when dealing with art and mathematics [2].

Polyhedra

"We owe a large number of *editions principes* of scientific texts – both in their medieval versions from Arabic both in the Humanistic versions translated directly from the Greek – to the cultural world and to Venetian typographers".

Maccagni thus writes in his catalogue of the exhibition "La Scienza a Venezia tra Quattrocento e Cinquecento. Opere manoscritte e a stampa." (Science in Venice between XV and XVI century. Manuscripts and printed works)[3]. This suggests a deep link between mathematical sciences and Venice in the Renaissance.

Such a link concerned also the artists, some of whom were among the most important mathematicians of the time. Among these were Albrecht Dürer and the manifesto of his interest for geometry represented by the engraving *Melencolia I*. In it, a magic square contains the date of the composition: 1514. While in Bologna, Dürer probably became acquainted with mathematician Luca Pacioli, a disciple of Piero della Francesca, a great artist and distinguished theoretician of perspective who authored the well-known treatise on regular solids entitled *De Quinque Corporibus Regolaribus* [6].

Pacioli incorporated Piero's treatise on regular solids in his famous 1509 book *De Divina Proportione* that was also published in Venice. It can be seen how the Renaissance history of geometry is strictly tied to Venice! The *De Divina Proportione* owes much of its fame to the fact that the sixty tables of geometric solids were "facte e formate per quella ineffabile mano sinistra a tutte discipline mathematici accomodatissima del prencipe oggi fra i mortali, pro prima fiorentino, Leonardo da Vinci" (made by that ineffable left hand well-versed in all mathematical disciplines of the prince now among the mortals, earlier Florentine, Leonardo da Vinci).

A portrait of Pacioli was completed towards 1498–1500, possibly by Jacopo de Barbaris, author of a famous map of Venice. The painting is particularly interesting, because in the top left corner it depicts a model – perhaps a glass model, a detail of non-secondary importance, since we are in Venice – which corresponds to table XXXV of the *De Divina Proportione*. It was exactly in Venice that Dürer arrived during his first trip to Italy in 1494–95. If Piero della Francesca is considered by historians of mathematics to be "the painter-mathematician and artist *par excellence*... the best geometrician of his time", Dürer is considered "the best mathematician". The Venetian period is certainly of great importance in Dürer's artistic and geometric development.

Another visitor to Venice, although many years earlier, was Paolo Uccello. Among the marble tarsias inside St. Mark's Basilica, there are two geometric ones that Muraro attributes to Paolo Uccello, who would have realised them while he was in Venice in 1425–1430 [8]. Of the two solids, the first is located on the floor at the entrance of the Basilica, while the other – less famous – can be found in a restricted area in the nave. The two solids have the same shape: two stellated dodecahedra: dodecahedra such that a regular pyramid has been placed on each of the faces. It was Lucio Saffarô[1] [9] who noticed their presence: himself an artist with a great interest in mathematics, he worked on polyhedra

[1] *Lucio Saffaro died in November 1998.*

for many years. When, in 1970, he noticed the polyhedron, Saffaro thought it unbelievable that no other mathematician had noticed it before. Later on, he found out [10] that the polyhedron was mentioned on page 88 of a work by German historian S. Günther that had been published in 1876 [11]. This was a surprise because the mathematical discovery of the stellated dodecahedron was attributed to Johannes Kepler, who – in his 1619 treatise *Harmonices Mundi* and therefore many years after Paolo Uccello – describes a solid that he calls *stellarum duodecim planarum pentagonicarum* [12].

In 1986, the image of Paolo Uccello's stellated dodecahedron was chosen as symbol of the "Biennale di Venezia" exhibition dedicated to the theme *Art and Science* [13]. Saffaro later noticed that, on the floor of the chapel of St. Pantalon, in the homonymous church, there are two marble tarsias that look exactly the same, although damaged by the floor restoration: they represent the other type of stellated dodecahedron appearing in Kepler's treatise. The author of the two tarsias is so far unknown, but it may well be Paolo Uccello himself.

The Ducal Palace was the great work of Venice at this period, itself the principal effort of her imagination
John Ruskin [14]

Symmetry

"Symmetry is a vast subject, significant in art and nature. Mathematics lies at its root, and it would be hard to find a better one on which to demonstrate the working of the mathematical intellect". This is how, in 1951, famous mathematician Hermann Weyl concludes his volume on symmetry written on leaving the Institute for Advanced Study in Princeton. He also wrote:

> One may ask if the aesthetic value of symmetry depends on its vital value: Did the artist discover the symmetry with which nature according to some inherent law has endowed its creatures, and then copied and perfected what nature presented but in imperfect realizations; or has the aesthetic value of symmetry an independent source? I am inclined to think with Plato that the mathematical idea is the common origin of both: the mathematical laws governing nature are the origin of symmetry in nature, the intuitive realization of the idea in the creative artist's mind its origin in art; although I am ready to admit that in the arts the fact of the bilateral symmetry of the human body in its outward appearance has acted as an additional stimulus. [15]

Weyl's book is rich in examples taken from the artistic field. It is on these examples that he builds a mathematical theory of symmetry groups. One by one, Weyl describes the various types of symmetrical transformations. One of the first is translation symmetry, exemplified by the Ducal Palace in Venice. A visit to Venice always leaves traces evoking memories and images even after many years!

It should be noted that there may exist symmetrical structures that are part of wider structures that are not symmetric; conversely, symmetrical structures may have asymmetric elements inserted in them. The very concept of symmetry has been changing over the years and the meaning of this term essentially depends on the area of discipline and on the scientific field in which it is utilised.

To illustrate how the same symmetrical elements may be regarded differently, one could take as an example the Ducal Palace itself. While Weyl stresses its rather evident elements of symmetry, Ruskin instead underlines its asymmetric character. Talking about the (seaward) façade of the Palace, he notes that the two windows on the right are lower than the other four and sees this layout as one of the most remarkable examples of a daring sacrifice of symmetry in honour of functionality, which he considers the most noble aspect of gothic art. They are therefore asymmetric elements inserted in a symmetrical structure.

The Grammar of Ornament. The Victorian Masterpiece on Oriental, Primitive, Classical, *Medioeval and Renaissance Design and Decorative Art* [15] – published in 1856 by Owen Jones – is a great encyclopaedia of decorative patterns. It also contains Byzantine patterns and therefore the mosaics and (in)tarsias that are part of the decoration in St. Mark's Basilica: in particular, it illustrates a dozen of the most interesting ones found on the floor. The layout of these patterns on the floor of the Basilica – as can be only noticed from a map, since most of them are covered up to protect them from visitors – is also symmetrical with asymmetric elements. It goes without saying that, when one exits the Basilica, one finds the image of apparent symmetry of Piazza San Marco (St. Mark's Square).

The human spirit always makes progress, but this progress is spiral.
Madame de Stael

Spirals

Every Venice visitor soon understands that the shortest distance between two points is not always the segment that unites them. In Venice, the shortest route is always winding, broken, a labyrinth. "There exists no natural labyrinth structure in which the work of man has gradually superimposed itself in such a way that it became a kind of Incan reading" [16]. According to Giuseppe Sinopoli, the manifestation of the sacred in Venice is sealed by the supreme symbol that represents it, determines it, specifies it and stabilises it: the natural double spiral of the Canal Grande around which the city has gradually grown.

It is near the point where the two spirals touch that the Department of Mathematics of the Ca' Foscari University is situated: this department is famous for its mirror hall, where Titian's teleri – now at the Metropolitan Museum in New York – used to be located.

Toward the end of the double spiral on the Canal Grande is the Basilica of Santa Maria della Salute. On 22nd October 1630, the Venetian Senate made a vow to God to erect a church in this town and to dedicate it to the Virgin Mary with the title "Santa Maria della Salute" (Saint Mary of Good Health) [17]. The town was being ravaged by the plague, the most recent of a series of Venetian pestilences and the most violent in terms of mortality [18]. Among the many projects put forward, after some disagreement, Baldassarre Longhena's was eventually chosen. In its final construction, the dome is supported by modillions in the shape of enormous Archimedes's spirals or large yoyos. Going up inside the dome, from the gallery one can see the floor of the basilica and note that it is composed of families of intertwined spirals that move farther and farther away from the centre. The floor of the basilica is remindful of another typically Venetian object concerned with the intertwining of many spirals: this is what Rosa Barovier Mentasti calls "the most important invention of the Sixteenth Century in the field of glasswork". In 1527, Filippo Catani asked for a 25-year exclusive right to sell "twist bands": it is the official birth certificate of this sophisticated technique in Murano, a technique that allows the creation of parallel bands, twisted in variable spiralling shapes, made of milkscab glass – an opaque white glass similar to fine porcelain – or coloured glass.

The human spirit always makes progress, but this progress is spiral – so is reported to have said Madame de Stael. In Venice, it is sufficient to board a *vaporetto* (steam-driven boat) that travels along the Canal Grande to get away from or closer to the centre of a double spiral: the Rialto Bridge. If the structure of the city is labyrinthine, the double spiral of the Canal Grande has the function of avoiding getting lost in the labyrinth. If – as Teilhard de Chardin wrote – the spiral is life, Venice need not worry about its future, as it is built around a double spiral.

So, as you roam around these labyrinths,
it is never clear if you are pursuing a goal
or running away from yourself,
if you are the hunter or the prey.
Iosif Brodskij [19]

Labyrinths

"I was therefore somewhere in the Venetian labyrinth. I knew one thing: its topology is absolutely unique and is not found in any labyrinthine construction of the past" [20]. There is no doubt that even people living in Venice have trouble acquiring an absolute certainty as to the best route to choose. If the aim is to reach St. Mark's Square from faraway areas, the difficulty is increased by the ever-increasing complexity of the routes and choices to be made. Naturally, we are talking of the terrestrial labyrinth: in fact, in Venice there are two labyrinths, a terrestrial one and an aquatic one that is made of canals. It is not at all obvious that the two labyrinths are complementary; on the contrary, when proceeding on foot, it is a good idea never to trust the direction of a canal and the opposite also applies. In this way, the city realised – many centuries in advance – the modern idea of separating the two types of traffic constituted by people and means, in this case aquatic means.

It is evident that the two levels of the labyrinth touch in many points and that it is possible to proceed for a while on water and then continue on land, by using a gondola-ferry. Sinopoli stresses that "the entrance and the exit from the labyrinth, in the case of the Venetian lagoon, is not only a way to see the world, or an idea that takes shape from the recognition of symbolic representations in the lagoon city, but it is also a real and dynamic process that takes place within the natural labyrinth as well as the one built by man". Mathematical models are being studied for this dynamic process, so as to foresee the behaviour of the ecosystem of the lagoon and to intervene to solve its problems, such as pollution, tides and the phenomenon of high water, when the natural labyrinth invades the man-made one and erases it. At that time, the city becomes one water labyrinth, the great water spiral dividing the city increases in size and merges with the lagoon.

"How beautiful the world would be if there were a procedure for moving through labyrinths" says William of Baskerville talking to Adso of Melk; these are the two protagonists of *The Name of The Rose* [21]. As anyone who has read the book or seen the film based on it will know, William was facing the problem of deciphering the shape of the library labyrinth. The two would have had a much tougher job if they had arrived in Venice without a map, possibly at night, and had had to reach the Ducal Palace, for instance! So much for Ariadne's thread!

Some mathematicians studied mazes with no junctions to classify them, while others studied labyrinths with junctions [22] and found mathematical rules for completing them, at least in some cases. In the last few years, computer programs have been written that allow traversing a maze with junctions. Given the small dimensions of today's laptops, these programs could be live-tested in Venice. This would certainly be a good test, even though the method I prefer for going around the city is that of wandering around the streets trying to understand which way to go based on reference points. Using this approach, many errors are made, as one often has to backtrack, but only this way does one get the impression of exploring a labyrinth that is always new and that seems to be there only for us.

Bibliography

[1] James J (1909) *Italian Hours,* Illustrated by Joseph Pennell, Heinemann, London
[2] Emmer M (1993) *La Venezia Perfetta,* Centro Internazionale della Grafica, Venice
[3] Maccagni C (ed) *La Scienza a Venezia tra Quattrocento e Cinquecento. Opere Manoscritte e a Stampa,* Venezia 3–15 Ottobre 1985
[4] Pacioli L (1494) *Summa de Arithmetica, Geometria, Proportioni et Proportionalità,* Venice
[5] Lane FC (1978) *Storia di Venezia,* Einaudi, Turin, pp 168–169
[6] Davis MD (1977) *Piero della Francesca's Mathematical Treatise,* Longo Editore, Ravenna
[7] Pacioli L (1509) *De Divina Proportione,* Venice
[8] Muraro M (1955) *L'Esperienza Veneziana di Paolo Uccello,* Atti del XVIII congresso Internazionale di Storia dell'Arte, Venice
[9] Saffaro L (1976) *Dai Cinque Poliedri all'Infinito,* Annuario EST – Mondadori, Milano, pp 473–484
[10] Saffaro L (1989) Anticipazioni e Mutamenti nel Pensiero Geometrico, in Emmer M (ed), *L'occhio di Horus: Itinerari nell'Immaginario Matematico,* Ist Enciclopedia Italiana, Rome, pp 105–116
[11] Günther S (1876) *Vermischte Untersuchungen zür Geschichte der Mathematischen Wissenschaften,* Lipzig
[12] Kepler J (1619) *Harmonices Mundi Libri V,* Linz
[13] Macchi G (ed) (1986) *Spazio,* Catalogo della Sezione, Biennale di Venezia, Venice
Emmer M (1987) *Computers,* film, *Art and Mathematics* series, 16 mm, 27 minutes. Film 7 Int, Rome. The film was partly shot in the *Space* section of the Venice Biennial (Biennale di Venezia) in 1986
[14] Ruskin J (1885) *Stones of Venice,* John B. Alden Publisher, New York. Quote from vol 2, Chapter VIII
[15] Weyl H (1952) *Symmetry,* Princeton University Press
[16] Sinopoli G (1991) *Parsifal a Venezia,* Consorzio Venezia Nuova, Venice, p 22
[17] A.S.V. *Archivio di Stato di Venezia,* Senato Terra, reg 104, ff 363v–365; Difesa 49–50
[18] Niero A (1979) Pietà Ufficiale e Pietà Popolare in Tempo di Peste, in *Venezia e la Peste 1348/1797,* Marsilio Editore, Venice, pp 287–293
[19] Brodskij I (1991) *Fondamenta degli Incurabili,* Adelphi, Milan
[20] Sinipoli G, *Parsifal a Venezia,* Consorzio Venezia Nuova, Venice, p 22
[21] Eco U (1983) *The Name of the Rose,* Translation from the Italian by William Weaver, Secker & Warburg, London, p 178
[22] Phillips A (1989) *Topologia dei Labirinti.* In: Emmer M (ed) *L'Occhio di Horus: Itinerari nell'Immaginario Matematico,* Ist Enciclopedia Italiana, Rome, pp 57–67
[23] Emmer M (1987) *Labirinti,* film, Arte e Matematica series, Film 7, Int Rome
[24] Phillips A (1993) *The Topology of Roman Mosaic Mazes.* In: Emmer M (ed), *The Visual Mind,* MIT Press
[25] Rouse Ball WW, Coxeter HSM (1974) *Mathematical Recreations & Essays,* 12th Edition, University of Toronto Press, Toronto

Santa Margherita

LUCIANO EMMER*

Translated from the Italian by Emanuela Moreale

The motorboat that leaves from Santa Lucia station and heads for St. Mark's Square does not travel along the "Canal Piccolo" anymore. Now, in order to see it, I have to get off at the San Tomà or Ca' Rezzonico stops. I prefer the former, as it gives me the opportunity to arrive at it through the narrow Venetian *calli* (streets) I was used to. I arrive in *campo* San Pantalon, just past the bridge, and there it is, standing just in front of me: it is a small deconsecrated church with a small and discreet façade and a wooden door with two wooden beams nailed across it.

I wonder what may still be found in there: perhaps the two columns that flanked the small apse, the small rose window high up and the two curtained windows. The rows of chairs will not be there anymore: they will have been removed and destined for other uses (perhaps in one of those small circuses that travel around the countryside) instead of waiting to rot from the humidity. The rest will be a desolate ruin.

That small church was the most important place of my childhood: the Santa Margherita cinema. I have fed on the films from the mythical era of the mute film since I was a child: then they were my daily bread.

I used to spend whole hours in that dark hall to see these films again and again; I would move from the first to the last row, or sit in the only row up on the balcony, where may once have been an organ. I wanted to make those magic images mine from all the points of view so as to never miss even a single second of magic.

For several years, there was no distracting piano music: only the mechanical noise of the projector – a locomotive in my fantasies. The noise in the hall was only broken once in a while by those trying to make sense of the subtitles. I would only read them absent-mindedly, as I understood the story perfectly from the images themselves.

It was at the age of three that I saw my first film. I remember the scene perfectly: a large tree – perhaps an oak – in the middle of a clearing; two men in black trousers and wearing open white shirts; a man dressed in dark clothes gives

* *The text is an extract from the book: Emmer L (1996) Delenda Venezia, Centro Internazionale della Grafica, Venice. With permission of the editor.*

them two guns; next to them two godfathers: one of them approaches the man on the left and whispers something to him; white text on the black background interrupts the scene. I was not yet able to read, but my aunt satisfied my curiosity: the text read "aim for his belly button – his gun is unloaded".

The small cinema was my natural habitat: home and school were simply a boring duty to me. Had it been possible, I would have moved to and lived in the dark of the cinema hall: the screen was my room window and it opened onto the undertakings of Max Linder, Buster Keaton, Larry Semon and Charlot. They were my companions, those that – I hoped – would descend into the hall, just like in the "Purple Rose of the Cairo", and drag me into the screen to join in their adventures.

I already snubbed "serious" cinema. I did not even collect the stickers of Greta Garbo, Ramon Navarro or Rodolfo Valentino. If I ever came into possession of one, I always exchanged it straight away for one of my favourite heroes while playing penny up, winning by tossing my coins closest to the wall on the cobbled surface.

I never disowned them either: later on I added to my admired list two of the greatest – the cruel witnesses of the stupidity of our century: Stan Laurel and Oliver Hardy.

I would enter the cinema leaving outside the light of the early afternoon and leave it in the shadows of the evening or even the darkness of the night (careless of the rebuke that was expecting me at home), while my aunt – my accomplice – waited for me in the dairy nearby. There we would have several *cialdoni* (large wafers) with cream and cinnamon sprinkled on them.

Outside the cinema, in *campo* Pantalon, the warm perfume of the *folpi* (chopped polyps in their own juice) kept in a container topped by a copper lid would await me. But I would not be tempted, partly because of financial reasons. The only stall that would have attracted me would have been one that offered white cream cakes, just like the ones that my heroes threw in each other's faces. It wasn't that I was a glutton: I was tempted to try out that cake throwing with an unpleasant teacher or friends that put on airs. My revolutionary impulses were frustrated: no one ever sold such cakes at the corners of the narrow Venetian streets.

My relationship with cinema started in Venice: the frequent visits to the small theatre, made out of a deconsecrated church, determined in my subconscious the decision to start telling my stories in a film.

When I left Venice for good, the saddest goodbye was that with the warm niche of the Margherita cinema: while leaving, I thought I would lose my interest in watching my heroes' adventures on the screen. This was indeed the case. Thrown about between Milan and Sardinia, I lost the habit of going to the cinema and, for a while, I betrayed cinema for the theatre.

My new impact with cinema happened again in Venice: the Film Festival had just been born.

Of course, this could not have taken place in my favourite movie theatre, among the few unstable wooden chairs; the chosen site was the lit fountains garden at the Excelsior Hotel in the Lido. To the visitors of the newly-born Festival, Venice represented an option, just like the ruins in Pompei for a visitor to Naples. But, apart from the insult to my town, the worst was the ugliness of the films presented. Fascism – with an excess of "nationalism" that Italy was not used to – had banned American cinema from our cinema theatres and replaced it with the cinema of the regime.

The incident happened during the premiere of the film "Scipione l'Africano", a "colossal" film by Carmine Gallone (that began filming with the help of Mussolini, who possibly regretted not being called to interpret the role of the protagonist).

I was sitting in one of the front rows of the fountain garden with a friend who shared my passion for cinema: Stefano Vanzina. A Roman friend of his was with us: he was one of those cynical and civil people who are becoming extinct. Behind us, separated by a large corridor, the first V.I.P. row was reserved to the authorities. Galeazzo Ciano and Count Volpi, both wearing a white tuxedo, were sitting next to each other.

Our sector (reserved for common people) would turn noisy to stress the absurdity of some scenes. In order to assert the value of the film, the two authorities used the kind of language that is employed by expert film critics: "Good sequence" – would say Galeazzo Ciano feigning competence – and Volpi would agree in silence. The third time the Fascist party official spoke his words aloud, Steno's friend – acting in a calm and detached fashion – turned towards Galeazzo Ciano and said aloud "What the hell are you talking about – brilliantined b***?" (It is well known that the Fascist party official followed the fashion of straightened hair covered in brilliantine). There followed a moment of embarrassment, some movement among the ranks of the plain-clothed police and then the speaker left, certainly to avoid scandals among the audience.

Venice witnessed my first documentary film: the war had switched off the lights of the open hall at the Lido. The Festival had moved to a cinema near St. Mark's Square. Out of all of its headquarters, this was the poorest, but it had an advantage: when you go out of it, you found yourself immersed in the soft blue light of the darkening.

When, at midnight, I boarded the ferry to spend the night at the Lido (Venice hotels were full of Fascist and Nazi party officials), I turned to look at it. The city was getting farther and farther away; the lights that usually bare the Doges' Palaces were off; the border between the water of St. Mark's dock and the stones of the Riva degli Schiavoni was undefined. I had newly found my childhood's town: dark, solitary and silent.

I felt as if I was abandoning it in its sleep for the second time and forever: perhaps, during the night, planes would awaken it by releasing their Bengal lights before carrying out a terrifying bombing. I woud then be a powerless witness of its end.

It was pointless to try and escape my Circe: at the end of the war, I allowed myself to be seduced once more by the first autumnal fogs that turn sky and water into one single sheet wound around the whole city. For the first time, I arrived from the sky and at the last light of the day; here it was, underneath me, immersed in the amniotic liquid of the lagoon, smaller, like a small Inca head and swarming with fireflies in the dark that was about to cancel it out of my view, even before I landed at the airport of San Niccolò at the Lido.

I stayed for two months to shoot three documentaries: I had fallen into the trap of that cynical romanticism that chooses Venice as its ideal setting of forgone passions and mournful groans?

Venice is a cheerful and discreet city that – with the labyrinth of its streets – helps you get rid of nuisances and leaves you alone to think. A few years later, I spent a very happy morning with Parise: he was to talk about the work of art he loved the most in front of a cine camera (TV cameras were not yet available). He had chosen St. Mark's Square and its pigeons. Hence, there was talk of sending them to sleep, taking them away or even eliminating them so as to prevent them from damaging the square – as if the square could even exist without them.

We sat at the café Lavena and, without even the comfort of a little orchestra, we were given the sad news: it was the last day Kiffels were going to be warmed up.

Do you not know this gastronomic delicacy? They are mini croissants made with almond flour: they are eaten warm, not only for their taste, but also to fully appreciate their delicious fragrance.

I do not believe in recovery or restoration: another small piece that made up the mosaic of my childhood was being lost.

The opportunities to go back to Venice have been getting less and less frequent: when we meet, a kind of mutual understanding, of solidarity, is established between us, as between old friends who have emigrated to Patagonia.

When I am in town – she realises it – she is finally inhabited. The others, the strangers, are not inhabitants: this would be like considering it normal to see chickens in a church or priests in a chicken run.

She has asked me many times: “Why don't you write a story about me?”. I could not refuse one can see the trees better when standing outside the forest, as Shakespeare seems to say in Macbeth; away from Venice, I got to know it better in my memory.

I have written a story that started many years ago and ends today, with her destruction. And what if, during the reconstruction of the final scene of the film – to make it more realistic – the city was really annihilated? I would never find Venice again, but I would have the comfort of knowing it was freed from her parasites.

I get a suspicion: in these times of “necrofilms”, the opportunities to broadcast the end of the glorious city reconstructed with a graphic computer would be multiplied.

In order to avoid witnessing this, I would find refuge in her ruins and carry out a rite: the rite of *consòlo:* the long lunch to honour the deceased – with never-ending courses of extremely refined Venetian dishes.

Mathematics and Music

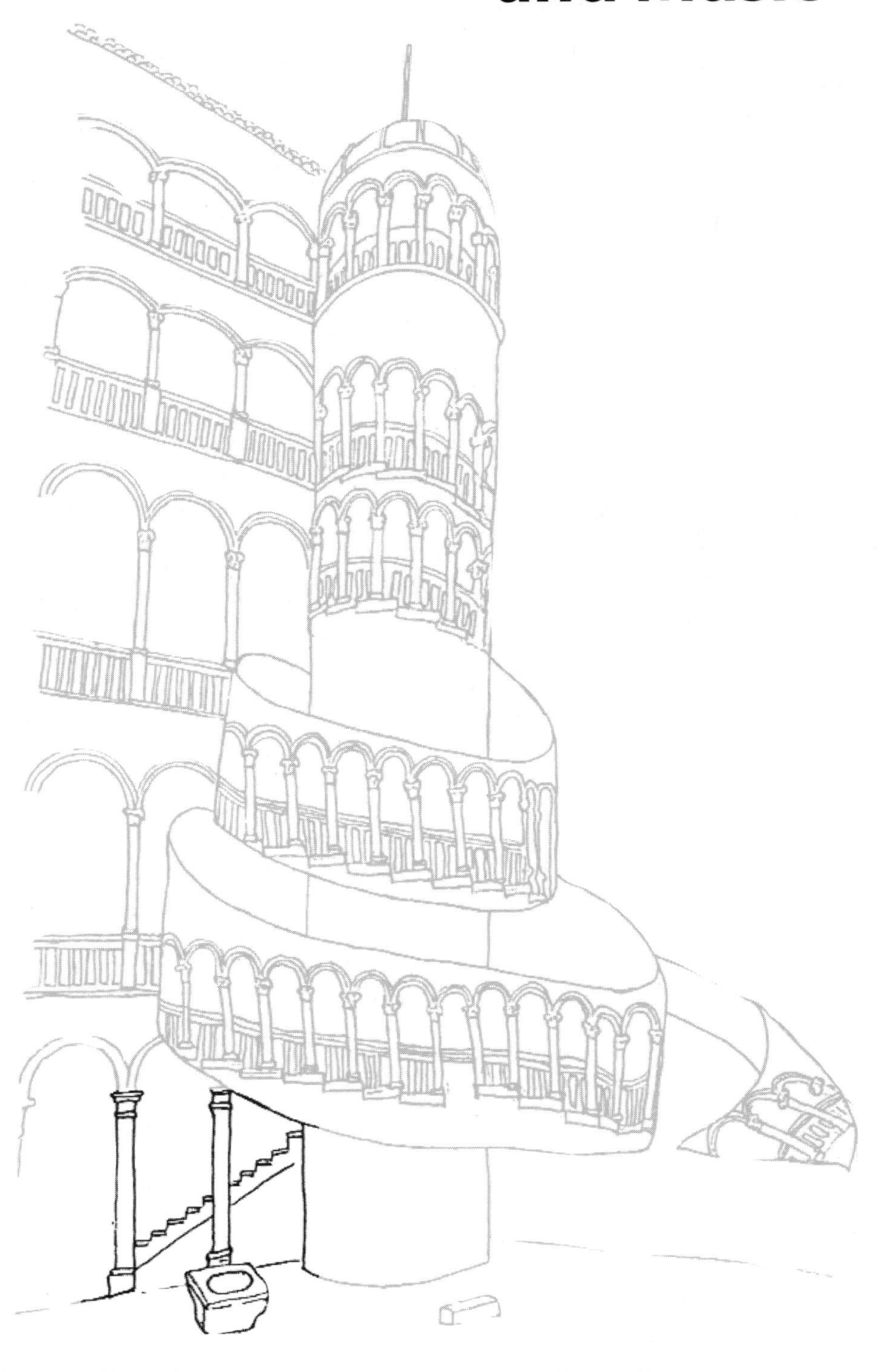

Mathematics & Cultures, Mathematics & Musics: a Model for Bronze Instruments

LAURA TEDESCHINI LALLI

Translated from the Italian by Emanuela Moreale

I consider it a mistake to make the Theory of Consonance the essential foundation of the Theory of Music, and I had thought that this opinion was clearly enough expressed in my book. The essential basis of Music is Melody. …and, secondly, that I should consider a theory which claimed to have shown that all the laws of modern Thorough Bass were natural necessities, to stand condemned as having proved too much[1].

The essential basis of Music is Melody. Harmony has become to Western Europeans during the last three centuries an essential and, to our present taste (1870), indispensable means of strengthening melodic relations, but finely developed music existed for thousands of years and still exists in ultra-European nations, without any harmony at all[2].

If the scale is the imitation of sound in its horizontal and successive dimension, chords are its imitation in the vertical dimension, in simultaneity. If the scale is analysis, the chord is sound synthesis. […] This (major triad) chord is undoubtedly similar to the fundamental sound, although not more than, for instance, Assyrian drawings of the human figure are similar to their models[3].

From Sound to Music: … Mathematically?

There are many cultures and there are many variations in music. There are also many phenomena of hearing organisation in time that are thought by some to be musical and by others to be non-musical (rituals, functional and so on). There are even cultures in which there is no activity denoted by a word corresponding to our "music"[4]. Nowadays, it is clear that, as soon as terms such as "musical sounds" are used – particularly if in contraposition to other perceptible and significant sounds that are believed to be "non musical" – a culturally-determined choice is thereby made.

1 *Helmoltz H (1954) On the Sensation of Tone, preface to the third German edition, Heidelberg 1870. English version, Dover, New York 1950.*
2 *ibid.*
3 *Schoenberg A (1922) Manuale d'Armonia, Universal Edition, vol. Cap. IV, Vienna 1922.*
4 *Giannattasio F (1993) Il Concetto di Musica, Nuova Italia Scientifica, Roma 1993.*

This is why we believe it useful to carry out an investigation into the possible existence of consolidated mathematical models that may have favoured or reinforced some descriptive categories in music. A particular model, for instance, may obviously highlight or ignore selected objective elements in the acoustic signal. While the need to limit the interpretation applicability of such models has been pointed to from the musical field as well as that of musicology, we believe that it is not only possible to discuss them in mathematical terms, but actually useful in clarifying the interactions between research areas when quantitative methods are used.

We are interested in understanding how mathematics has been used as a reinforcing element of cultural choices that are quite far from it and not dependent on it, such as the cultural choices found in the musical field. In particular, we would like to discuss – using mathematical language and methodology – how only those sounds that satisfy quite stringent conditions of periodicity, but are quite clearly defined in mathematical terms, acquired "musical" status in acoustics manuals and even in the common sense. We believe that one of the problems is that the chosen mathematical model, in itself privileges the description of simple periods. This is possible when, in moving concepts from one field to another – in our case our "mathematical modelling" of sounds – some characteristics of the model remain undiscussed, as if they were data of the phenomenon.

In particular, we would like to provide suggestions to relativise the choices of models at the basis of the analysis of the sound, particularly for the discretisation of the range of frequencies consistently and coherently used in a given repertoire that is usually termed "musical scale".

This is why we "invented" a (mathematical) theoretical musical instrument that does not comply with the conditions typical of current (mathematical) descriptive models. Then, if we imagine that a people would play these instruments century after century, it is quite natural to enquire which of the objective characteristics of the sound of these instruments would become structurally important, in what sense the continued use of these instruments – instead of others – favours a possible and coherent musical system.

One of the fundamental bases of the studies aimed at rationalising the description of musical activity is found in the study of the structure of "sound", that is of small timescale musical phenomena. It is hoped that this will give indications also on the mechanisms of perception and memory, as they are so deeply connected with any musical practice. Many mathematical approaches to music assume that, if the "sound" is well understood, then there are strong indications as to the temporal structure that unravels in a piece that is perceived to be "musical". For instance, structurally important notes in a musical piece (such as those that form an interval of a fifth or an eighth) would correspond to particularly important frequencies in the sound spectrum and the major chord in our tradition would be a reproposition in the relationships among the lowest spectrum frequencies.

Of course, only few really seem to believe such slightly prescriptive descriptions of what can be defined as "musical". Already Helmoltz among the physicists and Schoenberg among the musicians warned against drawing excessive conclusions from their studies on sound, as in the passages quoted at the beginning of this article. It may be worth noting that, while Helmoltz's studies are documented in scientific books, Schoenberg's are mostly in musical pieces, i.e. sounds.

The key idea that will guide us is a definition of music – which we will quote later – for which repetition of temporal patterns in slightly different sonic contexts is crucial to learning its structure and for the development of the capacity to recognise it. By recognising – in this case – we mean learning to repeat some characteristics in another context. This is certainly what we can do when we tune a musical instrument or our voice: we are able to extrapolate and repeat a temporal characteristic that is to date not totally understood and whose study presents remarkable degrees of complexity. We know that the "pitch" is linked to the fundamental frequency of sound, but we also know that it is a complex function of it. It is not determined by a mere principle of resonance, but – as an operation of perception and reconstruction – it is rich enough to allow ambiguity and musical "puns" that are nowadays studied in psycho-acoustic experiments and deliberately exploited by western composers. Among the first, we will list the classical experiments by Shepard in the 1960s and the beautiful music developed by James Tenney, his colleague at Bell Laboratories[5]. In oral tradition music, much of the same ambiguities are exploited in the *Pasi But But*, chant of native Taiwanese[6]. More recently, William Sethares systematically investigated the role of the structure of sound of instrumentals from a given culture, in the perceptual categorization of pitches. [7]

A mathematical model submits to mathematical treatment some aspects of a phenomenon in order to uncover previously unknown or unclear links. The model consists in identifying some quantitative variables and in the choice of an equation that connects these variables, thus describing in mathematical terms an aspect of the phenomenon at hand. Of course, it should be remembered that possible consequences – deriving from formal manipulations of the chosen variables – only concern the aspects undergoing such treatment: any other link and conclusion is to be discussed again, ensuring that the chosen model does not already have in itself the conclusions as hypotheses. On the other hand, it is neither clear nor simple thing to discuss which hypotheses are necessary and which can be made more relative.

In this article, we are studying the linear part of an operator of vibration. We will first consider a consolidated model in literature, the "vibrating string" and then discuss the characteristics that have been used to describe other vibrating objects but that are instead its own. Further, we will discuss two of the elements of the model, its one-dimensionality and its elasticity and will notice how they separately influence the specific spectrum characteristics. We set forward two more models, changing the vibrating string in either direction.

[5] *James Tenney "For Ann (rising)", December 1969 (version with digitally-generated sound. Artifact Recordings, FP001/ART1007).*

[6] *Le Voix du Monde, CD III (Polyphonies), Les Chants du Monde, CMX 3741010.12.*

[7] *William A. Sethares "Tuning, Timbre, Spectrum, Scale", Springer Verlag 1998.*

This work was originally purely theoretical. Later on, we asked ourselves if there existed a people educated for centuries by music played on rigid rather than elastic instruments, and we replied affirmatively. In fact, there is at least all the "Musical Culture of the Gong and Chimes" in South-East Asia that has fascinated western composers for the quality of its sound and its complexity for over a century. We will therefore report the first results of a field study carried out at the gongsmiths' who create the bronze instruments on the island of Bali; these results confirm some of the results obtained theoretically, and not known previously.

This research is the result of several separate collaborations. It has been a pleasure to work with so many people filled with curiosity and imagination. The study of the vibrating bar with non-standard boundary conditions was carried out in collaboration with Silvia Notarangelo. The field research in Indonesia, as well as all the evaluation from an ethnomusicological viewpoint, has taken place through a continuous dialogue with ethnomusicologist Giovanni Giuriati. The first measurements of Bali instruments were taken on the *gamelan gong-kebyar* in possession of the Embassy of the Republic of Indonesia in Italy, on the *gamelan gender wayang* belonging to Giuriati and in the museum of musical instruments of the Accademia SSTI of Denpasar, with kind permission from director Dr. I Made Bandem. We would also like to thank master gongsmiths and tuners in Bali for their hospitality and for sharing their secrets with us.

Changing "Music", Changing Mathematical Model

Definition

Strictly for the purposes of this paper, we will define "music" to be a collective meditation on sound, coded in the activity of composition of sounds in time.

Any mathematical approach to sound starts with a mathematical model. The mathematical model at the basis of classical approaches is that of a "vibrating string", that is the differential equation:

$$\frac{\partial^2 u}{\partial t^2} = c^2 \frac{\partial^2 u}{\partial x^2}$$

where x is the position on the string, t is the time and the function aimed for is the shift $u(t,x)$ describing the string vibration. c is a constant of the problem.

Let us consider which constituting elements can be changed in this equation and what influence this would have on the type of vibrating object that will make our new "instrument".

This equation can also be obtained as limit of a difference equation describing a chain of harmonic oscillators, that is a chain of material points separated by "springs". In this way, we are highlighting the fact that – when thinking of a vibrating object as a mathematical "vibrating string" and therefore choosing this equation – we are thinking that the force that connects two near-by elements is, from a mathematical viewpoint, of an elastic type, just like in a "spring" or "harmonic oscillator". It can also be said that, in this case, the object resists tension and the term "elastic object" is used. Moreover, this equation – when variable x varies in $\mathbb{R}$, is a one-dimensional equation in space (the string length).

The equation of the vibrating string has been a rather important mathematical model and not only used as the basis of classical treatises in theory of sound: it underpins many theories, also theories that have no musical applications. In the 1999 edition of "Matematica e Cultura", the contribution by Piergiorgio Odifreddi[8] goes back over some of the routes that the vibrating string has allowed or inspired, stressing how – historically – the existence of this model has influenced western musical practice (through theoreticians) and how it still influences the development of recent physics theories today. In other words, more complex models for conceptually distant phenomena organise themselves around this model.

From Sound to Music: "Archetypal Instruments" and Their Spectrum

In terms of vibrations, air columns behave just like vibrating strings. We can therefore think of air columns and vibrating strings as constituting the "archetypal instruments" of our music, i.e. of our musical culture. (This is basically true in Europe and finds a wide systematisation in XVII and XVIII century orchestras for "learned" music or written art music).

In order to change music, from here, we can go in two directions: we can either vary x in $\mathbb{R}^2$ – that is think of the instrument as two-dimensional, as an "elastic membrane" – or assume that the vibrating object is not totally elastic, but resists bending and torsion, i.e. that it is a *semi-rigid* object.

Given these conditions, we try to assess which types of "music" can be deduced, i.e. we try to come up with a whole, coherent vibration system and understand which temporal characteristics of this music would be different from a music (theoretically) based on ideal strings and air columns.

Helmoltz's idea is that the human ear behaves as an analyser of "frequency spectrum" that is able to distinguish two sounds on the basis of how they can be approximated by overlaying simple vibrations, such as sinusoidal vibrations. This hypothesis is not as strange as it may seem and is based on the physical principle of resonance. While we are not aware of any experiments that have seriously tested this hypothesis, we keep it as a starting hypothesis, but ensuring that we make it explicit. The idea is to initially make few, but qualitatively very important changes with respect to the basic model, so as to proceed systematically to the discussion of the possible consequences.

On the other hand, such types of experiments on the ear are hardly feasible until we learn more on the interaction mechanisms between perception and memory. In the last century, Helmoltz claimed that ear experiments are simply not possible.

[8] *Odifreddi P (1999) Il Clavicembalo Ben Numerato, in: Emmer M (ed.) (1999) Matematica e Cultura 2, Atti del Convegno di Venezia 1998, Springer-Verlag Italia, Milan.*

Linear, One-Dimensional Elastic Music

The first characterisation of the solutions of the vibrating string is their "harmonic spectrum": in fact, they can be approximated well by overlaying simple vibrations whose frequencies are in whole number ratios to each other. On the physical side, a harmonic spectrum allows easier measurement by resonance, as it assumes that the basic frequencies are separated by a fixed distance and by its multiples.

A people that is exposed to mathematical string and air column sounds – that is sounds that behave just like vibrating strings – for centuries, would develop a subtle sensitivity for the "harmonic" spectra and frequencies in rational ratio. They would become able to align these frequencies, so as to identify as "similar" two sounds when many of their components are in resonance.

When two simple vibrations have similar – but not identical – frequencies, their overlaying gives rise to the characteristic phenomenon of interference "beats". Our people of listeners of string and air columns players would thus develop a social sanction against interference beats, as a sign of possible evaluation errors of nearby/similar frequencies and therefore of identification, comprehension and reproduction of the basic alphabet.

Our "learned" music – that is our music written in a learned environment and whose description has been the most systemic and rationalised – is a music based mainly on strings and air columns. The beautiful story narrated above is therefore plausible, but written "a posteriori", after it has happened: nearby/similar sounds in this sense are labelled "consonant" in our culture.

The story as narrated, though, now plausibly puts forward censorship towards interference beats as a cultural phenomenon (towards objective characteristics), rather than a physiological-aesthetic phenomenon.

Let us see what would happen with types of music based on other instruments.

Elastic – but Two-Dimensional, Square – Music

If the vibrating object is square and the restoring force continues to be elastic, the differential equation is the same, but x varies in $\mathbb{R}^2$, and the spectrum of its vibrations is given by two harmonic series and their combinations. The frequencies that a people used to square instruments would choose as comprehensible and easily reproducible would be, first of all, the first ones of the series:

$$f_{n,m} = \sqrt{n^2 + m^2} \qquad n, m = 0,1,2\ldots$$

for instance,

$$f_{0,1} = 1, f_{1,0} = 1, f_{1,1} = \sqrt{2}\ldots$$

A "square" music would therefore select as recognisable by the ear and reproducible in various instrument, for instance, an irrational ratio between frequencies such as $\sqrt{2}$: a frequency that in the western music of the last few centuries has been purposefully avoided as difficult to match and therefore unstable: it is used only following or together with others. A trace of this sanction can be found in the name *diabolus in musica*, indicating the tritone interval (three tones, augmented fourth or diminished fifth interval in our scale). The tritone, in an ideally tempered western scale, would in fact separate two frequencies in $\sqrt{2}$, ratio, by dividing symmetrically in halves the octave interval.

This speculation is possible: so much so that, in some compositional music research sectors, work is being done in this direction. For instance, at the IRCAM laboratories in Paris, sounds that respect various types of spectra are synthesized and the intervals they prescribe unravelled.

Also in this direction, are the ground-breaking experimental studies of existing musical systems, which are contained in the book by W.A. Sethares already mentioned, of which we became aware after writing this paper.

However, we think that the consequences affect not only on musical intervals but also the very principle of "resonance" as main basis for musicality of sound and of a sound system. In particular, if we accept that irrational frequency ratios are founding, the sanction against interference beats is drastically reduced, as they become a necessary, recognisable and controllable feature.

A Semi-rigid, One-Dimensional "Musical Instrument"

The Equation

If we are correct in thinking that the object that is made to vibrate is semi-rigid, the force that reestablishes the equilibrium is not of an elastic type: the object offers resistance to bending and torsion and its vibrations satisfy a fourth order differential equation, of which we search solutions of the type $u(x,t)=h(x)\,g(t)$:

$$\frac{\partial^4 u}{\partial x^4} + m^2 \frac{\partial^2 u}{\partial t^2} = 0$$

Here, again, x varies in $\mathbb{R}$, that is the spatial variable is one-dimensional so the equation that ties the variables, the mathematical model, goes under the name of "bar".

First Consequence

There are no waves that travel at constant velocity and with unmodified shape. A people educated by ear with such instruments would develop a subtle discerning of spatial relations, even before the subtle discerning of ratios of frequencies, for instance the initial deformation producing the sound and its propagation depending on the material. The vibration waves passed on to the air through contiguity would be different in different points of the instrument and would change rapidly in time:

a people whose ear is used to the sound of semi-rigid instruments would have a subtle spatial awareness and perhaps would not privilege listening to music in which spatial information is cancelled out of a desire for uniformity. This reminds us of commercial packages to carry out Fourier analysis (decomposition of sound in sinusoidal components of periods that are in simple ratios between themselves) that necessarily involve time-averages, without making this evident. Software packages are measurement tools and it is advisable to know that the relationship between what is read on a computer and a spectrum that is variable in space and time is very … averaged.

Second Consequence

There are many possibilities of boundary conditions. For a vibrating string – we have not mentioned it – the boundary conditions are that the ends must be fixed. A mathematical bar is also generally thought to have fixed ends, i.e. to have its ends locked onto a support. However, when – as is reasonable to think in the case of semi-rigid objects – we consider that the bar rests on one of its points and its ends are free to vibrate, one can speak of "beam".

If the bar is resting on another object, its vibrations encounter an obstacle in its suspension point (resting point); it therefore becomes crucial for the structure of its vibration to express what is meant by "resting point". From a modelling viewpoint, a resting point can be described mathematically in various ways, depending on whether it is thought that some vibrations stop in it, whether it is still and also whether the deformation flattens in its vicinity. It is therefore possible to give different boundary conditions for the differential equation in that point, all of which are reasonably worthy of being called "suspension points"[9].

In order to clarify our ideas so as to impose "conditions of musicality" on the construction of the instrument, we locate the origin of the coordinates in the centre of the bar and we denote the following:

- l is the half-length of the bar ($x = l$ and $x = -l$ coordinates at the ends)
- a is the coordinate of the suspension point
- $h(x)$ is the shift function (at a fixed time)

We propose the following conditions as possible modelling definitions of the "suspension point":

1) $h(a) = h(-a) = 0$ and $h''(l) = h''(-l) = 0$ in the case of fixed suspension point and ends free to rotate

2) $h'(a) = h'(-a) = 0$ and $h'''(l) = h'''(-l) = 0$ in the case of humps in the suspension points and cutting force nil at the edges

3) $h'(a) = h'(-a) = 0$ and $h''(l) = h''(-l) = 0$ in the case of maximum vibration points in suspension points and ends free to rotate

Looking for particular solutions to the differential equation of the type $u(x,t) = h(x)g(t)$, the partial differential equation above splits into two ordinary differential equations as follows:

$$g''(t) + \mu^2 k g(t) = 0$$
$$h^{IV}(x) - k\,h(x) = 0$$

the former with initial conditions in time, the latter with boundary conditions to be chosen among the three described above, and k, a constant.

9 S. Notarangelo, *Sul problema delle scale musicali extra-europee: vibrazioni di piastre percosse (Bali)*, Tesi di laurea in Matematica, Università degli Studi di Roma Tor Vergata, 1996–97.

We want to define "musical instrument" a system with many keys of different lengths, but all with the same vibration structure, at least within the solutions considered here. They would have therefore all the same musical "colour" and reinforce the musical culture of our imaginary people, by repeating the temporal structure of sound.

We therefore wonder how difficult it is to calculate the suspension point for the whole system of keys (instrument), so that they vibrate ideally with the same vibration structure, even if they have different length. In mathematical terms, we wonder about the dependency of the spectrum of the vibration operator on the boundary conditions and the length of the key.

"Musicality" and Consistency Conditions

We have said that, for the purposes of this study, by musicality conditions we mean the consistency of the vibration spectrum from key to key when the length changes. In other words, we refer to the possibility of building an instrument made of several keys, all characterised by the same sound colour, at least as far as the operator studied so far is concerned and excluding non-linear and decaying phenomena. This exclusion is important not only from the mathematical viewpoint, but also because the whole perception of the colour and quality of sound is in fact deeply tied to these strong temporal aspects. The research and measurements by H. Fletcher are nowadays in this direction.

For the moment, out of methodological reasons, we will privilege the information that reaches the ear from the spectrum of the operator of vibration: the theoretical instrument is to be compared with the vibrating string model that we are reevaluating and that is studied in similar conditions.

We will then ask ourselves if and what existing instruments are built in this way.

We must therefore assume that the spectrum is repeated from key to key of the instrument and that the people of players and listeners therefore learns to decontextualise some characteristics so as to repeat them in a different temporal context, that is "to abstract".

As to the first "instrument" that we talked about – the one imagined as if built by (fixed border) elastic rectangular membranes – this condition consists in maintaining the ratio between the two sides of the rectangle. In fact, it is these two linear dimensions that determine the double series of overtones and, if their ratio is preserved, the entire spectrum simply rescales perfectly starting from another fundamental, much as in the vibrating string. Possible irrational ratios would then be exactly retained.

As to the vibrating string, the spectrum does not change with the length of the string: we therefore require that this spectrum information be consistent (i.e. the same) across keys in our "beam" instrument.

If we consider essential that the spectrum of the vibrations of a musical instrument be harmonic, one of the few compatible boundary conditions for the beams is that $l = a$, that is such that its suspension point coincides with the end of the bar. (We notice that musical treatises often assume this to be true in musical instruments known as glockenspiel, small tuned metallophones. Most of the actual metallophones have bars neither clamped nor resting at their edges.[10])

Let us continue the game and let us impose the condition that the system of different-length bars always vibrates with the same structure. A people used to this type of vibration would be able to recognise this structure from its sound. (Later, to change music, we can decide, in particular, that it should vibrate consistently with a non-harmonic spectrum).

The condition that we found, such that a beam vibrates with the same spectrum – independently of its length – is that the ratio between the length of the bar and its suspension point is kept constant:

$$l = b\,a, \qquad \text{with } b \text{ constant on all keys}$$

We remind readers that, in this simplified model, we are assuming that x varies in $\mathbb{R}$, that is we are only studying the transversal vibrations of the beam-bar.

Field Research: South-East Asia Gong-Chime, Smith Craftsmen

At first measurement of the plates that compose the vibraphones of our own western symphonic orchestras, it would seem that the ratio conditions that we found do not hold, neither as rectangles nor as longitudinal beams.

At this point, it is natural to wonder if there exist a people musically used to instruments that – unlike ours – are not based mostly on vibrating strings and air columns or their modifications, but rather on instruments in which the rigidity element is predominant. This people should also have been playing these instruments for centuries.

Ethnomusicologists talk of "gong and carillon cultures" when referring to South-East Asia. Travellers visiting that area of the world often write of "the tingling music" in their diaries, as if this was part of the landscape[11].

In the vast insular South-East Asia, in particular, there is a strong predominance of bronze instruments. The Java and Bali *gamelan* orchestras, for instance, consist of dozens of metallophones – both plates/slabs and circular-framed ones – and variable-sized gong systems. It is known that Debussy was fascinated by and spent whole days studying the music and instruments of the Javanese gamelan at the 1889 International Exhibition in Paris.

The description of the musical scales in use in Java and Bali is a deep problem that is as yet not completely understood by ethnomusicologists. The music here is passed down orally and has an extremely sophisticated tradition. The used pattern of intervals varies across villages and ensembles, but only within certain limits. The very intervals of fifth and octave are seriously brought into question[12].

[10] *There are two exceptions: one traditional and one recent; the mbiras (or "thumb pianos") of Zimbabwe, and the Fender Rhodes electric pianos. Thanks to W. Sethares for the puntualization.*

[11] *Lewis N (1951) A Dragon Apparent, Eland, London.*

[12] *Deutsch W A, Fodermayer F, Altezza vs Frequenza: sul Problema del Sistema Tonale Indonesiano, in: Carpitella D, 1989 ed., Ethnomusicologica, Seminari internazionali di etnomusicologia 1977–1989, Quaderni dell'Accademia Musicale Chigiana 43, Accademia Chigiana, Siena.*

We went to see the gongsmiths that forge the bronze musical instruments. From the moment in which the point of suspension is measured and drilled through, the bars are tested for sound, held tightly with two fingers positioned on the just-made hole, which testifies to its importance in the judgement as to the quality of the sound. The whole operation is checked by ear by each of the craftsmen, as well as by the master gongsmith coordinating work in the whole workshop. It is, in fact, the master gongsmith that accurately measures the point in which the hole should be created. We visited each of three craftsmen's workshops and – although in different ways, in three different phases of the forging process, with different measuring techniques – in each case we noticed that the suspension point is calibrated at ¼ of the length of each bar[13]. According to our numerical simulations, a 1:4 ratio (that is, in our notation, $l = 2a$) is not only easy to measure geometrically, but it also guarantees a non-harmonic spectrum.

And what about interference beats? We have already mentioned that, if the spectrum is non-harmonic and if this is essential in the subtle discerning by ear, interference beats should be a characteristic feature and should not be avoided.

In Bali, instruments are always forged, tuned and played by the twos. In each couple of metallophones, there is one that is slightly lower. The two instruments, played together, form interference beats that are sought for and controlled by a master tuner who establishes their speed. This speed is accurately determined and controlled by slightly lifting or lowering the tuning of the plates by means of filing. The master tuner also finishes the tuning of the whole orchestra, by checking the whole range of interference beats obtained through several instruments playing at the same time. Orchestras are characterised by a glittering aura of vibratos in interference beats, whose subtle controlling gives life to sound[14].

This aura of interference beats that is so essential to Bali music is called "Ombah". Without Ombah, the gamelan is simply not ready to be played.

[13] *Tedeschini Lalli T (1999) Forgiare I Suoni, video, Laboratorio Audiovisivo del Dipartimento e Spettacolo dell'Università di Roma, Rome.*

[14] *Hood KM (1995) Angkep-angkepan. See General section. In: Ndroje Balendro: Musiques, Terrains et Disciplines, textes offerts à Simha Arom, edited by V. Dehoux et al, Peeters, Paris, p.323–38.*

Numbers in Asian Music

Trân Quang Hai

Translated from the Italian by Emanuela Moreale

The ancients saw music as an application of cosmic algebra. The Chinese, in their science, considered only the qualitative aspect of the Numbers that they manipulated as signs and symbols. Among the three functions of numbers, the distinction between cardinal and ordinal use is less essential than the distributive function. Thanks to this quality, numbers provide the function of uniting a set, of grouping.

The ratios expressing relationships among musical sounds have a correspondence in all other aspects of an event. The governance of these relations allows comparisons between musical harmony and all other harmonic classifications: colours, shapes or planets.

Theory

The number 5 represents the 5 elements, the 5 movements. "Five" evokes the 5 senses, the 5 organs that are a coagulation of breaths (as wind instruments).

Element	EARTH	METAL	WOOD	FIRE	WATER
Notes	Gong	Shang	Jiao	Zhi	Yi
European Notes	FA/F	SOL/G	LA/A	DO/C	RE/D
Organ	Spleen Stomach	Lungs Small Intestine Large Intestine	Gall Liver	Heart Intestine	Kidneys Bladder
Colour	Yellow	White	Blue-Green	Red	Black
Planet	Saturn	Venus	Jupiter	Mars	Mercury
Emblem	Phoenix	Tiger	Dragon	Bird	Turtle

Function	Emperor	Minister Public Services (objects)	People	Services	Products (Military/Religious))
Number	5–10	4–9	3–8	2–7	1–6

The pentatonic range has a centre (GONG) surrounded by four notes assimilated to the four directions in space (SHANG-West, JIAO-East, ZHI-South, YI-North).

It is therefore the law of numbers that governs the proportions of the musical edifice. Sounds, just like numbers, obey the stimuli of attraction and repulsion. Their successive or simultaneous order – i.e. melodic and harmonic movements of the chords respectively – allows for an apparent structure of musical form. Sounds are ordered, paired up and make up structures that evoke real and imaginary worlds.

The root note (also known as tonic or keynote) constitutes the basis and the centre: it is the reference point that allows the construction of the musical edifice.

In the example where the tonic is the FA, the DO note will maintain a fifth interval with its tonic. The very note DO will play a role of fourth in a range of SOL. Our usual DO (Ut) is a tonic in the construction of the classical scale model.

The first and basic tube (Huang Zhong) reproduces the FA (Gong tonic). This FA is close to the FA sharp of the physics scale with its 708.76 vibrations per second. This generator tube represents the Central Palace around which the other elements gather. Yellow is the emblematic colour of the centre: it evokes the Sun, centre of the Sky or heart of the flower. It is reserved to the Emperor, central individual on Earth.

The 12 LÜ or Musical Tubes

We will not insist on the well-known legend of the discovery of the LÜ scale at the hands of Ling Lu (Linh Luân), a music master at the time of the famous Emperor Huangdi (Hùynh Dê, 2697–2597 before our era). In the solitary valley of the mount Kouen Louen (Côn Lôn), at the western borders of the empire, he found bamboos of the same thickness and obtained the fundamental sound, the HUANGZHONG (Hoàng Chung, the Yellow Bell) by blowing in one of the canes after cutting it in between two knots. He obtained the complete LÜ scale thanks to the sound made by a male and a female phoenix.

The LÜ scale would basically correspond to the modern chromatic scale. The absolute pitch of the fundamental sound, the HUANGZHONG, changes according to the dynasty. We will choose the FA, just like Louis Laloy, a French missionary specialist in Chinese music at the beginning of the nineteenth century. The LÜ scale will then be:

HUANGZHONG (Hoàng Chung, the Yellow Bell):	FA
TALÜ (Dai Lu, The Great Lyu):	FA#
TAIZU (Thai Thô'c, The Great Iron of Arrow):	SOL
JIAZHONG (Gia'p Chung: The Narrow or Still Bell):	SOL#

GUXIAN (Cô Tây, Ancient Purification):	LA
ZHONGLÜ (Trong Lu, cadet Lyu):	LA#
RUIBIN (Nhuy Tân, Beneficial Fertility):	SI
LINZHONG (Lâm Chung, The Bell of the Woods):	DO
YIZE (Di Tac, The Same Rule):	DO#
NANLÜ (Nam Lu, The Lyu of the South):	RE
WUHI (Vô Xa, the Imperfect):	RE#
YINGZHONG (Ung Chung, The Bell of the Eco):	MI

But, when composing their melodies, the Chinese did not use the scale of the 12 sounds thus obtained – starting from the tonic note, the HUANGZHONG, through a succession of fifths. Instead, they were content with the five degrees that form the pentatonic scale GONG, SHANG, JIAO, ZHI, YI (Cung, Thuong, Giôc, Chuy, Vu). In the SHI JI (Su Ky: Historical Memoirs), Xi Ma Tian (Tu Ma Thiên), has given the dimensions of the tubes that create the notes of the Chinese pentatonic scale:

$9 \times 9 = 81$ (placing 81 millet grains one next to the other gives a length corresponding to that of a bamboo cane giving the fundamental sound of the Yellow Bell HUANGZHONG).

In their cosmogonic system, the Chinese determined relations based on the law of numbers. These relations are also valid in the music world. The 12 musical tubes, or LÜLÜ, are the basis of this music theory. They generate in turn, in a rhythmic proportion, both by diminishing by a third and by increasing by a third: generation therefore happens through the action of the Three.

According to the type of operation carried out, two generations of tubes are created. The inferior generation provides a shorter tube, whose sound is more acute than the other one and length is reduced by a third. The superior generation provides a longer and deeper sound than the previous one, with respect to which it has been increased by a third.

The inferior generation is the result of a multiplication by two thirds, that is the inverse of a fifth. So, the first tube measuring 81 produces the second shortest tube: $81 \times 2/3 = 54$. Therefore, from FA (81) and DO (54) there exists a fifth.

The superior generation is the result of a fourth, since a 4/3 ratio is what characterises it. Therefore, the second tube (54) produces the third tube: $54 \times 4/3 = 72$. From DO (54) to SOL (72, longer and therefore deeper), there exists a fourth.

FA	first tube	81		
DO	second tube	54	81 × 2/3	
SOL	third tube	72	54 × 4/3	
RE	fourth tube	48	72 × 2/3	
LA	fifth tube	64	48 × 4/3	
MI	sixth tube	42	64 × 2/3	
SI	seventh tube	57	42 × 4/3	
FA sharp	eight tube	76	57 × 4/3	inversion
DO sharp	ninth tube	51	76 × 2/3	"
SOL sharp	tenth tube	68	51 × 4/3	"

RE sharp	eleventh tube	45	68 × 2/3	"
LA sharp	twelfth tube	60	45 × 4/3	"

The GONG (Cung) note = FA.
2/3 of 81 give 81 × 2/3 = 54; this note is known as ZHI (Chùy): DO
4/3 of 54 give 54 × 4/3 = 72; this note is known as SHANG (Thuong): SOL
2/3 of 72 give 72 × 2/3 = 48; this note is known as YI (Vu): RE
4/3 of 48 give 48 × 4/3 = 64; this note is known as JIAO (Giôc): LA.

Edouard Chavannes quoted and commented Lyus's passage in appendix II, p. 636, of the third volume of his Historical Memoirs: the HUANGZHONG produces the LINZHONG; the LINGZHONG produces the TAIZU; the TAIZU produces the NANLÜ; the NANLÜ produces the GUXIAN and so on. Another part is added to the three parts of the generator to create a superior generation; one part is taken from the three parts of the generator to create an inferior generation. The HUANGZHONG, the TAIZU; the JIAZHONG, the GUXIAN, the ZHONGLÜ, the RUIBIN belong to the superior generation; the LINZHONG, the YIZE, the NANLÜ, the WUYI, the YINGZHONG belong to the inferior generation.

According to LIU PUWEI, the tube whose length is 4/3 of the generator tube belong to the superior generation and provides the fourth inferior, that is the low octave of the fifth of the sound of the generator tube. The tube whose length is 2/3 of the generator tube belongs to the inferior generation and provides the fifth of the sound of the generator tube.

These five notes correspond – according to SIMA QIAN, quoted by Maurice Courant – to "the 5 LÜ" HUANGZHONG, TAIZU, GUXIAN, LINZHONG and NANLÜ, the only ones whose measure is expressed in whole numbers starting from the basic 81:

GONG (81):	Huangzhong:	FA
SHANG (72):	Taizu:	SOL
JIAO (64):	Guxian	LA
ZHI (54):	Linzhong	DO
YI (48):	Nanlü:	RE

This scale would have been used under the Yin dynasty (1776–1154 B.C.). According to Maurice Courant, the heptatonic scale – obtained by adding the two complementary or auxiliary notes BIEN GONG (Biên Cung) and BIEN ZHI (Biên Chùy) to the pentatonic scale – "existed at least a dozen centuries before the Christian era". These auxiliary degrees are obtained by pushing the succession of fifths up to the seventh, starting from the fundamental sound. If we give the GONG (Hoàng Chung) degree the pitch of the FA, the BIEN GONG will have the pitch of a MI and the BIEN ZHI that of a SI.

Therefore, the cycles obtained starting from the fifths present analogies with the movements of the sun (seasons), of the moon (moon months), but also with planets mirroring in the organs. The Chinese associate them with a symbolic colour that has a therapeutic value.

Symbolism in Musical Instruments

In Chinese mythology, there are eight instruments destined to let resound the eight forces of the compass rose. Each instrument has a sonorous body made of a different material, which determines its peculiar character.

1. The sound of SKIN corresponds to the North
2. The sound of CALABASH-GOURD corresponds to the N-E
3. The sound of BAMBOO corresponds to the East
4. The sound of WOOD corresponds to the S-E
5. The sound of SILK corresponds to the South
6. The sound of EARTH corresponds to the S-W
7. The sound of METAL corresponds to the West
8. The sound of STONE corresponds to the N-W

In the North, and at the SKIN, the drums are eight. The GOURD (or CALABASH), in the North-East, has the peculiarity of consisting of a series of 12 Liu, some YIN, some YANG. The instrument allowed making four different sounds at the same time.

The sound of BAMBOO, in the East, was produced by sonorous tubes (Koan Tse). There were three types of these, all of which had 12 tubes: deep, medium and acute sounds, corresponding to the Earth-Man-Sky triad). Later, they evolved towards a separation between yin and yang tubes that constituted two distinct and complementary instruments.

The sound of WOOD was represented by various instruments of which the Tchou – shaped like a bushel and named after Ursa Major – would initiate the concert in the same way as the Ursa indicates the beginning of the day or of the year with its position.

The sound of SILK was produced by stringed instruments known as Qin. These were five-string zithers that originally had a rounded top part representing the Sky and a flat front part representing the Earth. They had five strings to represent the five planets or the five elements.

The sound of EARTH was produced by instruments made of clay that gave a GONG – that is the tonic FA – as deep sound as well as four more tones (SHANG; JIAO; ZHI; Yi). In the West, the sound of METAL was rendered by twelve copper and tin bells that gave the twelve semitones of the LÜ. The sonorous STONES situated in the North-West were assigned to ceremonies that evoked the Sky, establishing a spiritual link thanks to the pure quality of their sounds.

Yoga

In Yoga, there exist 7 chakras corresponding to 7 vowels, 7 sounds or pitches, 7 overtones and 7 points of the human body. The author has carried out experimental research in the presence of overtones in Yoga. The result of his three-year study was presented at the International Congress of Yoga in France in 2002.

According to his research, the fundamental of voice should be at 150Hz.

Number	Name of Chakras	Location	Overtones of Hz	Vowels	Number
1	Mulâdhâra	coccyx	H n° 4	U	600Hz
2	Svâdhishthâna	genitals	H n° 5	O	750Hz
3	Manipûra	navel	H n° 6	Ö	900Hz
4	Anâhata	heart	H n° 8	A	1200Hz
5	Vishuddha	throat	H n° 9	E	1350Hz
6	Ajnâ	between eyebrows	H n°10	AE	1500Hz
7	Sahasrâra	top of head	H n°12	I	1800Hz

Instruments

The lute in the shape of a Vietnamese moon – DAN NGUYET or DAN KIM – was conceived in a harmonious style and in a totally empirical way. Every part of this instrument can be divided into three parts: the two deep strings (0.96 mm) and the high string (0.72 mm) have a vibrating length of 72 cm.

The instrument measures 108 cm overall; the sound board 36 cm; the thickness of the resonance chamber is 6 cm. The bridge is 9 cm long and 3 cm high. The decorative part of the instrument, located opposite the bridge, measures 12 cm in length. The wooden pins measure 12 cm and there are 9 ring nuts measuring 3 cm in width.

The Vietnamese monochord DAN DOC HUYEN (unique stringed instrument) or DAN BAU (gourd instrument) is the only musical instrument in the world using Pythagorean theory to create overtones by the division on harmonic knots of the unique steel string of the instrument into 2, 3, 4, 5, 6, 7, 8 equal parts corresponding to the series of overtones. (In fact, if the open string is tuned in C, the overtone 2 will be C an octave higher than the pitch of open string, 3 = G, 4=C 2nd octave higher, 5= E, 6=G, 7=Bb, 8= C 3rd octave higher.)

Chinese tradition has it that – over 5000 years ago – Emperor DU XI asked his lute maker to produce a zither. This imperial instrument was to be based on the relationship between the Sky and the Earth. It therefore had the length of three thuocs, six tâcs and one phân, so as to match the number 361 representing the 360 degrees of the circle plus the centre, that is unity and multitude. The height of the zither was eight tâcs and its bottom four tâcs, to match the eight half seasons and the four cardinal points or four seasons, that is space and time. Its thickness, two tâcs, bore the emblem of Sky-Earth. The twelve strings vibrated similarly to the twelve months of the year, while a thirteenth string represented the centre.

This story shows the domination of the Number and its application in Chinese thought.

Musical Notations

Around 1911, a musician playing the Chinese fiddle, called LIU Thien Hoa, adopted the cipher notation for writing musical scores proposed by Jean Jacques Rousseau in 1746, perfected by Pierre Galin (1786–1821) and later made popular by Aimé Paris and Emile Chevé (1804–1864). The number 1 corresponds to the root note – whatever the tonic (DO, RE, FA, SOL, LA) – and the five main degrees correspond to the five number (1,2,3,4,5). The 6 and 7 represent intervals.

The cipher notation also appears in Indonesian music: it is found in the musical scores used in the GAMELAN.

Numbers are found in the notation used for Tuvan and Mongolian throat singing. The harmonics are numbered according to frequencies starting from the root note. A cipher notation was proposed to write musical scores for split-tone singing. Ted Levin and the author have used cipher notation to transcribe Tuvin and Mongolian folksongs.

Frequency Analysis Using a Sonagraph

The sonagraph, an instrument for measuring spectra, allows pushing split-tone singing beyond basic and experimental research. Since 1970, many people – including Emile Leipp, Gilles Léothaud, Trân Quang Hai, Hugo Zemp in France, Gunji Sumi in Japan, Ronald Walcott in the United States, Johan Sundberg in Sweden, Graziano Tisato in Italy, Werner Deutsch and Franz Födermayer in Austria – have utilised the sonagraph or similar types of instruments to increase the precision of the harmonics produced by split-tone singing thanks to hertzian spectroscopy.

An examination of sonagrams shows the diversity of styles in the split-tone singing of Tuvans, Mongolians, Tibetans and of the Xhosas in South Africa. It allows to establish a classification of styles, to identify the number of harmonics and better understand the how and why of vocal techniques – something that has so far been impossible. The sonagraph allowed Trân Quang Hai to carry out introspective experimental research in overtone singing (with one tonic and two independent partials or harmonics) as well as in "harmonic drone" with variation of the tonic – DO (harmonic 12), FA (harmonic 9), SOL (harmonic 8) and DO octave (harmonic 6 to create the same harmonic pitch). Other experiments by Trân Quang Hai that have produced interesting results are those on the realisation of different spectra starting from ascending and descending singing scale (normal voice, overtone voice with parallel harmonics, overtone singing with opposite moving harmonics between drone and harmonic melody). Basing himself on self-analysis, his research, whose originality lies in its experimental character, has led him to high-light the link between the harmonic drone and the tonic melody, which is the opposite of the initial principle of traditional overtone singing. Moreover, he has interwoven the two melodies (tonic and harmonic); explored overtone singing with one tonic and two independent partials or harmonics and highlighted the three harmonic zones on the basis of the same tonic sound.

Using the sonagraph and other medical systems of analysis (laryngoscopy, fibroscopy, stroboscopy, scanner), the author has been able to propose a new dimension of undertone study with which to develop undertones or subfundamentals by creating F-2 (an octave lower than the sung fundamental), F-3 (an octave + a fifth lower than the fundamental) or F-4 (2 octaves below the sung fundamental). This new aspect of research on undertones has attracted a number of researchers, namely Leonardo Fuks (Brazil), Johan Sunberg (Sweden), Masashi Yamada (Japan), Tran Quang Hai (France), Mark Van Tongeren (The Netherlands), and a few overtone singers from the Western world (Steve Sklar from the USA, Bernard Dubreuil from Canada).

This article is a brief introduction to the use of numbers in various areas of Asian music. It represents merely the beginning of a study that will continue in the future.

Recommended Reading

Amiot J M (1780) *Mémoires sur la Musique des Chinois Tant Anciens Que Modernes,* vol. VI de la collection, Mémoires Concernant les Chinois, Paris, 185pp

Deutsch W A, Födermayer F (1992) Zum Problem des zweistimmigen Sologesanges Mongolischer une Turk Völker, Von der Vielfalt Musikalischer Kultur, Festschrift für Josef Kuckerts (Wort und Musick 12), Verlag Ursula Müller-Speiser, Anif/Salzburg, Salzburg, pp 133–145

Gunji S (1980) An Acoustical Consideration of Xöömij, *Musical Voices of Asia,* The Japan Foundation, Heibonsha Ltd, Tokyo, pp 135–141

Kunst J (1949) Music in Java, Its History, Its Theory and Its Technique, Second Edition, Translated from the Danish Original by Emil Van Loo, vol 1, 265 pp, vol 2, 175 pp, Amsterdam

Laloy L (1910) *La Musique Chinoise,* Editions Henri Laurens, Paris, 128 pp

Leipp E (1971) Considération Acoustique sur le Chant Diphonique, *Bulletin du Groupe d'Acoustique Musicale,* 58, Paris, pp 1–10

Leothaud G (1989) Considérations Acoustiques et Musicales Sur le Chant Diphonique, *Le Chant Diphonique,* Dossier, Institut de la Voix, Limoges, 1:17–43

Picard E (1991) *La Musique Chinoise,* Editions Minerve, Paris, 215 pp

Rousseau JJ (1979) Dissertation Sur la Musique, *Ecrits Sur la Musique,* Editions Stock, Paris

Sunberg J (1977) The Acoustics of the Singing Voice, in Scientific American 236, USA, pp.82–91

Tisato G (1990) Il Canto degli Armonici, Nuove Tecnologie e Documentazione Etnomusicologica, *Culture Musicali,* 15, 16, Rome

Tongeren M Van (2002) Overtone Singing / Physics and Metaphysics of Harmonics in East and West, 1st edition, Fusica Publishers, 1 accompanying CD, Amsterdam, 271 pp

Trân Quang Hai, Guilou D (1980) Original Research and Acoustical Analysis in Connection with the Xöömi Style of Biphonic Singing, *Musical Voices of Asia,* The Japan Foundation, Heibonsha Ltd, Tokio, pp 162–173

Trân Quang Hai, Zemp H (1991) Recherches Expérimentales Sur le Chant Diphonique, *Cahiers de Musiques Traditionnelles:* VOIX, Ateliers d'Ethnomusicologie/AIMP, Genève, 4:27–68

Trân Quang Hai (1995) Le Chant Diphonique: Description Historique, Styles, Aspect Acoustique et Spectral, EM, Annuario degli Archivi di Etnomusicologia dell'Accademia Nazionale di Santa Cecilia, Roma, 2:123–150
Trân Quang Hai (1995) Survey of Overtone Singing Style, EVTA (European Voice Teachers Association, Documentation 1994 (Atti di Congresso) Detmold, pp 49–62
Trân Quang Hai (2002) A la decouverte du chant diphonique , in G.Cornut (editor), Moyens d'investigation et pedagogie de la voix chantee, Symetrie (publishers), 1 accompanying CD-Rom, Lyon, pp 117–132
Walcott R (1974) The Chöömij of Mongolia – A Spectral Analysis of Overtone Singing, *Selected Reports in Ethnomusicology 2* (1), UCLA, Los Angeles, pp 55–59
Zemp H, Trân Quang Hai (1991) Recherches Expérimentales Sur le Chant Diphonique (voir Trân Quang Hai, Zemp H)

Discography

This selected discography considers only cd.

Tuva

Shu-De. Voices from the Distant Steppe, Realworld CDRW 41, London, U.K., 1994
Tuva / Tuvinian Singers and Musicians, World Network 55.838, Frankfurt, Germany, 1993
Tuva – Echoes from the Spirit World, Pan Records PAN 2013 CD, Leiden, the Netherlands, 1992
Tuva: Voices From the Center of Asia, Smithsonian Folkways CD SF 40017, Washington, USA, 1990

Mongolia

Jargalant Altai/Xöömi and Other Vocal Instrumental Music From Mongolia, Pan Records PAN 2050CD, Ethnic Series, Leiden, Hollande, 1996
White Moon / Tsagaan Sar/ Traditional and Popular Music from Mongolia, Pan Records PAN 2010 CD, Leiden , The Netherlands, 1992
Mongolian Music, Hungaroton HCD 18013–14, collection UNESCO, Budapest, Hungary, 1990
Mongolie: Musique et Chants de Tradition Populaire, GREM G 7511, Paris, France, 1986

Siberia

Chant Epiques et Diphoniques: Asie Centrale, Sibérie, vol. 1, Maison des Cultures du Monde, W 260067, Paris, France, 1996
Uzlyau: Guttural Singing of the Peoples of the Sayan, Altai and Ural Mountains, Pan Records PAN 2019CD, Leiden, Hollande, 1993

Vietnam

Tieng Dan Bâu / Thanh Tâm, (The Sound of the Monochord / Thanh Tâm), Dihavina, Hanoi, Vietnam, 1999
Vietnam: Dreams and Reality / Trân Quang Hai & Bach Yên, Playasound PS 65020, Paris, France, 1988

Filmography

Le Chant des Harmoniques (English Version: *The Song of Harmonics,* 16 mm, 38 minutes. Authors: Trân Quang Hai and Hugo Zemp. Realisation: Hugo Zemp. Coproduction CNRS Audiovisuel et Societé Française d'Ethnomusicologie, 1989. Distribution: CNRS Audiovisuel, 1 Place Aristide Briand, F-92195 Meudon, France

Mathematics and Medicine

Matematici

Mathematics, Clinical Decisions and Public Health

CARLA ROSSI, LUCILLA RAVÀ

Translated from the Italian by Emanuela Moreale

The Contribution of Mathematics to the Understanding of Medical Problems

The contribution of mathematics to a better understanding of medicine-related problems (clinical medicine, epidemiology, diagnostics, public health and so on) is analogous to the contribution of mathematics to other technical-scientific disciplines (such as engineering, physics, biology) and socio-economic ones. In other words, mathematics has served as an instrument built and developed to adequately solve practical problems [1]. In fact, mathematics helps see problems in an integrated way. Doctors' mathematical background should give them a better angle on medical problems and help them see medical problems in a wider context. It is then possible to evaluate each alternative decision in a quantitative way by introducing concepts such as uncertainty and probability in the analysis of medical problems.

The concepts of uncertainty and probability constitute a central element in a series of considerations aiming to clarify the need for mathematical mental models to better comprehend problems in some way related to medical decision-making.

We would like to emphasise that we are not going to talk about the contribution of mathematics to medical (experimental) research, but rather we will attempt to highlight the cultural contribution of mathematics: by this we mean the development and fusion of ideas coming from different fields as opposed to the idea of specialised erudition.

The most common error is that of failing to recognise that mathematical reasoning is the natural continuation and enrichment of intuitive capacities, whose essence is accessible and indispensable to all of us, independently from their knowledge of formal tools. Therefore – except for some routine or simple cases – medics should turn to other experts, while their own knowledge and intuitions will remain central to solving the problem if they are expressed and communicated at an adequate scientific level. It is likely that – if a remarkable mismatch is found between the results of the experts' work and the medics' intuition – the medics are actually right and the model used by the experts needs to be reviewed and modified. However, it is always necessary to think over the facts and actual data: the help of mathematics – of its language made of symbols and formulae – lies exactly in expressing facts and real data in a way that is neater, clearer and therefore less susceptible to distortions. Mathematics is of help not because it adds something, but, on the contrary, because it guarantees not adding anything extraneous.

It is important for non-mathematicians to learn to look at facts and data without distortions, without letting detail hide what is essential. A mathematical *forma mentis* provides a critical attitude that is attentive in facing everyday problems in any field and, in particular, in those situations in which there are decisions to be made under conditions of uncertainty.

All real medical problems are affected by uncertainty. It is therefore important to be able to understand the degrees of probability and to communicate one's own evaluations appropriately. It is essential to be able to evaluate – through the probability of possible effects – the advantage given by requesting a specific piece of information deemed useful in order to reduce uncertainty and adequately evaluate a conditional probability. This highlights the deep difference between using probability-grounded quantitative decision criteria and criteria of a qualitative and emotionally intuitive type. The latter would seem to suggest, for instance, that before taking a certain decision, one should gather all the potentially accessible information or – in any case – at least those that can be obtained, ignoring the cost of this operation and any risk involved as well as any delay caused by this operation. The rational criterion, instead, consists in evaluating probabilistically the cost and advantage of all pieces of information, of each test, of anything causing delay in the decision-making. This way, every Euro spent on obtaining information or carrying out tests or avoiding delays always results in an expected gain that is at least equivalent in magnitude. Let us see – concretely, with some examples – in what way and in what measure this type of mathematical *forma mentis* becomes necessary to apply rational criteria in taking decisions, both at an individual (clinical decision affecting one patient) and collective (public health decisions) level.

Clinical Decisions: Decision Trees and Algorithms

A diagnostic (or prognostic) process is an inferential procedure based on probabilistic and statistical models. In general, it can be seen schematically as a classification problem in which the aim is to assign an individual to one of the possible diagnostic (or prognostic) classes when this cannot be observed directly for various reasons. However, we can suppose that there exist some events or measures linked to the diagnostic class (with probability distribution depending on the class they belong to) through some (known) model and that such events or measures can be observed directly. The observation of the latter allows carrying out – through the use of the same model as inferential key – a probabilistic evaluation of the individual belonging to each possible class taken into consideration. A general classification problem can be seen schematically as follows.

Let us assume that there are k distinct populations $A_1, A_2, \ldots, A_k$, (in the case under consideration, possible morbid cases, according to which the patient under exam should be classified) and one or more quantities X that can be measured and whose (statistical) distribution in the i-th population $(i = 1,2,\ldots,k)$ is $F_i(x)$ [X represents the result of a measurement – or lab analysis, for instance – having different statistical behaviour across the various populations].

The goal is to specify a behavioural rule $A(x)$ determining how to decide what population an individual presenting a given value of X (indicated as x) should be attributed to.

To solve this problem, an appropriate probabilistic schema can be used.

Let $P(A_i)$ be the proportion of individuals belonging to population A_i out of the entire population, that is what is technically known as the prevalence of A_i (that, for simplicity, is supposed to be known).

Let $F(x/A_i)$ be the probability distribution of X in the population A_i.

Given an observed value x, the posterior probability of the population A_i can then be calculated by using Bayes' theorem.

Bayes' theorem – in the case under consideration – provides an evaluation of the probability with which an individual with measured quantity x belongs to the *i-th* population (that is, in our case, that the individual is affected by a certain disease).

It is then possible to establish "reasonable" behavioural rules to carry out the classification, such as, for instance:

- $A(x) = A_i$ if $P(A_i/x)$ is the maximum value of the posterior probability calculated by applying Bayes' Formula, that is if A_i is the most likely class for an individual with observed measure x;
- $A(x) = A_i$ if $P(A_i/x)$ is the maximum value of the posterior probability calculated by applying Bayes' Formula and if, in addition, if $P(A_i/x) > h$, where h is a predefined value.

In the former case, the assignment rule minimises the probability of classification error given by $1- P(A_i/x)$. Such criterion is known as "Bayesian classifier".

The latter case is clearly a bounded variant of the former and attributes an observation to a certain population only when the posterior probability is above a certain pre-determined critical threshold, i.e. is when the probability of an incorrect classification is less than $1-h$.

It is, however, evident that Bayesian classification does not distinguish between the different error types that can be made through an incorrect assignment. In order to clarify things, without losing generality, let us consider the case in which there are only two possible classes: A_1 and A_2. In this case, the only classification errors consist in assigning to A_1 a patient that belongs to A_2 and, inversely, assigning to A_2 a patient belonging to A_1.

It is possible that the consequences deriving from the two types of errors may have different degrees of importance; it is then necessary to introduce some mechanism to keep track of the differences. In fact, in medical applications, it is often necessary to avoid some types of errors more than others. For instance, in the case of the diagnosis of breast cancer, it is considered to be more important to avoid false positives – that is the classification of a healthy patient as an ill patient – rather than false negatives, since a false alarm can result in a mastectomy. In other situations, instead, it may be more important to avoid false negatives. For instance, in the case of cervical cancer, it is essential that no positives be missed.

If the different types of erroneous classification have different consequences, such diversity can be taken into account through the introduction of an appropriate "loss function".

This function depends on the class chosen for the patient and on the class the patient really belongs to – which is not observable. So, we define $L(A_i, A_j)$ as the loss corresponding to the error of assigning to A_i a patient that actually belongs to A_j. The general properties that such a function should have are quite obvious:

$$L(A_i, A_i) = 0 \text{ and } L(A_i, A_j) > 0 \text{ for all } i \text{ different from } j$$

Moreover, the function is identified to within multiplication by a constant. The consequent classification rule is based on the minimisation of the expected loss function $EL(A_i/x)$ that is obtained by calculating the weighted mean of the loss function with weights given by the posterior probabilities of the membership classes:

$$EL(A_i / x) = \sum_{j=1}^{k} L(A_i, A_j) P(A_j / x)$$

Further technical details and application examples can be found in [2].

In order to start the classification process, it is necessary to specify the sources of uncertainty and to spell out all the information on the problem at hand that are already in our possession (or that can be acquired later).

From this schema, the a priori information can be derived: this is not linked to the situation of a particular patient, but to all the population potentially interested in the classification. It is constituted by:

- possible classes: it is necessary to specify the set of "diseases";
- probability distribution of these classes: it is necessary to know or be able to estimate the prevalence[1] of these diseases in the reference population;
- types of events or measures that can be linked to the classes: it is necessary to take into consideration all indicators and measures that can be altered by the presence of one or more of these diseases;
- probability distribution of such types of information within the specified classes: it is essential to know or estimate the frequency distribution or probability of the various events or measures specified above in the various diseases. These constitute the known model that allows evaluating of the posterior probabilities of the classes by using Bayes' formula.

Medical decision making, therefore, usually takes place under conditions of uncertainty and is based on "a priori" (available at the beginning) information and further information that is acquired dynamically as it becomes available during the process. The correct combination of all information allows minimising

1 *The prevalence of a disease in a population indicates the probability that a randomly-chosen individual belonging to the population is affected by that disease. It also corresponds to the proportion (usually estimated from the statistical frequency) of the disease in the population of interest.*

the level of uncertainty. It is important to identify the most relevant information for the problem at hand. This can be done by carrying out preliminary statistical analyses starting from the data available on already-classified patients and by collecting results of similar observations reported in the literature. Moreover, physicians with expertise in a relevant area may be consulted.

It is therefore up to the medical practitioner to request the most appropriate information at each moment in time. Non-essential information does not help get at better conclusions: on the contrary, it delays the decision process, with all possible negative effects.

The dynamic process of acquisition and classification (diagnosis, choice of therapeutic response and so on) can be modelled through techniques collectively known as "decision trees and algorithms" [2]. A decision tree is the descriptive instrument of a decision process; a decision algorithm is the strategy applied in traversing the tree. The next section will consider a concrete example.

A Diagnostic Problem

Let us suppose that a doctor needs to decide whether an individual that wants to donate blood – and who does not present any suspect symptom and declares not to have been exposed to HIV risk – is affected by the HIV virus (which causes AIDS). Let us also suppose that the physician attempting to make a diagnosis has at his or her disposal not only general information on the prevalence of this disease in the population the patient belongs to, but also the Elisa test[2] (a more complex situation is analysed in [2]). The possible diagnostic paths are described in the graph (decision tree) in Figure 1.

First of all, let us consider the symbols utilised in constructing the tree.

- There are nodes identified by rounded rectangles and circles (ellipses), arches (directed segments – represented by arrows – originating in a node and terminating in another node). The graph should be traversed in the direction indicated by the arches, that is in the direction pointed to by the arrows.
- The nodes having no arches originating in them are terminal nodes (in the graph in Figure 1, there are 2 terminal nodes represented by squares corresponding to the two possible diagnoses).
- There are two types of nodes: decision or choice nodes (squares) and chance or event nodes (circles). Terminal nodes are always decision nodes.

An arch originating from a decision node indicates the action chosen by the individual who makes the decisions (the decision-maker). An arch originating from a chance node indicates an event that is not under the control of the decision-maker, in that it depends on the result of observations, tests, measurements or other, whose outcome is not known in advance. For instance, in the case at

2 *The Elisa test is a type of blood test that gives either a positive or negative result that – with a certain probability to be specified later – indicates the HIV status of the individual.*

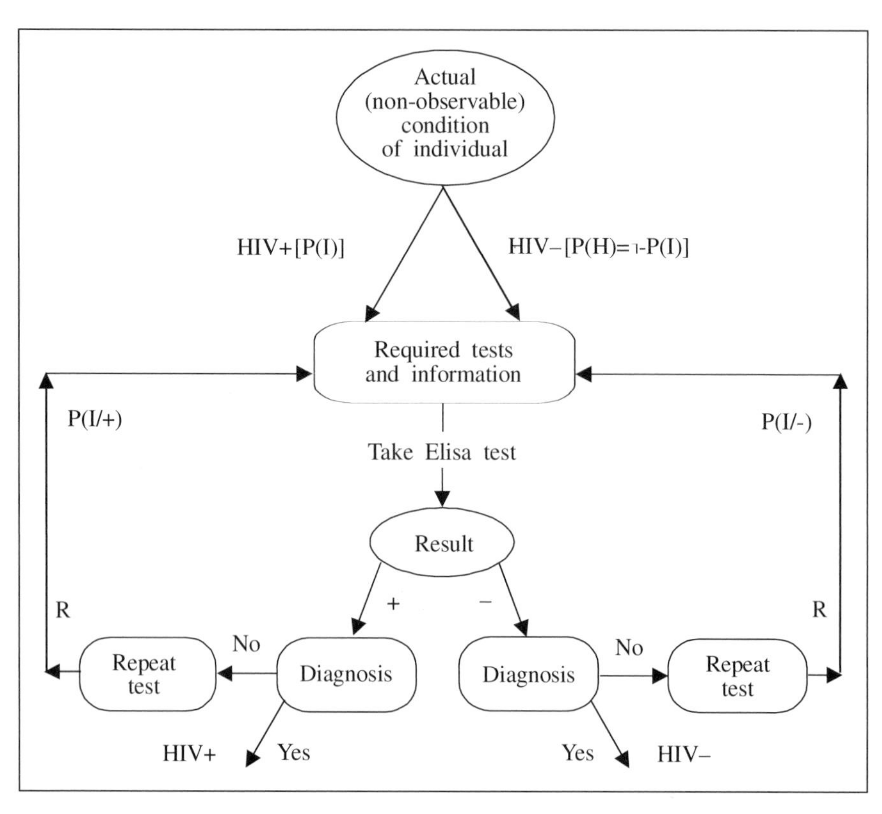

Fig. 1. Decision tree illustrating the possible (hypothetical) diagnostic paths relating to the classification of a blood donor. Rounded rectangles represent phases in the decision process under the control of the decision-maker (decision nodes). Circles (ellipses) represent phases that are independent of the will of the decision-maker (chance nodes). Arches represent the decision-maker's actions originating from a decision node and events originating from a chance node. The arch marked "R" indicates the decision to repeat the Elisa test. In this case, the a priori probability of this new test is the posterior probability of the previous test.

hand, from the chance node relating to the test result, the right branch will be followed if the result is negative; otherwise, the left branch will be chosen.

It is possible to have obligatory decision nodes (only one arch coming out of a node). In the case at hand, we are supposing that the Elisa test is going to be taken in any event at the beginning of the diagnostic process.

The tree described above, although specific and simplified, shows the general workings of a medical decision process made up of the following phases:

- problem description: in our case, the individual's membership of a certain population with given prevalence of the disease;
- formulation of diagnostic hypotheses: in our case, HIV infection, HIV positive or HIV negative;

- choice of data to be gathered to discriminate between two hypotheses in an "optimal" way, that is by minimising the probability of error: in our case, the Elisa test and its possible repetition.

It can be noted that, even if this particular example is limited to the diagnostic process, other processes – such as therapeutic, experimental or prognostic – can be represented in trees that are similar in construction and interpretation.

It should be observed that, in some cases, a descriptive representation of a decision process may be sufficient and that an evaluation of the various strategies represented in the corresponding tree might not be required. This is certainly the case when there is only one single arch coming out of each node (this means that all choices are obligatory when the result of chance nodes are known). This case can be called "deterministic" and, although not very realistic, constitutes a reference point for real cases in which the diagnosis is probabilistic. In fact, the strategies employed to traverse a decision tree (decision algorithms) are based on choices relating to requests for information that aim to reduce uncertainty and "lean" – so to speak – towards the ideal deterministic case, by producing the maximum probability difference between the alternatives being considered as possible diagnostic choices.

Let us go back to the example under consideration so as to introduce rules to calculate probabilities relating to the possible decision paths.

By $P(I)$ – and $(P(H) = 1-P(I))$ respectively – we will denote the proportion of ill and healthy individuals in the population, i.e. the prevalence of the disease in the population: in other words, the probability that an individual chosen at random from the population is in fact ill. Moreover, let $P(+/I)$ and $P(+/H)$ be the probabilities of obtaining a positive Elisa test for an ill and healthy individual respectively. Finally, $P(+)$ is the probability of obtaining a positive test, independently of the status of the individual under consideration. We can then write the following relationships between the various quantities introduced (where the $\cap$ symbol indicates the intersection between events and generates the event which is true if and only if both the events are true):

$$P(I \cap +) = P(I/+)P(+) = P(+/I)P(I) \tag{1}$$

$$P(H \cap -) = P(H/-)P(-) = P(-/H)P(H) \tag{2}$$

$$P(+) = P(+/I)P(I) + P(+/H)P(H) \tag{3}$$

$$P(-) = P(-/I)P(I) + P(-/H)P(H) \tag{4}$$

As a consequence of (1), (2), (3) and (4), the following relationships are derived (by applying Bayes' Theorem):

$$P(I/+) = \frac{P(+/I)P(I)}{P(+)} = \frac{P(+/I)P(I)}{P(+/I)P(I) + P(+/H)P(H)} \tag{5}$$

$$P(H/-) = \frac{P(-/H)P(H)}{P(-)} = \frac{P(-/H)P(H)}{P(-/I)P(I) + P(-/H)P(H)} \tag{6}$$

Let us now suppose that $P(I)$ is equal to 0.006 and that the test gives a positive result in 99% of the cases in which the individual is I and a negative result in 99% of the cases in which the individual is H. Formally, we can express this through the quantities:

$$\text{Sensitivity} = P\,(+/I) = 0.99;$$
$$\text{Specificity} = P\,(-/H) = 0.99;\ P\,(-/I) = 0.01;\ P\,(+/H) = 0.01.$$

If we now use Bayes' Theorem (formula (5)) to calculate the probability that an individual, chosen at random from the population and whose test result was positive, is I, we obtain: $P\,(I/+) = 0.374$.

In other words, given the error margin implied in the type of test and the "low" prevalence of infection in the population (a priori probability), one can expect that only 37% of the individuals who test positive are in fact ill.

What is the best choice to be made in such a case? Can we attempt a diagnosis or should we gather more data?

It is evident that it is not appropriate to attempt a diagnosis, and that it is much more reasonable to obtain more data. In cases such as the one at hand, the usual procedure involves retaking the same test, independently, after a while.

If a new test is taken independently of the first, all that is needed is to retraverse the same tree, replacing the a priori probability for the first test $P(I)$ with the posterior probability for this test, $P(I/+)$, because the latter has now the role of a priori probability for the second test.

If we now assume that the second result is also positive, we get:

$$P(I/++) = \frac{P(+/I+)P(I+)}{P(+/I)P(I+) + P(+/H+)P(H+)}$$

Because of the hypothesised independence between the two test results, we can also write:

$$P(I/++) = \frac{P(+/I)P(I/+)}{P(+/I)P(I/+) + P(+/H)P(H/+)}$$

By carrying out the calculations, we get: $P(I/++)=0.983$, a value that allows making a diagnosis. For completeness, we will also calculate the probability $P(I/+-)$, the probability in the case in which the second test is negative. The result is: $P(I/+-)=0.006$. As can be expected, the negative result in the second case has cancelled out the effect of the first test by reproducing the value of the a priori probability.

We would also like to point out that, if the first test turns out negative, it is best to make a diagnosis without requesting a further test, as in this case the probability $P(H/-)$ is in fact very high (greater than 0.9999) and therefore requesting a further test uselessly delays an almost certain diagnosis.

As can be seen, the information required allows the evaluation of probabilities that are useful for the diagnostic process. Given the crucial importance of requesting the optimal information, that is the importance of using an "optimal" decision algorithm, we will now add a few general considerations on this topic.

Value and Cost of Information

There are different types of information. Some information is immediately available at the moment of the first visit and can be considered to have zero cost: for example sex, age and symptoms. Other types can be requested and must be evaluated on the basis of their potential informational content and their cost. The latter quantity must take into account all types of cost: economic, time required before the result is available and related risks (such as invasive procedures and possible complications). The cost function must be estimated by experts: they request the information and – at the end of the decision process – proceed to classify the patient. The cost function is based also on the loss function defined earlier. In order to decide whether to request a test or not, it is therefore necessary to evaluate both the potential contribution of the test result to the diagnosis and the cost of the test itself [3–5]. It is possible to carry out a precise probabilistic evaluation by using quite technical mathematical rules (explained in detail in the articles quoted earlier): these evaluations are essentially based on the variation of the expected loss function.

We will limit ourselves to a few general considerations on the evaluation of the possible strategies for traversing a decision tree or decision algorithms, using another example linked to possible therapeutic choices. In this case, too, the example is simplified, because the cost-benefits of the various choices are not taken into account and the optimisation is carried out only on the target function that, in this case, is given by the probability of survival after ten years.

Analysis of a Therapeutic Decision Process

A decision algorithm is a set of precise rules that allow determining the "optimal way" to traverse the branches of a certain decision tree, i.e. choosing the optimal decision strategy. Let us consider the case of a patient with an already diagnosed first-stage lymphoma: what is now needed is a therapeutic decision such that – independently of any cost considerations – the patient's survival chances after ten years are maximised. Figure 2 illustrates the decision tree; leaf nodes also indicate the values of the various survival probabilities at ten years.

At decision node 1, the choice is based on surviving surgery in case of lymphadenectomy and on the subsequent survival function at ten years. It is obvious that surgical removal is the best choice, since the risk of death due to surgery is very low (0.5%), while the mortality at ten years if no surgery is undertaken has a probability as high as 100%. At decision node 2, the decision to be made relates to whether or not to undertake radiation therapy. The survival rate at ten years is 97% for treated patients, while it is only 10% for non-treated patients. Therefore, the optimal decision at the second node is to carry out the treatment.

Other examples of decision trees and algorithms are illustrated in [2].

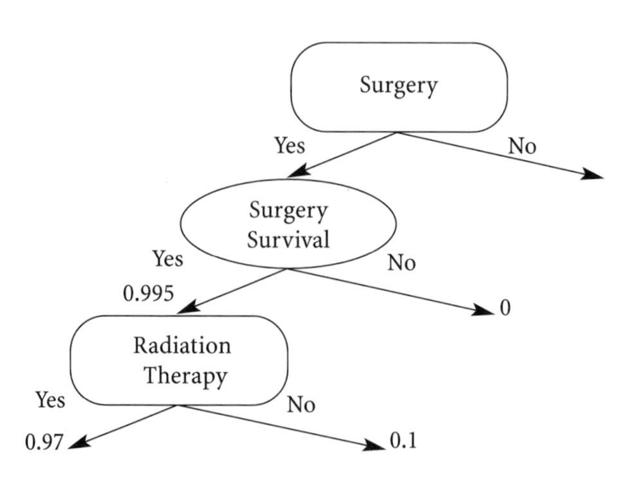

Fig. 2. Example of decision tree with binary choices and strategy aiming to maximise survival probability at ten years. In the example at hand, surgery – in spite of its 0.5% likelihood of death – is the optimal one-step decision for the chosen target function: in fact, after ten years, patients who have undergone surgery have a survival rate of at least 10%, while patients who have not undergone surgery have a survival rate of 0%. If the best strategy is chosen among the two-step options, then surgery together with radiation therapy is the optimal choice, as this gives a survival rate of 97% for treated patients against a survival rate of just 10% for non-treated patients at ten years. Overall, the maximum survival rate can be reached by combining surgery with radiation therapy and is about 96.5%.

Public Health Decisions: Scenario Analysis and Use of Forecast Models

Scenario analysis has been widely used in the field of economics in forecast studies analysing the impact of decisions with the aim of optimising economic policy.

The group of methodologies constituting such type of analysis were born out of the necessity to explore potential future developments of a phenomenon and to point out the critical points for the decisions implicating choices that may eventually modify such developments.

The keywords in scenario analysis are: aims, target, decision, policy-oriented research and cost-benefit analysis.

The set of methodologies consists of some fundamental steps:

- Basic problem analysis: the nature and fundamental aspects of the problem at hand are described in terms of its past, present and future trends;
- Development of a conceptual (mathematical) model: the elements and relationships between elements of relevance to this problem are described through a model;
- Development of exploratory scenarios: "zero" scenarios (making no interventions in the natural development of the phenomenon) and "what-if" scenarios;
- Development of strategic scenarios: these are characterised by a predefined desired final outcome and are about determining the correct strategies leading to such situation.

A peculiar aspect of scenario analysis consists in the various forecasts obtained through the mathematical model not being presented in an abstract form, but rather in a decision context. In general, the conceptual schema is represented in graphical form through a block diagram such as the one in Fig.3. Here, within the block model, one can introduce the details relating to the compartments and fluxes that describe the situations of interest, as explained in the example that we will deal with later.

It is a set of relations tying together quantities that are useful to describe the phenomenon to be studied: these are usually called "variables" and divided into "input variables" or data and "output variables" or forecasts.

Input variables are supposed to be known, whereas output variables become known only once the data has been entered, when they are calculated on the basis of relationships in the model itself.

The coefficients appearing in the relationships of a model are known as model "parameters". One can distinguish between "fixed parameters" (tied to the basic scenario) and "scenario parameters" that are usually varied to simulate the various situations of interest.

The meaning of the input variables can either be the same as that of the output variables or be different. For instance, in an epidemic model, the input variables can represent the number of healthy and/or ill individuals at the beginning, while the output variables can represent the same quantities at a later time. In a genetic counselling problem, the input variables can be the parents' genotypes and the output variables the probabilities of the various possible genotypes for a possible offspring.

The problems that are tackled through the use of a mathematical model are of two main types:

- direct problems: these consist in generating forecasts on the basis of the relationships, data and values chosen for the parameters;
- inverse problems, on the other hand, consist in utilising the data (known or predetermined forecasts obtained outside the model) and possibly known parameters to estimate the value of the other (unknown) parameters.

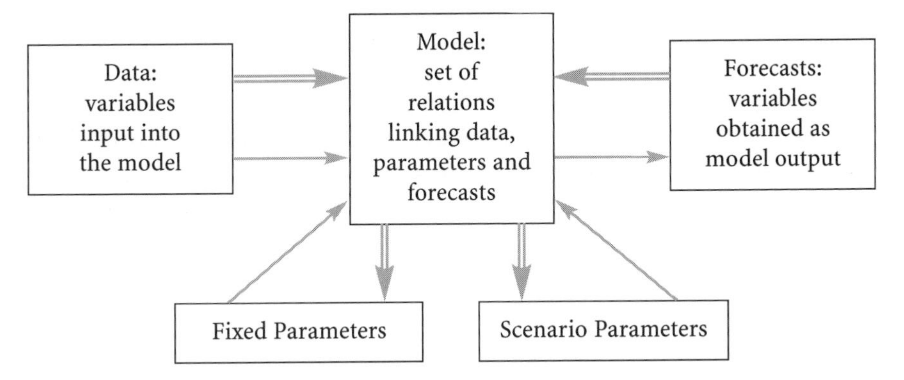

Fig. 3 General diagram of a scenario analysis model.
: Flows relating to "direct" problem; : Flows relating to "inverse" problem

Exploratory scenario analyses belong to the former type, except the estimate of the basic scenario, which usually belongs to the latter type, at least as far as the estimate of some parameters is concerned. The construction of strategic scenarios, instead, belongs to the latter type, since exploratory scenarios are generated from the model by varying the values of the parameters and simulating the corresponding values of the forecasts (direct problem).

On the other hand, strategic scenarios solve the problem of what parameter values allow a certain strategic result to be obtained in terms of wanted conditions on the values of the output variables (inverse problem). Once the appropriate value is identified, it is also necessary to identify what actual actions are necessary to obtain it.

In order to better clarify what explained above, we are going to use a concrete example that takes into consideration forecast models and estimates of the size of the AIDS epidemics in Italy. We will consider only the basic scenario providing the actual estimate of the size of the phenomenon and we will not take into consideration alternative scenarios that are explored in detail in the articles quoted earlier.

How to Use Mathematical Models to Estimate the Size of the HIV/AIDS Epidemics in Italy

In order to plan and channel efforts in the area of prevention and care of the HIV infection and the acquired immunodeficiency syndrome (AIDS), it is necessary to have reliable estimates of the size and dynamics of the epidemics. Obtaining such estimates requires the use of powerful statistical-mathematical instruments. In fact, since the HIV infection is generally characterised by a long asymptomatic phase and a long incubation period[3], studies based on observation are not adequate to provide information as to the dynamics of the epidemics. More appropriate types of studies are those based on dynamic compartmental models and on Back-Calculation models: these allow estimating and foreseeing the incidence and prevalence of HIV and AIDS, by concentrating the attention on the different aspects of the epidemics. The Mover-Stayer model is a dynamic compartmental model developed to study the HIV/AIDS epidemics among the general population: it is a useful tool, particularly for carrying out scenario analyses. On the other hand, Empirical-Bayesian Back-Calculation (EB-BC) allows reconstructing the dynamics of an epidemic for categories at risk. Such classes of models present different types of advantages and contraindications: this suggests that the two methods should be used together, so as to achieve greater flexibility in the model to be used for scenario analyses as well as greater robustness and accuracy in the estimates.

[3] This is the time interval between the moment when infection takes place and the onset of the first AIDS symptoms

Table 1. Compartments of the Mover-Stayer model corresponding to the stages of HIV and AIDS infection defined by lymphocyte CD4 levels.

Latency Period	HIV infection compartments	AIDS compartments
	Y1	-
$CD4 \geq 500$	Y2 and Z1	-
$200 \leq CD4 < 500$	Y3 and Z2	-
$CD4 \simeq 200$	Y4 and Z3	A1
$100 \leq CD4 < 200$	-	A2
$50 \leq CD4 < 100$	-	A3
$CD4 < 50$	-	A4

The Mover-Stayer Model

The Mover-Stayer model (MS) [6–7] is a dynamic compartmental model allowing the study of the HIV infection and AIDS epidemics and the forecast of future epidemic scenarios on the basis of a comparison between the simulated data derived from the model and observed data. According to such model, the initial population $X(o)$ is divided into two groups that differ in behaviour: $(1\text{-}S(o))X(o)$ or "movers" – the individuals at higher risk because of their behaviour – and $S(o)X(o)$ or "stayers" – those individuals who are not involved in the spread of the HIV infection. Since the MS model was developed for the study and the spread of an epidemic in the general population, when reproducing an epidemic, only general risk behaviours are taken into account, without distinguishing between types of individuals (such as drug addicts, homosexuals or heterosexuals). As a matter of fact, it would be impossible to tell whether a drug addict caught the infection through sexual intercourse or by using infecting syringes; therefore, estimates and forecasts would not be very reliable.

Infected individuals are classified in 4+3 compartments on the basis of the level of CD4[4] lymphocytes (4 Y compartments, for individuals with risky behaviour, and 3 Z compartments, for individuals with safe behaviour). Similarly, also ill patients (AIDS cases) are classified in 4 compartments on the basis of the level of CD4 lymphocytes (Table 1). This allows writing the model equations both in the deterministic version (differential equations) and in the stochastic version (point processes). A graphical description of the model is given in Fig. 4.

For an efficient use of the model – particularly in terms of robustness and identifiableness – most biomedical and epidemiological parameters should be predefined on the basis of the data derived from ad hoc studies or from the literature *(fixed parameters).* On the other hand, only a limited number of socio-behavioural parameters – that cannot be determined through direct field studies and

[4] CD_4 lymphocytes are the immune system elements attacked by the HIV virus; therefore, their level is an indicator of the progress of the infection. A low level indicates a greater damage to the immune system.

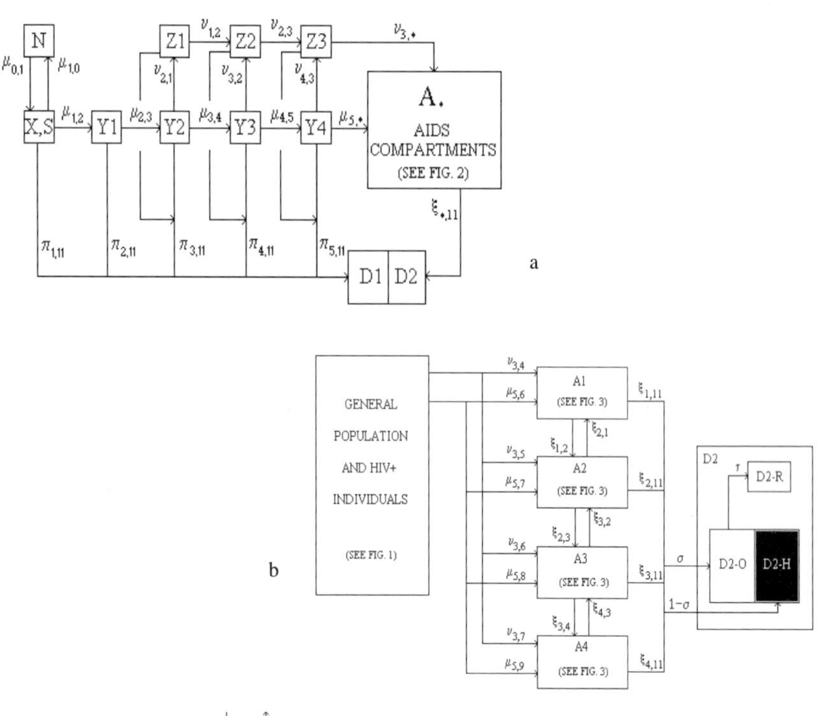

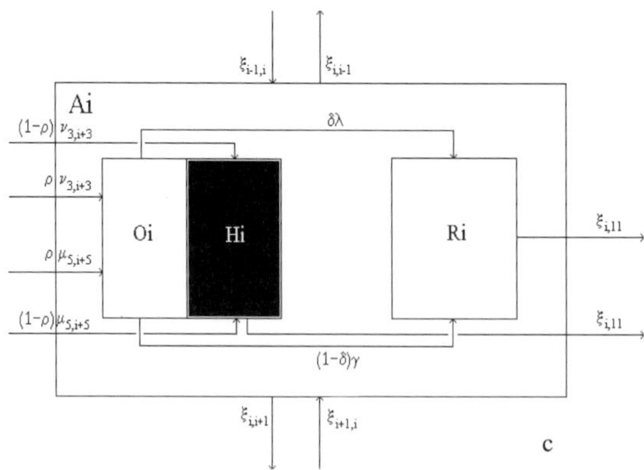

Fig. 4. a–c. Diagram of the compartments of the Mover-Stayer model and description of the dynamics of such compartments. **(a)** Compartments of non-infected individuals and individuals infected with HIV; compartments of individuals affected by AIDS and corresponding transition rates here indicated globally as A and. ν_3. μ_5. $\xi_{\cdot 11}$, (see Fig. 4(b) for more details).
(b) Compartments of individuals with AIDS. Generic compartment A_i (i=1,...,4) is described in detail in Fig. 4 (c). Compartment D_2 represents AIDS deaths, of which σ% % observed and (1-σ)% hidden; t is the rate of delay in notification. **(c)** Description of the i-th compartment of individuals with AIDS, A_i. Compartments O_i, H_i e R_i represent observed, hidden and notified individuals respectively. Individuals coming from compartments $\mathbf{Y}_4$ and Z_3 entering compartment A_i, are constituted by a proportion (1-ρ) of hidden individuals. Modifications in the pathological status are represented by exchanges with nearby AIDS compartments (A_{i-1} and A_{i+1})) and are regulated by transition rates $\xi_{\bullet\bullet}$. Diagnosed ill individuals move over to notification compartment R_i at times distributed according to a mixture of two exponentials, with rates λ and γ, and mixture proportion d and (1-d)

certainly produce behavioural perturbations (indeterminism) – should be estimated by the model *(characteristic parameters* or *scenario parameters).*

The characteristic parameters of the model are the initial proportion of stayers $S(o)=S_0$; the macro infectivity rat $\mu_{1,2}$, incorporating both the infectivity and probability of contact between infected and susceptible individual – and the initial year of the epidemic T_0 (more precisely, T_0 is the instant in which the prevalence of the infected people in the population reaches the value of 1 in a million of inhabitants). Such parameters are estimated by comparing observed data (usually provided by national surveillance systems) and that simulated by the model (forecasts) by using the method of the maximum likelihood or the minimum χ^2, – observed data (usually provided by national surveillance systems) and that simulated by the model (forecasts). A software application called TOVAIDS has been developed as an application of the MS model: this software has a simple user interface that allows simulating the progress of the HIV/AIDS epidemic both under the basic scenario and under possible alternative scenarios taking into consideration the effect of pre-AIDS therapies and/or prevention campaigns. This program is available from the Università degli Studi di Roma Tor Vergata website, at the URL: http:// www. mat. uniroma2.it/~rava/tovaids_home.html.

Empirical Bayesian Back-Calculation

Back-Calculation is a class of statistical methods that – initially proposed as a method to come up with the most conservative estimate of the HIV/AIDS epidemic [8] – were soon utilised also to estimate and foresee the incidence and prevalence of HIV and AIDS infection [9,10]. The idea at the basis of each of these methods is that of reconstructing – according to some distribution of the incubation period – the number of individuals that would have had to have become infected with HIV previously so as to give rise to the observed AIDS incidence. Then – by applying the distribution of the incubation period to the HIV infection curve – it is possible to foresee the incidence of AIDS in the short term.

Let *A(t)* be the total number of diagnosed AIDS cases by time *t, h(s)* the HIV infection rate at time *s* and *F(t)* the distribution of the incubation period. Then the convolution equation

$$A(t) = \int_{o}^{t} h(s)F(t-s)ds \tag{7}$$

– through the distribution of the incubation period – ties the HIV infection rate to the incidence of AIDS. The preceding equation is justified by observing that an individual receiving an AIDS diagnosis within time *t* must have necessarily got infected at time *s*, where $s<t$, and have undergone an incubation period of length smaller than *t-s.* Therefore, the fundamental idea of the Back-Calculation is that of using an instance of *A(t),* that is actual data on the incidence of AIDS, an estimate, usually external, of *F(t)* and equation (7) – or an equivalent equation derived from it – for the purpose of obtaining information concerning past HIV infection rates, *h(s)* $s<t$.

The Back-Calculation method presented here, the Empirical Bayesian Back-Calculation (EB-BC), [11–13], is implemented in the S-plus language, in a discrete framework, and is based on an empirical Bayesian approach. Such method approximates the HIV infection curve *(h(s))* by a constant step function with a relatively small number of steps representing infection rates, while the lengths, usually annual, represent the infection intervals. Infections occur in each time interval j $(j = 1,\ldots,J)$ according to a Poisson process of intensity g_j. It is then assumed that the incidence of AIDS in each diagnosis interval i $(i=1,\ldots,I)$ is independently distributed according to a Poisson distribution of average $m_i=S_j Z_{i,j} g_j$, where $Z_{i,j}$ is the probability that an individual infected in time interval j develops AIDS in interval i. Re-parametrising the infection parameters g_j to ensure that the estimates are non-negative, the fundamental equation of the Back-Calculation assumes the following form:

$$E(Y_i \mid b) = m_i = \sum_{j=1}^{J} =Z_{i,j}\, g(b_j)$$

where b_j are the infection parameters to be estimated.

With this method, it is necessary to use a constrained optimisation technique to avoid obtaining estimates of the HIV incidence relating to two consecutive time intervals that otherwise could turn out to be extremely variable. Therefore, smoothing constraints are introduced and – following empirical Bayesian theory – they are assigned the role of a priori distribution for the parameters to be estimated b_j.

EB-BC utilises estimates of the distribution of the incubation period in the form of multistage Markov models, in which the first stages are infection stages – usually identified by different CD4+ lymphocyte levels – while the following stages (at least one) are AIDS stages and, if necessary, the last one is that of death from AIDS [14–16]. The graphical form of the model would not be very different from that relating to compartments of type Y and A in Fig. 4.

This method therefore allows the introduction into the model of some covariates and the study of the progression of the epidemic, while taking into account a variety of external information that may be available, ranging from demographic data to the date of diagnosis of the HIV positive status.

Application of the MS and EB-BC Models to the HIV/AIDS Epidemic in Italy

Both the MS and EB-BC models have been applied to the study of the HIV/AIDS epidemic in Italy [17]. We will now illustrate part of the results obtained by using the quarterly data on AIDS diagnoses, as notified to the "Centro Operativo AIDS" (COA) of the Istituto Superiore di Sanità during the period 01/01/1981 – 31/03/1998 according to the definition of AIDS established by the U.S. Center for Diseases Control (CDC) in 1993 [18]. Data excluded from the analysis are pediatric cases (independently of the type of transmission), cases due to hematic transmission and nosocomial cases, since the MS model is a transmission model through contacts of various nature between an infected individual and a healthy individual that is susceptible to the infection. The analysis was therefore carried out among the following five subpopulations: homosexual and bisexual males, heterosexual males and females plus male and female drug-addicts. It should be noted that, since these five groups of individuals represent almost the totality of the population, we will from now on refer to their union as the "total population".

Because it is still difficult to provide a precise evaluation of the efficacy of the AIDS treatments introduced in 1996 (known as high efficacy antiretroviral therapy or, more commonly, triple therapy), the results shown here were obtained only on the bases of AIDS cases diagnosed by 31st December 1995.

According to the estimates obtained through the use of the MS model, the epidemic started during the third quarter of 1978: the initial proportion of stayers was 0.9985 and the $\mu_{1,2}$ rate equal to 0.0000116. The estimated value for S_0 implies an estimate of the size of the initial group at risk (at the end of the third quarter of 1978) of 90,000 individuals, which is consistent with other available epidemiological information. It should be noted that the MS model works on a non-closed group, in that it takes into account demographic dynamics as people enter or exit the compartment of individuals susceptible to infection and this is essential to obtain an estimate of the endemic phase of the epidemic. In fact, if the group under consideration were closed – as many dynamic models applied to groups of homosexuals described in the international literature are – the epidemic would necessarily extinguish itself out of a total saturation effect of the individuals at risk, without ever giving rise to an endemic phase. Instead, the MS model takes into account demographic dynamics and allows forecasting a stabilisation of the epidemic on the endemic level after 2001, with a quarterly incidence of HIV infections of around 300 cases. Such results correspond to a threshold value of $S_0^*=0.9997759$ [6] – obtained by simulating the epidemic over a period of over 100 years – which implies a stable endemic phase with 2.25 over 10,000 susceptible individuals with risky behaviour in the general population. This threshold value allows estimating the average infectivity time as being around eleven years. The average incubation time leading to the development of the disease is instead around twelve years and the median time is less than fifteen years.

Contrary to the case of the MS model, the EB-BC was carried out separately for each of the five sub-populations under consideration. For each of them, in the EB-BC, the progression of AIDS was reproduced through non-reversible four-stage Markov models (with three HIV stages – identified on the basis of decreasing levels of CD4+ lymphocytes – and one AIDS stage) with transitions allowed only between consecutive stages; the parameters for such models were determined on the basis of data relating to the Italian population. The incidence and prevalence of AIDS were estimated in the hypothesis that, starting in the third quarter of 1997, new HIV infections occur with a rate that is equal to the last estimated rate.

Figures 5 and 6 show the (annual) HIV and AIDS incidence curves for the whole population that were estimated using the MS and EB-BC models. As can be seen, the two incidence curves – in spite of differences in the final stages – have rather similar shapes. This provides cross-validation for the two estimation methods, especially seeing that the underlying hypotheses on rates for the last few years are not corrected by available data. In particular, since the version of the MS

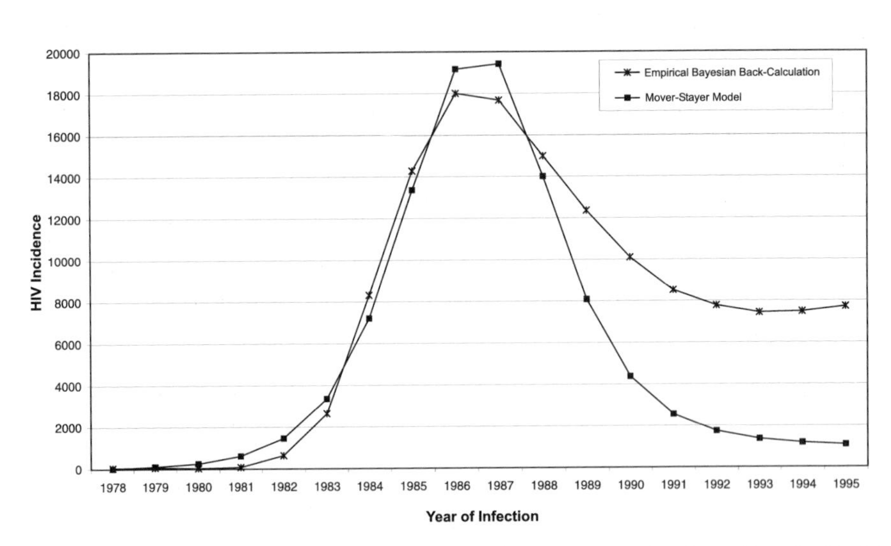

Fig. 5. HIV incidence curve for the total population estimated using MS and EC-BC models.

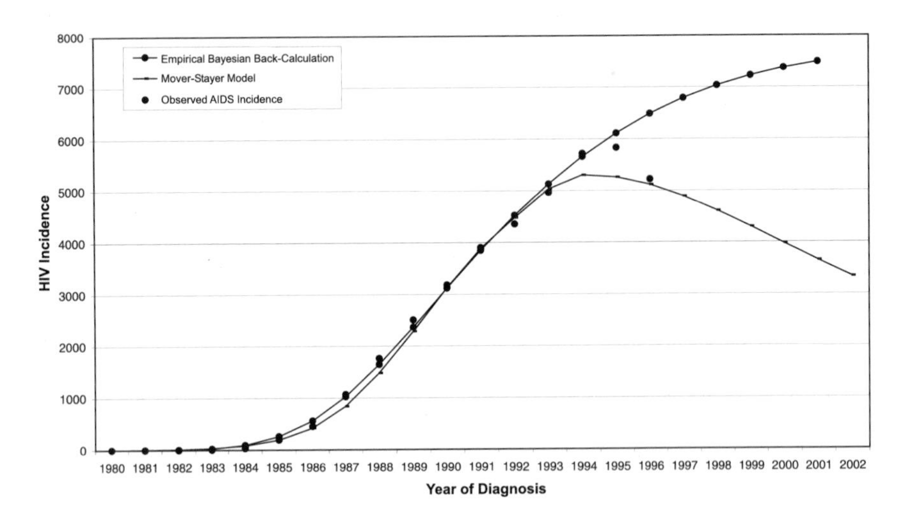

Fig. 6. AIDS incidence curve for the total population estimated using MS and EC-BC models.

model used in this case does not take into account any changes in risk behaviour (following prevention campaigns, for instance), the low right tails of the curves estimated by this model are due to a saturation effect of the epidemic. On the other hand, methodological assumptions underlying EB-BC (concerning the choice of a priori distribution and therefore the smoothing constraint, which are not dealt with here) justify the rather high level of the right tails of the curves obtained with this method. It is also necessary to take into account the fact that –

due to the long incubation period, a limitation of both these methods – the estimates relating to the most recent periods never prove very accurate and reliable and, in any case, are always characterised by a rather high variability. This phenomenon turns out to be particularly evident in the case of EB-BC in which, therefore, the more limited dipping in the relative curves is induced more by the amount of uncertainty than by a real underlying trend.

The two curves relating to the incidence of AIDS are even more similar to each other, being more influenced by the data. Moreover, such curves show a good adaptation of the various models to the observed data, also taking into account the possible fluctuations of such data due to various causes that cannot be traced back to a purely random Poissonian fluctuation foreseen by the standard deviation.

Figures 7 and 8 show the HIV and AIDS incidence curves estimated through EB-BC for each of the five sub-populations. It seems clear that, during the first decade, the epidemic reached its maximum diffusion among male drug users. On the other hand, since 1991, heterosexual females have had the highest HIV incidence and represented the sub-population in which the epidemic has been spreading the fastest. Nevertheless, the highest incidence and prevalence values are still found among male drug users, as – in the first period of the epidemic – the incidence of HIV in this group had been far higher than in the other groups. It can also be noted that, for each sub-population, the incidence and prevalence curves show their peaks around the years 1986/87 and 1990/91 respectively and have now reached a *plateau*, except for the sub-population of heterosexual females, among whom the epidemic is still growing.

The obtained results, partially reported in this article, clearly show that the HIV/AIDS epidemic in Italy shares many of its characteristics with these epidemics in other Mediterranean countries. In particular, since the beginning of the epidemic, most infections have taken place among drug users, whereas homosexual and bisexual men have always been – and still are – a group in which the incidence of HIV infection is limited and decreasing. At present, while the epidemic is in a phase of reduction or at least stability in the main groups at risk, it shows a fast growing increase in its diffusion among heterosexual females, mainly due to their contacts with infected drug users.

The MS and EB-BC models, just like other dynamic and Back-Calculation models, have been developed to tackle several types of problems relating to the objectives of this study. Therefore – although both provide estimates and projections on the incidence and prevalence of HIV and AIDS in the short and midterm – they present different characteristics and potentials; their utilisation, whether separate or joint, is determined in each case by the objectives of the study.

One of the interesting characteristics of the MS model is that the modelling approach allows carrying out corrections on the data to account for delay in notification and undernotification when estimating the parameters. This allows, on the one hand, a unitary view of the process of comparison between forecast data and actually observed data, with the bonus of greater coherence and elegance in the mathematical modelling; on the other hand, greater accuracy in the estimate process based on solving just one "inverse problem". It is well known how critical

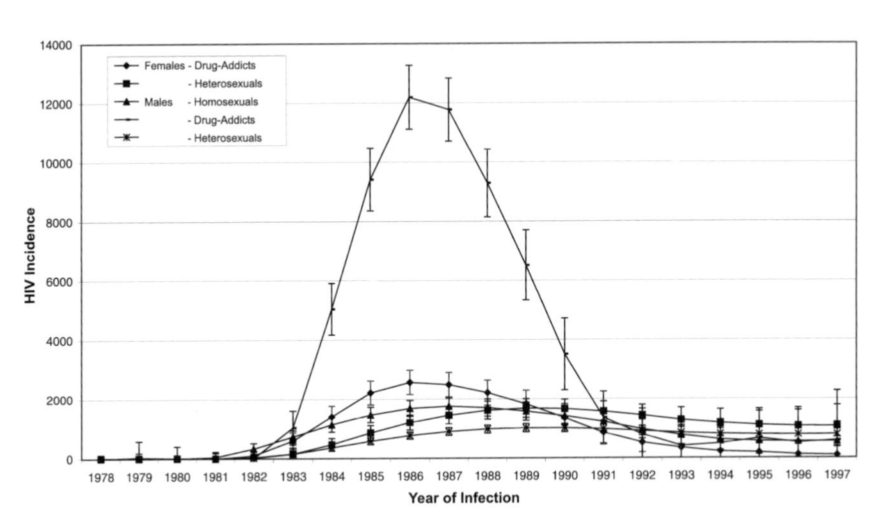

Fig. 7. HIV incidence curve by transmission category estimated using EC-BC.

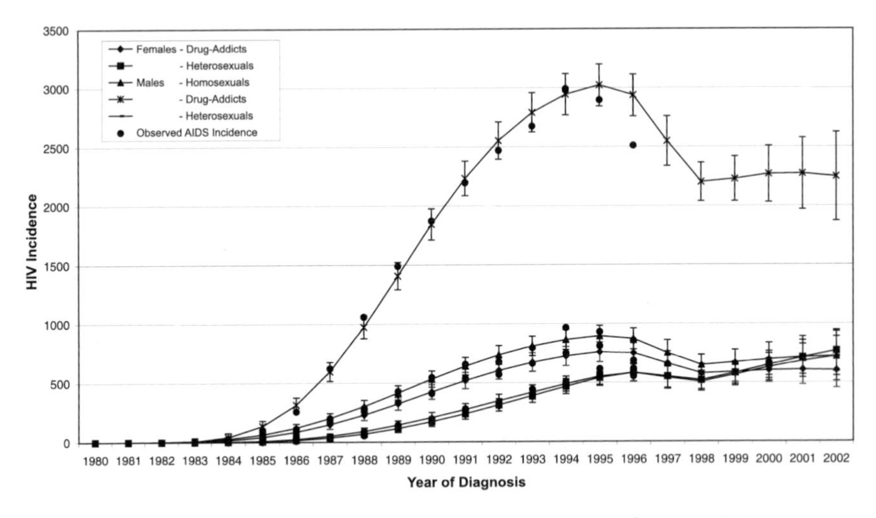

Fig. 8. AIDS incidence curve by transmission category estimated using EC-BC.

the solution to inverse problems is, both from the point of view of the instability of the estimates and the problems of ill-conditioning. The usual reconstruction method of the theoretical data prior to the estimate procedure of the actual parameters involves the solution of at least two consecutive inverse problems: reconstruction for under-reporting and reporting delay. Both of these problems produce uncertainty in the reconstructed data and this has repercussions in the final estimate of the characteristic parameters of the model. The method of theoretical generation of the observed data through a dynamic model, in which parameters are estimated by solving just one inverse problem (instead of three), allows – for instance through the determination of likelihood intervals – limiting and better controlling uncertainty in the final estimate of the model parameters. Moreover, it makes it rather easy to frame the likelihood procedure within the Bayesian approach to estimation problems and in particular the context of the Bayesian inductive method and not the various Bayesian adhockeries (empirical Bayesian methods, conjugate classes and so on).

Moreover, the MS model allows carrying out analyses that go under the name of *"What if scenario analyses"* in which the question is asked, for "what kind of reduction in the incidence of HIV should we expect from the free distribution of condoms in schools for x months?". In fact, the MS model allows the simulation of the progression of the epidemic starting from a more or less complex hypothesis of the interaction between the population susceptible to infection and the infected population. It therefore also allows evaluating the impact of any intervention on the susceptible population.

However, such type of models – just because they are based on the interaction between at least two population groups – easily become too complex to be treated adequately and can give rise to instability problems in the estimates. This is particularly the case as the number of interactions between the various groups (drug users, heterosexual males and females, possibly stratified also by age or type of behaviour) grows.

The Back-Calculation method used in this study, EB-BC – thanks to its flexibility – allows estimating the progress of the epidemic in the various risk groups and carrying out scenario analysis, taking into consideration the possible peculiarities of the epidemic under study and of the multiple variations in the natural history of the HIV infection. These variations have repeatedly occurred in the past and are still taking place now: they are caused, for instance, by changes in the definition of AIDS and, mainly, by the introduction of new therapies and prophylactic treatments. Luckily, EB-BC multistage incubation models allow obtaining an estimate of the prevalence and incidence of both HIV and AIDS for each of the infection stages as well as the cumulative incidence of HIV infection (i.e. the total number of infected individuals since the beginning of the epidemic, including possible deaths from AIDS). Because of this, EB-BC results constitute a useful basis for studies aiming to develop and evaluate the efficacy of new treatments and prophylactic and therapeutic interventions.

On the other hand, the main disadvantage of Back-Calculation methods lies in the fact that they are estimation procedures that do not model the interaction between susceptible and infected people. Therefore, they do not allow estimating the impact of interventions on the population at risk before infection and, even less, to carry out scenario analysis for the a priori evaluation of preventative intervention (which is, however, possible with the MS model).

An interesting alternative to the disjoint use of these two method consists in using Back-Calculation to estimate the size of the epidemic in the various groups making up the population of interest – as this would not require a dynamic model of all the possible interactions – and a simple dynamic model for the general population (or for individual groups). This latter would need to allow robust estimates of the few parameters in use and, at the same time, carrying out scenario analyses also on the group of susceptible people, in particular for the evaluation of possible prevention campaigns [17].

Bibliography

[1] de Finetti B (1996) L'Apporto della Matematica alla Comprensione dei Problemi Economici, *Metra,* vol V, 3:355–366

[2] Rossi C, Ferranti E, Cugini P (1994) Alberi e Algoritmi Decisionali, *MEDIC,* 2: 271–279

[3] de Finetti B (1964) Teoria delle Decisioni, in: *Lezioni di Metodologia Statistica per Ricercatori.* Istituto di Calcolo delle Probabilità e Istituto di Statistica, Università di Roma, Roma, pp 89–161

[4] Fabi F, Rossi C (1989) *Bayesian Decisions in Medical Diagnosis and Prognosis,* Atti del Secondo Congresso Nazionale dell'Associazione Italiana di Statistica Medica, Pavia, pp 323–336

[5] Rossi C (1992) A Bayesian Consulting System in Medical Clinics, in: Barrai I, Coletti C, Di Bacco M (eds), *Applied Mathematics Monographs,* Giardini Editori e Stampatori in Pisa, pp 222–243

[6] Rossi C (1991) A Stochastic Mover-Stayer Model for HIV Epidemic, *Mathematical Biosciences,* 107, pp 521–545

7] Rossi C, Schinaia G (1998) The Mover-Stayer Model for the HIV/AIDS Epidemic in Action, *Interfaces,* 28 (3), pp 127–143

[8] Brookmeyer R, Gail H G (1986) Minimum Size of the Acquired Immunodeficiency Syndrome (AIDS) Epidemic in the United States, *Lancet* (2), pp 1320–1322

[9] Brookmeyer R, Gail HG (1988) A Method for Obtaining Short-Term Projections and Lower Bounds on the Size of the AIDS Epidemic, *JASA* (83), pp 301–308

[10] Rosenberg P, Gail MH, Pee D (1991) Mean Square Error Estimates of HIV Prevalence and Short-term AIDS Projections Derived by Backcalculation, *Statistics in Medicine* (10), pp. 1167–1180

11] Heisterkamp SH, Downs AM, van Houwelingen JC (1995) Empirical Bayesian Estimators for Reconstruction of HIV Incidence and Prevalence and Forecasting of AIDS. I. Method of Estimation, in: *Quantitative Analysis of HIV/AIDS: Development of Methods to Support Policy Making for Infectious Disease Control,* PhD Thesis, University of Leiden, pp 65–98

[12] Heisterkamp SH, van Houwelingen JC, Downs AM (1999) Empirical Bayesian Estimators for a Poisson Process Propagated in Time, *Biometrical Journal,* 41 (4), pp 358–400

[13] Downs AM, Heisterkamp SH, Brunet JB, Hamers FF (1997) Reconstruction and AIDS Prediction of the HIV/AIDS Epidemic Among Adults in the European Union and in Low Prevalence Countries of Central and Eastern Europe, AIDS (11), pp 649–662

[14] Longini IM, Scott Clark W, Byers RH, Ward JW, Darrow WW, Lemp GF, Hethcote HW (1989) Statistical Analysis of the Stages of HIV Infection Using a Markov Model, *Statistics in Medicine* (8), pp 831–843

[15] Longini Jr IM (1990) Modeling the Decline of CD4+ T-lymphocyte Counts in HIV-infected Individuals, *Journal of AIDS* (3), pp 930–931

[16] Hendriks JCM, Satten GA, Longini IM et al (1996) Use of Immunological markers and Continuous-time Markov Models to Estimate Progression of HIV Infection in Homosexual Men, *AIDS* (10), pp 649–656

[17] Pasqualucci C, Ravà L, Rossi C, Schinaia G (1998) Estimating the Size of the HIV/AIDS Epidemic: Complementary Use of the Empirical Bayesian Back-Calculation and the Mover-Stayer Model for Gathering the Largest Amount of Information, *Simulation,* 71, 4, pp 213–227
[18] COA (Centro Operativo AIDS) (1998) *Sindrome da Immunodeficienza Acquisita (AIDS) in Italia:* Aggiornamento dei Casi Notificati al 30 Settembre 1997, Istituto Superiore di Sanità, Roma

The Radon Transform and Its Applications to Medicine

ENRICO CASADIO TARABUSI

An argument in favour of the practical – as well as cultural – utility of socalled "pure" mathematical research is the sheer number of problems originally introduced and studied without considering their applications and only later successfully utilised in other contexts. A well-known example of this is that of ellipses: they were studied by Apollonius in the third century B.C. and used centuries later in astronomy by Kepler. An even more significant case, in many respects, is that of the Radon transform. A totally natural object studied since the beginning of the century with satisfactory results, the Radon transform has nevertheless remained unknown for over half a century even to applied mathematicians – who repeatedly "rediscovered" it independently together with partial results – and finally became popular starting from the 1970s thanks to a revolutionary application such as tomography in medicine. A curious and noteworthy fact is that physicist A. M. Cormack and engineer G. N. Hounsfield were awarded the Nobel Prize in Medicine for the invention of the CAT scanner – an application essentially based on a mathematical result!

What the Radon transform is can perhaps be explained with a "game". Let us suppose that on each square of a chessboard, for example a 4×4, there is a small heap of rice grains. Just like in the game of chess or Battleships, we shall denote each square with a letter (A to D) identifying the column and a number (1 to 4) identifying the row. We want to determine the number of grains in each heap and this number can vary from square to square. Let us suppose that we cannot determine it directly, but we are allowed to obtain information about the grain total for each column, each row, and each diagonal (not only diagonals of length 4, such as A1-B2-C3-D4 and A4-B3-C2-D1, but also those of length 3, such as A2-B3-C4, or of length 2, such as A3-B4, or even of length 1, such as A4). To work out the value of the 4×4=16 unknowns (the number of grains in each heap) we therefore have at our disposal 22 data items (4 row totals, 4 column totals and 7 diagonal totals in each direction). Are we in a position to solve the problem? More precisely:

1) Are the data sufficient to determine the values of the unknowns, or are two or more different rice grain distributions possible that give the same collected data and therefore cannot be distinguished from one another on the basis of the knowledge at our disposal? Are we in a position to identify precisely such possible indeterminateness?
2) If the data are sufficient, how can we proceed to determine the values of the unknowns? In general, even if they are not, how can the maximum knowledge possible of such values be obtained?

3) Are the data items independent of one another, or can any of them be derived from the knowledge of the others, in which case it provides us with no additional information? (If obtaining each data item has a certain cost associated with it, we might decide not to obtain those that are "superfluous").
4) How can we verify that the data are plausible – that is, that they really correspond to a situation such as the one described above – and they are not randomly generated data or the results of measurement errors instead (for instance if the number of grains is very high)? If the data may contain errors and the measurements cannot be repeated, how can we still proceed to reconstruct the values of the unknowns and minimise the error? (At this stage, it may be useful to have more data than strictly needed – see question (3).)

These are only some of the questions that can arise from this simple situation.

As to question (2), for instance, we always know what the grain numbers of the heaps in the corner squares (A1, A4, D1, D4) are, since each of them is a diagonal of length 1 and this information is available directly as collected data. It is evident that we can immediately work out the total number of grains present on the chessboard by simply adding up the row totals, or the column totals or else the diagonal totals in one direction or those in the other. This answers question (3): for instance, the total of column A can be obtained by adding up all row totals and then subtracting the totals of columns B, C, and D. At the same time, this shows how ensuring that the number of data items is not smaller than that of unknowns does not automatically guarantee a positive answer to question (1).

Inquiring readers will have identified several different distributions that give the same collected data. For instance, one is obtained by putting one grain in each of the squares A2, B4, C1, and D3 and leaving all other squares empty; another is the mirror image of the former distribution, with a grain in A3, B1, C4, and D2 and no other grains elsewhere. In either one, each column or row total equals 1; the total of every diagonal of length 3 or 2 also equals 1, while all other diagonals have no grains. However, with a bit more work, one can determine that this is "essentially" the only possible indeterminateness of the type discussed in question (1): in fact, any two distinct distributions giving rise to the same collected data are such that one can be obtained from the other by removing the same number of grains from each of the squares A2, B4, C1, and D3 and adding the same number of grains to those already present in each of the mirror squares A3, B1, C4, and D2; or conversely. In particular, since we do not admit negative numbers of grains, the data set collected from a distribution that has zero grains in, say, A2 and A3 cannot be collected from any other distribution (in fact, no grain can be removed from A2 and A3) and this shows how data can – in some cases – univocally determine the values of the unknowns.

Going back to question (2), we have already talked about corner squares. A procedure to obtain the values of the central squares could be the following: for instance, to have the grain count in B2, the datum of column B is added to that of row 2 and diagonals A3-B2-C1 and C4-D3; then the totals for diagonals A3-B4, C1-D2, A2-B3-C4, and B1-C2-D3 are subtracted and the rest is divided by 3. Indeed the contribution of B2 occurs three times with a positive sign and none with a negative sign, while the contribution of each of the other squares either does not occur at all or it occurs once with a positive sign and once with a negative sign. The cancelling out of the terms relating to other squares is made possible by the use of values relating to lines that do not contain B2 (in our case, diagonals C4-D3, A3-B4, C1-D2, A2-B3-C4, and B1-C2-D3). Other procedures to obtain the value of B2 are possible, but – similarly to the one detailed above – none of these can make exclusive use of lines containing B2 (that is column B, row 2 and diagonals A3-B2-C1 and A1-B2-C3-D4). We leave to the willing reader the task of finding a procedure that allows the determination of the values in the eight midside squares. We shall recall here that such values are generally not unique, because of the indeterminateness described in the previous paragraph.

As to question (3), we have previously noticed how the equality of the four grand totals (by rows, by columns, and by diagonals in either direction) provides three relations among data that must be satisfied. These can be used to "acquire" three fewer data items or, as suggested in question (4), to decrease data collection errors. If, for instance, the sum of the row totals is greater than the sum of column totals, the former are reduced and the latter are increased (following a standard procedure that we shall not describe in detail here) so as to make the totals agree. The reconstruction procedure on these modified data generally gives more accurate values for the unknowns.

The "game" described above and the ensuing problems are in fact simply matters of linear algebra (precisely, they concern the solvability of systems of linear equations), but we have used a more geometric terminology, so as to introduce the objects that intervene in the Radon transform in its general meaning: a base space (the chessboard), an unknown function (the grain distribution) over the elements in this space (the squares); some subspaces (columns, rows, and diagonals) over which the totals – that is the *integrals* – of such function are known. Let us now replicate the game on a 1000×1000 chessboard having the same borders as the original one: the squares will now be minuscule and the grains microscopic. Let us assume that all the information we have of the unknown distribution of grains is their column totals, row totals, and 45° diagonal totals, as well as totals of lines at other given angles. Let us finally imagine enlarging the chessboard by replicating it indefinitely in each direction, so that those same small squares tile up the entire Cartesian plane. This means that we shall have infinitely long rows, columns and diagonals. We shall therefore have to impose some restriction on the grain distribution to exclude the possibility of the corresponding totals being infinite, for instance that the overall total amount of grains on the chessboard is finite.

This last scenario provides a satisfactory approximation (or, more precisely, a discretisation) of the problem that J. Radon introduced [1] and resolved as follows:

a) If – on the Cartesian plane – the integrals along all straight lines of an unknown function satisfying certain conditions are known, it is possible to reconstruct the function itself (in each point of the plane) by means of an explicit formula.
b) Having assigned a value to each straight line on the plane, such mapping (that is such function over the set of lines) satisfies certain explicit conditions

if and only if there exists a function (satisfying the same conditions as in point (a)) of which, line by line, such value is the integral.

The *Radon transform* is the operation that allows going from the function on the plane (satisfying certain conditions) to the mapping of each line to the integral on it; the mapping is called *Radon transform* of the function.

In the chessboard example, the Radon transform lets us go from the set of number of grains on each square (that is from a function on the set of squares) to the set of row, column and diagonal totals (that can be seen as a function on the set of considered alignments). Statement (b) therefore gives necessary and sufficient conditions on a function over the set of lines for it to be the Radon transform of a (suitable) function on the same plane. Statement (a), instead, gives an *inversion formula* of the Radon transform, allowing expressing a function in terms of its transform, in the same way as we expressed the number of grains on a certain square in terms of the row, column, and diagonal totals.

It is clear that statement (a) answers questions (1) and (2) positively in terms of functions on the plane. The former would have had an affirmative answer also in the case of a 4×4 chessboard, had we allowed data to be available on square alignments different from those (0°, 45°, 90°, 135°) we considered. On the other hand, this would have made their definition less natural. As to the conditions mentioned in statement (b) and in relation to question (3), there are relationships among the values assigned to the various lines on the plane: the simplest (albeit not explicit in [1]) is probably the analogue of the one described for the chessboard, for which the overall grain total is obtained by adding the data of all the alignments parallel to a common direction, whatever this may be. Statement (b) also solves explicitly the first problem posed in question (4): similarly to the case hinted at in the case of the chessboard, this solution can be used as a basis for deriving a method for improving imperfect data before proceeding to reconstruct the function.

The so-called *dual Radon transform* allows passing from a function on the set of lines on the plane to the function assigning to each point the integral in the angular variable of the value of the former on the lines that contain the point. In other words, if each line is assigned a value, the dual Radon transform assigns to each point the average (up to a multiplicative constant) of the values on the lines through it. In the chessboard example, given a value to each row, column, and diagonal, the dual transform would assign to square B2 the average of the values of column B, of row 2, and of diagonals A3-B2-C1 and A1-B2-C3-D4. As can be seen, such a transform acts in the opposite direction of the "direct" transform, since – unlike it – it allows going from functions on the set of lines on the plane to functions on the plane itself. (The dual transform should not be confused with the inverse transform, although it approximates it sufficiently well for some purposes.) In [1] the equivalents of statements (a) and (b) for this transform are proved; in particular, a formula is exhibited that expresses an unknown function over the set of lines in terms of its dual Radon transform.

Let us now describe the working of the most famous application of the Radon transform: *CAT* or *Computerised Axial Tomography.* Before its introduction in the 1960s, the feature of X-rays of penetrating the human body more deeply than visible light was used only to observe it "in transparency", directly on a screen or by means of exposed photographic plates. Even though some tricks were available to improve its precision, such technique did not provide a real internal mapping of the areas of interest (particularly inside the cranium) – useful for identifying and precisely locating neoplasias and oedemas among others – to help at the diagnostic stage or with surgery, if required. On the other hand, CAT does provide this mapping indirectly, by employing a computer to process the attenuation data of single X-ray beams that cross the area of interest, as we shall explain next.

It is possible to emit a beam of X-rays so thin that it can be considered with good approximation to be located along a straight line. Let us imagine that – after pointing it towards the head of a subject – we can follow it along its route into the body. For each millimetre down the route – depending on the wavelength of the beam and on the chemical and physical properties of the tissue being penetrated – a greater or smaller part of the beam will be absorbed or deflected and the remainder will continue along the same straight line. Depending on the degree of attenuation caused by each linear millimetre of tissue being penetrated, the latter is said to be more or less *opaque* to X-rays of that frequency. On exit at the other end of the body, only a small part of the original beam will be left: this can be gathered and measured by a sensor. We therefore have at our disposal a measure of attenuation of the beam (the ratio between the intensity prior to and after crossing the body): the greater this attenuation, the more opaque the scanned tissue and/or the longer the route across it. In fact, it is easy to see that such attenuation (or, to be precise, its logarithm) is equal to the sum, millimetre by millimetre along the beam, of the opacity of the tissue being scanned at each point or, more precisely, to the integral along the line of the opaqueness function.

The whole process can be repeated with a beam aligned along a different line each time, so as to obtain the integral of the opaqueness function along it (in the simplified but realistic hypothesis that the opaqueness of the tissues in question is – at each point – isotropic, that is, independent of the direction of the beam hitting the point). If we do this with all the lines lying on the same axial plane, that is perpendicular to the vertical axis of the body, in accordance to what stated in (a) it is possible to use a computer to obtain the opaqueness function on that plane. In other words, it is possible to work out how opaque the tissue is at each single point of the plane. The function thus obtained can then be visualised, for instance on a screen, by representing a virtual section of the body at that axial plane and using grey tones so that the more opaque areas appear lighter (consistently with ordinary radiographic plates in which the bones, more opaque, appear lighter than the rest). Since an alteration of the tissue usually entails an alteration of the opaqueness (for instance, cancerous brain tissue is less opaque than healthy brain tissue), this visualisation is precious for the identification and precise localisation of anomalous tissue. Visualisation can further be improved by first inoculating the patient with an appropriate contrast agent, particularly if the reason of the investigation are physiological alterations due to occlusions or similar causes.

In practice, it is important to take into account several implementation problems. One of them is: there are infinite lines in a plane, therefore there are infinite data theoretically available, but only a finite number of data items can be measured and handled. The choice of how many and which lines to use must take into account various issues such as the ones outlined next. On the one hand, the dose of radiation administered to a patient must be limited (therefore the lines must not be "too many"); on the other hand, data interpolation on the missing lines must be satisfactory, or in any case, the reconstruction formulae of the unknown function must show a good adaptation to the real situation (thus the lines cannot be "too few"). As to the latter aspect, interesting Fourier analysis issues are shown in [2], together with a detailed survey of this and other problems in which mathematics is applied to tomography; in [3, §5] the ill-conditioned character of the problem of inversion – that is the instability of the dependence of the solution on the data – is discussed.

The procedure described earlier for a single axial plane is usually repeated for many parallel planes at small distance from one another, so as to gain virtual sections at several "heights" and obtain a mapping of the whole three-dimensional body region. Apart from representing the single sections as mentioned earlier, graphical computer algorithms allow the synthesis of the results in three-dimensional visualisation, such as oblique cross-sections or view of only the tissue characterised by a certain opaqueness (for instance, the skeleton), virtual navigation inside the body (for example, along a blood vessel, so as to locate occlusions or other anomalies), and so on. Further information on CAT and many other medical and non-medical applications of the Radon transform, along with detailed historical notes and a very large bibliography, can be found in [4]. A detailed description of CAT and NMR (Nuclear Magnetic Resonance, see below) are found in [5].

The basic problem studied by Radon is so natural to lend itself to numerous extensions and generalisations, some of which he himself proposed in his article. Instead of lines in the Cartesian plane, lines – or planes – can be considered in a three-dimensional space or in a Euclidean space of higher dimension; lines or planes can be replaced by curves or surfaces of some specified type (for instance: circumferences or spheres); also, the Euclidean space can be replaced by a non-Euclidean space (for example: the surface of a sphere, the Lobachevskii-Bolyai hyperbolic space). Alternatively, the environment can be a discrete space, just like in the chessboard example: see [6] for an up-to-date review on this. The study (or rediscovery) of these extensions has been greatly stimulated by the success of tomography, following which many of these have found application, often in medicine.

An example is the Nuclear Magnetic Resonance, or NMR, introduced by P. C. Lauterbur in the 1970s and whose purpose is essentially the same as that of CAT, with which it is often used in combination or as an alternative. NMR is a technique for mapping – not the X-ray opaqueness of tissues but – the concentration of atoms of a specific element, usually hydrogen, as this is very abundant in tissues, being a component of the water molecule. The subject is immersed in an appropriate combination of magnetic fields, some of which are oscillating, so as to provoke resonance with the precession motion of the rotation axes of the hydrogen nuclei that are located on a prescribed plane (or, rather, within a very small distance from this plane). The subsequent exponential process of relaxation occurs with different characteristic times along the components longitudinal and transversal to the direction of the precession axis. Such times – indicated with T_1 and T_2 – depend on the type of tissue in which water molecules are located, and differ in a usually relevant way between healthy and pathological tissue. For each of the two components, the intensity of the electromagnetic radiation (typically in the radio-frequency range) emitted by resonant nuclei is measured at some instant during relaxation and results from the sum of the contributions of each small area of homogeneous tissue on that plane. Each contribution is proportional to the number of hydrogen atoms present in that area and is counted with the higher weight, the longer the characteristic time for that tissue is (because, at acquisition, the relaxation will be almost over in the molecules with short characteristic time).

By repeating the procedure in different (parallel and non-parallel) planes, one obtains the integral along each of them of the function obtained by multiplying pointwise the hydrogen density by this weight that depends on the characteristic time (T_1 or T_2) of the tissue type. Such functions (one resulting from T_1 and the other from T_2) can be explicitly obtained by using an inversion formula of the Radon transform relative to the space planes and can later be used similarly to what was described for CAT: just like in the latter, the visualisation of some features can be improved by first inoculating a contrast agent. Alternatively – with more complex configurations of magnetic fields – it is possible to measure the integrals on lines, instead of planes, or directly the point-by-point density. The three alternatives have advantages and disadvantages concerning resolution, contrast, and so on, and are selected depending on the situation at hand.

As we observed when talking about the chessboard, one soon becomes persuaded that – in order to find out the value of the unknown function at a certain point of the plane – it is not sufficient to use the values of its Radon transform only on the lines that contain it, since cancellations are required. Much less obvious is that – in the case of lines in the plane – the values on all the lines are actually required, even those very far from the point in question (whose influence will be small but not null): the inversion is therefore called *non-local.* This phenomenon has important consequences for CAT since, if only one region of a certain axial plane is required to be mapped (for instance the spine in the thorax) without gathering the data of the beams that would pass far from it (so that the subject is irradiated less) – these data however, as just said, are indispensable for a precise inversion – it is necessary to introduce an approximation in the reconstruction formula and to make do with a less precise image.

From this viewpoint, the Radon transform on planes in space is better behaved: to find out the value of the unknown function at a given point, it is still insufficient to know the values on the planes containing it, but – this time – those on the planes sufficiently near it (precisely, whose distance is smaller than a predetermined constant, for instance a millimetre) are sufficient. The corresponding inversion is therefore called *local.* The Radon transform is intimately linked to wave theory: it is beyond our scope to describe it in detail, but we shall highlight that this difference between dimension 2 and 3 (and, in general, between even and odd dimension) corresponds to the *Huygens principle.* According to this principle, the circular wave produced in the plane by a punctual instantaneous event (for instance a rock falling into the water) is always followed by other concentric waves of lesser intensity, but this does not happen in space (which allows us to hear an instantaneous sound without successive "tails").

Numerous other diagnostic methods are based on tomography and make use of variations of the Radon transform. One of these is *Positron Emission Tomography* or *PET*, which maps the concentration of a radioactive substance, such as Carbon 11 or Oxygen 15, previously inoculated into the subject. In a random event whose probability of occurring is proportional to the concentration, the substance emits a positron that immediately collides with an electron and, annihilating itself, gives rise simultaneously to two gamma rays of the same known intensity in opposite directions. When two of these rays simultaneously hit two of the sensors that surround the subject, it can be concluded that the emitting particle is located along the line joining them. The number of such events, in a sufficiently long period of time, for each pair of sensors, provides an estimate of the number of radio-nuclide particles located along that line. Again by applying the Radon transform inversion one obtains the unknown function that – as explained earlier – is the concentration of the substance: this can be useful, for instance, to gain information about the metabolism. The situation can be slightly more complex if the attenuation is not negligible. In fact, gamma rays can, just like X-rays, be attenuated along their route before they exit the body of the subject, thus altering the measurement. However, the overall attenuation of the two rays of the same emitted pair depends only on the common line of propagation and not on the emission location, since each point on the line between two sensors is in any case traversed by one of the two rays. Therefore, the model gives rise to a multiple of the classical Radon transform and the inversion is basically the same.

In *SPECT*, or *Single Photon Emission Computerized Tomography,* a different type of radio-nuclide is used, such as Iodine 131: this emits single photons directly. The sensors surrounding the subject are endowed with collimators: this allows determining from which direction each photon hitting them comes. The photon count registered by a sensor in the unit of time provides an estimate of the number of radio-nuclide particles along the line identified by the corresponding collimator. Just like in the case of PET, the final step (not mentioning application details) consists in applying the inverse of the Radon transform. Unlike PET, however, if the attenuation is not negligible, it has a greater effect on the beams emitted from points that are farther away from the sensor in question, since they have to travel longer distances inside the body. Therefore, the acquired datum is not the simple line integral of the density function of the radio-nuclide, but rather of this multiplied by an exponential attenuation factor that depends on the route travelled by the photon. The result is the *attenuated* or *exponential Radon transform:* for this, too, formulae are available that allow its inversion.

More recent and less established is the *EIT*, or *Electrical Impedance Tomography*, whose aim is that of mapping the electrical conductivity inside the body. Electrodes are placed all around the chest – for instance – of the subject, on the same axial plane; (low amperage and therefore innocuous) current is applied to two of these electrodes in turn and the electric potential induced on the others is measured. We shall not describe the complex dependence of such potential (the data at our disposal) on the internal conductivity (the unknown); we shall instead limit ourselves to mentioning that it is strictly (and surprisingly) tied to the Radon transform on the Lobachevskii-Bolyai hyperbolic plane – although measurements are performed in Euclidean space!

Aside from diagnostic aid, the Radon transform and its variations have other types of medical applications. One of these is radiotherapy planning, the aim of which is that of irradiating a certain internal region of the body (typically a neoplastic formation) while hitting the rest as little as possible. Having fixed an axial plane, it is necessary to establish the intensity of the X-rays to be emitted (in one of the two directions) along each line lying in it. Each beam will hit the points along its trajectory with an intensity that will be more and more attenuated by the stretch already travelled, as already seen regarding PET and SPECT. Overall, each point in the plane will therefore be hit by beams passing in the various directions and the total radiation that it receives will be the sum (or, more precisely, the integral in the angular variable) of the residual intensities that the various beams hitting it have in such point of their trajectory. The irradiation function, defined on the points of the plane, is therefore the dual attenuated Radon transform of the "initial intensity" function defined on the set of lines. By prescribing the former on the basis of therapeutic necessities (for instance giving a null irradiation value to the points outside the targeted region and a sufficiently high constant value to the ones inside), the second function could be obtained with a direct application of the inversion formula. The function thus obtained, however, does not – in itself – have any physical meaning, since – in general – it assigns some lines negative intensity values: this corresponds to the intuitive fact that it is impossible to irradiate an internal region without irradiating also the rest of the body. With a standard procedure (that we shall not describe here), the function provided by the inversion formula is then modified so that all values are non-negative and the radiotherapy can be carried out using this new function. In this way, the area that is not the target of the irradiation will in fact be exposed to only a very small – the realistic minimum – amount of radiation.

It is worth concluding these notes with some hints on the odd series of discoveries and rediscoveries of the Radon transform leading to its establishment and universal recognition in the 1970s (further details can be found in the already quoted [4], as well as in [7, Part I] and [8] for applications and in the historical notes of [6] for theory). The earliest evidence goes back to an uncertain moment around the year 1900: in a 1906 article on crystallography, H. B. A. Bockwinkel attributes an inversion of the transform of planes in space (that – as we have mentioned – is "simpler" than that of lines in the plane) to unpublished work by his teacher H. A. Lorentz. This article, along with its generalisation to any dimension by G. E. Uhlenbeck in 1925, remains long unknown to mathematicians. In 1904–1906, H. Minkowski gives an inversion of the transform on great circles on the surface of a sphere (under the necessary condition that the unknown function has the same value on each pair of antipodal points, for these are indistinguishable through great circles). Further, in 1916, P. Funk provides an inversion that can more easily be extended to other situations. In his article [1] written the following year, Radon, stimulated also by conversations with G. Herglotz, inverts the transform of lines in the plane and its dual and extends calculations to any dimension by observing the difference between even and odd dimension which we mentioned earlier (but without quoting the works of Lorentz or Bockwinkel, which he is probably not aware of). He also states the general problem on an abstract space – of which he solves some particular cases – and quotes the results by Minkowski and Funk.

Later, P. Mader in 1927 and F. John, a student of Herglotz's, in some works starting from 1934, take up the subject again and improve the results in various ways: important contributions by the latter are mainly the recognition of the links with the Fourier transform and the new type of conditions that he proves one can replace to the originals in statement (b). In 1936, H. Cramèr and H. Wold – presumably unaware of preceding work – rediscover the transform to apply it to statistics problems: the inversion of the Radon transform is in fact called "Cramèr-Wold theorem" in statistics. It is necessary to wait for a 1952 article by A. Rényi (taken up again three years later by W. M. Gilbert) for recognition that the two are the same thing!

In 1940, G. D. Birkhoff solves the following question: by drawing in the plane only straight lines, each of which is uniformly "dark", is it possible to obtain any "drawing"? The reader will recognise that this problem is analogous to that stated for radiotherapy, but in absence of attenuation. By applying the inversion of the dual Radon transform (without attenuation), in general a function is obtained on the set of lines which takes also negative values; therefore the answer is "no" (for instance, it is impossible to draw a black solid circle on a white background), unless one admits the possibility of negative dark tones, that is, the possibility to "erase" uniformly along lines. It is mainly with John's 1955 book that the Radon transform (thus called for the first time) is popularised among mathematicians and soon many researchers start to take an interest in it. Among them are B. Fuglede in 1958, S. Helgason (to whom a substantial part of the theory developed up to this day is owed) starting from the following year, I. Gel'fand and G. F. Shilov from 1960 and many more thereafter.

The first known application of the transform to a problem involving experimental data goes back to 1936, when V. A. Ambartsumian uses it to obtain the velocity distribution of stars near the sun, thus solving a problem posed by A. S. Eddington earlier that year. In 1956, R. N. Bracewell (who also did not know about Radon's article) rediscovers it to obtain maps of the luminosity distribution of the lunar surface by using data gathered through the alignment of parabolic mirrors. The first medical application – basically a precursor to CAT – is by W. H. Oldendorf in 1961, followed by D. E. Kuhl and R. Q. Edwards in 1963 with pioneering work on PET and SPECT. In the same year, A. M. Cormack, too, rediscovers the Radon transform while planning his first prototype for a CAT scanner. Based on his articles, starting from 1968, more prototypes are patented and successfully tested by G. N. Hounsfield of EMI (Electrical and Musical Industries

– the producers of the Beatles' records!). As they spread, the era of tomography officially starts. The long series of rediscoveries of the Radon transform ends shortly after with the articles by I. N. Shtein, B. K. Vainshtein, and S. S. Orlov in 1972 and by C. M. Vest and Cormack himself the following year: these articles recognise the key role of the Radon transform within tomography.

Bibliography

[1] Radon J (1917) Über die Bestimmung von Funktionen durch ihre Integralwerte längs gewisser Mannigfaltigkeiten, *Ber Verh Sächs Akad Wiss Leipzig, Math-Nat Kl 69*, pp 262–277. English translation in [4, Appendix A]

[2] Natterer F (1986) *The Mathematics of Computerized Tomography,* Revised Reprint, Wiley, Chichester

[3] Talenti G (1978) *Sui Problemi Mal Posti,* Boll Un Mat Ital A (5) 15, pp 1–29

[4] Deans S R (1993) *The Radon Transform and Some of Its Applications,* Revised Reprint, Krieger, Malabar

[5] Prestini E (1996) *Applicazioni dell'Analisi Armonica,* Hoepli, Milan

[6] Helgason S (1999) *The Radon Transform,* 2nd Edition, Progr Math, vol 5, Birkhäuser, Boston

[7] Gindikin S, Michor P (Eds) (1994) *75 Years of Radon Transform,* Conf Proc Lecture Notes Math Phys (Vienna 1992), vol 4, International Press, Boston

[8] Mathematics: The Unifying Thread in Science, *Notices Amer Math Soc* 33, 1986, pp 716–733

Risk Analysis for Liver Surgery

Heinz-Otto Peitgen, Bernhard Preim, Dirk Selle, Dominik Böhm,
Andrea Schenk, Wolf Spindler

Introduction

Surgical procedures for the resection of pathologic lesions are often risky interventions. The risk concerns postoperative complications as well as the success of the operation in terms of its longterm effect. Postoperative complications may result if too much tissue is removed. The longterm effect depends on whether all lesions are detected and are removed entirely together with a saftey margin. Risk analysis is an important aspect in preoperative surgery planning. Currently, preoperative planning considers the general state of the patient as well as special images resulting e.g. from computer tomography (CT). The currently prevailing techniques for preoperative planning are highly subjective and errorprone as no reliable methods exist to evaluate the extent of anatomic and pathologic structures and to quantitatively analyze these structures.

In this paper we discuss how to assess the risk involved in surgical interventions by image processing and visualization techniques combined with quantitative analysis. In particular, we describe how resections in liver surgery can be planned taking into account the intrahepatic vessels, and the regions of liver parenchyma supplied by them. For this purpose, radiological CT data are analyzed in order to identify the relevant anatomic and pathologic structures.

Virtual Reality Techniques in Surgical Planning

In order to assess the potential of visualization and analysis in surgery planning, it is useful to briefly discuss how similar techniques are used in other areas. In the construction of cars or planes, for example, risk analysis is carried out more and more by means of Virtual Reality techniques. For this purpose, geometric 3D models are constructed and combined with a functional model, which describes e.g. the elasticity of the structures involved and the behavior under pressure. Such a combination of a geometric and a functional model can be used for the simulation of a crash and the analysis of the crash with respect to the pressure applied to the persons involved. Virtual Reality (VR) techniques, which consider the individual patient data are promising also for risk analysis in surgical procedures. The main difference between the previously described areas and the medical domain is the necessity to reconstruct a geometric model on the base of individual patient data. The accuracy of such a model is limited by the resolution and the inherent distortion of the image acquisition process, e.g. in computer tomography. In the following we briefly survey the use of VR-techniques in image diagnosis and treatment planning.

In the past, basic algorithms for the visualization, image processing and image analysis have been developed and refined to adapt to the peculiarities of medical image data. Examples are algorithms for edge detection, contrast enhancement, noise reduction, image segmentation as well as visualization techniques, like maximum-intensity projections and other volume rendering approaches.

For difficult diagnostic processes, like the detection of microcalcifications in mammograms or the detection of polyps in virtual colonoscopy[1], these basic techniques must be appropriately parameterized and combined in a certain manner. Such problems are much too difficult to be solved by trial-and-error. The customatization of algorithms for a special clinical question requires a deep understanding of the underlying algorithms as well as the influence and interaction of parameters. This difficult task can only be carried out by experts which are hard to be found in hospitals, except some leading-edge research institutions. Therefore many attempts are targeted at the combination and parameterization of basic tools and their integration in preconfigured and easy-to-use applications. Customization and integration of basic algorithms characterize the current stage of VR-techniques to support computer-aided diagnosis. The resulting tools lack the flexibility and general applicability of basic algorithms, like edge detection. Instead they are adapted to emphasize certain anatomic or pathologic structures relevant for the question at hand. Like traditional diagnostic methods, these computer-supported diagnosis tools are assessed in terms of their sensitivity and specifity concerning the detection of pathologic situations.

While the applications described above are concentrated on a well-defined precise question, like "Is there a malignant structure?" some new applications concentrate on the ambitious task of supporting therapeutic decisions. Therapeutic decisions are complex problem-solving tasks involving a number of possible questions which are related to each other. The support for such a process requires an intensive collaboration of radiologists, surgeons and developers. In this collaboration, developers must get a deep understanding how therapeutic decisions are made, which information and measures may be relevant for a decision, which persons are involved in the decision and how it is documented. On the other hand, radiologists learn how the parameters of the image-acquisition, e.g. the interslice distance, influence the results of the image processing. The refinement of protocols for the use of medical devices is thus triggered by the demands of image processing. The software which finally supports radiologists and surgeons in their decision about an appropriate therapeutic strategy is even more complex than the software for computeraided diagnosis.

1 *Microcalcifications in mammogram and polyps in colonoscopy are indications for cancerous changes.*

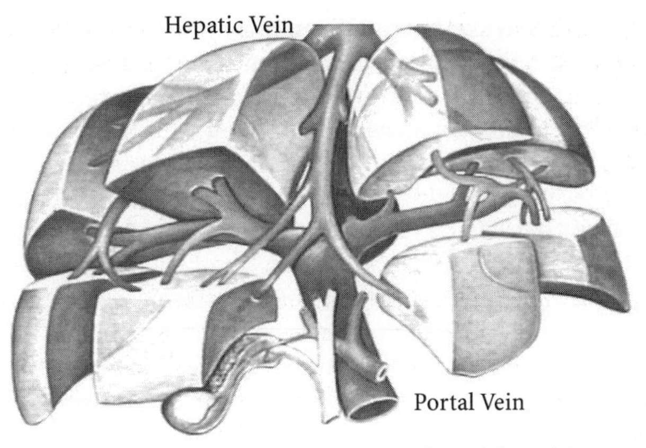

Fig. 1. Scheme of the functional portal liver segments, adapted from [5].

We call these *software assistants* and describe in this paper an example of a software assistant. This example concerns preoperative planning and risk analysis for oncologic liver resections.

Preoperative Planning in Liver Surgery

The central issue of modern liver surgery is to take into account the individual intrahepatic vessel and segment anatomy. Following the Couinaud model [1], the human liver is divided into different segments which are defined according to the branching structure of the portal vein and the localization of the hepatic vein. In Figure 1 a scheme of the functional liver segments as adapted from [5] is depicted. The smallest functional unit of the liver, a liver segment, is defined by the supplied territory of a third-order portal vein branch. Typically the hepatic veins proceed between the different segments (Figure 2).

For the planning of segment-oriented liver resections, it is desirable to make the individual branching pattern of the hepatic vessels available to the surgeon, to identify the patient's individual liver segments, and to localize hepatic lesions in relation to the vessels and segments. Based on such a visualization the surgeon can more precisely decide where to resect the liver tissue. As a simple tool a clip plane can be used to define the resection area. If the resection area is defined in advance quantitative measures can be determined to further improve the planning. The most important aspect of the quantitative analysis is the calculation of the volume to be resected compared to the remaining volume. A resection can be safely carried out if at least 30 percent of the liver volume remains intact. The calculation of distances, in particular distances between the planned resection area and the major blood vessels is important to assess the risk involved in the resection.

The preoperative planning of segment oriented surgery is challenging, since the liver segments are highly variable from patient to patient in shape, size and number. Furthermore, the boundaries between the segments are not straight and can not be localized by external landmarks. Applying simple schemes as shown in Figure 2 although very convenient for the daily radiological routine, is highly questionable from an anatomical point of view [2].

Image Processing for Liver Surgery Planning

The underlying clinical examinations for the diagnosis and planning of liver surgery are computer tomographic images which represent the whole liver slice by slice as shown in Figure 3 and Figure 4. A contrast agent applied to the patient provides a high vessel-to-tissue and tumor-to-tissue contrast revealing the relevant anatomic structures. The following subsections shortly describe the image processing steps to analyse these structures based on CT images [3]:

Image Segmentation of Liver and Tumors

The liver as a whole as well as the lesions are extracted from the clinical data sets with a segmentation algorithm derived from the watershed transformation known in mathematical topology. A segmentation can be performed interactively by only marking two points – one inside and one outside of the structure of interest in the image. Thus the boundary of an object can be computed and, if necessary, be refined with further interaction.

Extraction of the Intrahepatic Vessel Systems

For a clinical CT data set all voxels which belong to the intrahepatic vessel system can be specified by a threshold decision on their signal intensity (Figure 2). Since the vessels are highlighted by a contrast agent, this yields acceptable results. Better results can be obtained by using filter functions for noise reduction (blurring, median filter) and for background compensation (Laplace-like filters). The extraction of the vessels is based on a region growing algorithm, which aggregates all high intensity voxels of the intrahepatic vessel systems

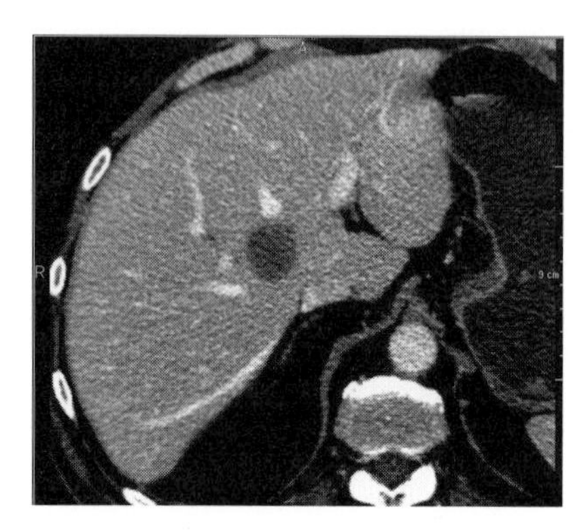

Fig. 2. 2D slice of a CT data set. The contrast agent reveals the vessels of the liver, which appear bright, whereas the tumor appears as dark spots.

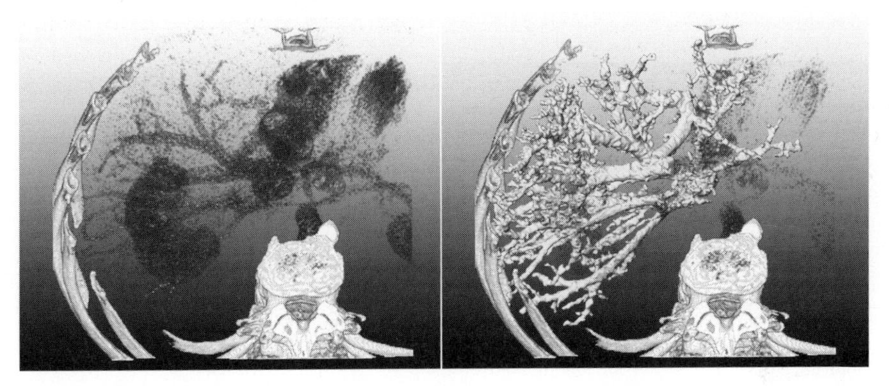

Fig. 3. *Left:* Transparent volume rendering of an unprocessed CT volume data set. *Right:* The intrahepatic vessel systems (portal venous system and hepatic vein fragments) which have been previously extracted are emphasized.

(Figure 3, right). The proposed method automatically suggests a threshold for an appropriate specification of the vessel systems and additionally provides a fast manual manipulation and visual control of the suggested threshold.

Analysis of the Intrahepatic Vessel Systems

Separating the extracted portal and hepatic vein and identifying the main portal subtrees is a prerequisite for the determination of the liver segments. To achieve this automatically, the structure and morphology of the vessel systems have to be analyzed. For this purpose we have chosen to "skeletonize" the vessels. Here a skeleton is defined as a one voxel width framework running along the medial axis of the branches (Figure 4, left) together with the local vessel radii associated with each point on the medial axis. Compared with the voxel-based shape representation of the vessels, the skeleton representation can be interpreted as a graph which provides an easier access to the geometry of the branches (medial axis and radius) and to structural information (ramifications). Using a model based approach, the portal vein and hepatic vein (fragments) can be identified automatically (Figure 4, left). The graph representation of the vessels is also the basis for user interactions such as selecting and coloring trees, subtrees and paths. Measuring the radius, length or volume of branches and their distance from tumors is supported as well. In Figure 4, right a hepatic vein fragment was removed and the main branches of the portal vein were colored automatically by determining the eight most voluminous portal subtrees.

Calculation of the Liver Segments

For the prediction of the liver segments based on an incomplete portal vein different models have been developed and tested: the Laplacian and the nearest neighbor approximation. Even though these methods differ very much in their

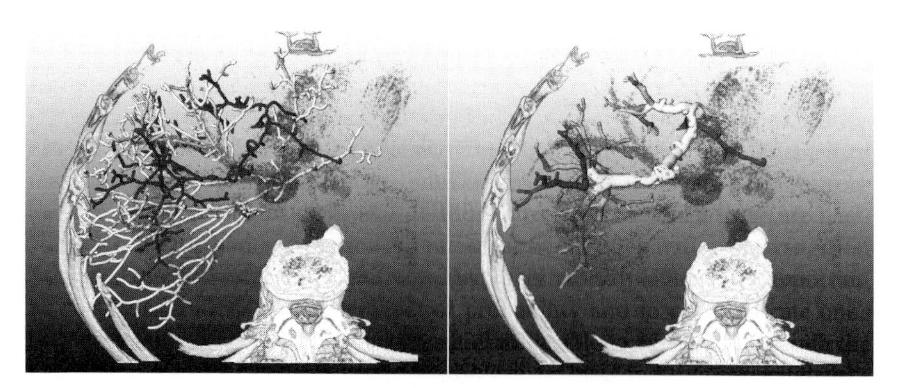

Fig. 4. *Left:* Skeleton of the extracted vessels from the dataset shown in Figure 4. A hepatic vein fragment is separated from the portal vein automatically.
Right: Automatic determination of the main branches of the portal vein.

mathematical formulation, they yield rather similar results. However, because the Laplace approximation is based on a fundamental equation of physics, it offers interesting venues for a scientific understanding of the prediction method. The nearest neighbor approximation is very attractive from a practical point of view, because of its conceptual simplicity and rather low computational complexity. The result of the segment approximation is shown in Figure 5, left, which is based on the portal branches in Figure 5, right. In Figure 5, right the same dataset with four embedded tumors is visualized. The transparent 3D rendering reveals the location of tumors in relation to the portal tree and the segments. The tumors are located in the segments I, V, VII and VIII with respect to the Couinaud scheme (see Fig. 5 right). The resection of these segments can be simulated by simply hiding a segment. Their volume and also the volume of the remaining intact liver parenchyma can be determined to estimate the risk of the operation. Also the length of branches and their distance to the tumors can be considered for planning operative vessel reconstructions. Algorithmic details and a careful evaluation of the image analysis steps may be found in a recent publication [3].

Evaluation and Results

The evaluation concerns two aspects: one is the anatomical correctness, in particular with respect to the estimation of liver segments. The second aspect concerns the clinical usefulness. Anatomical correctness of a method is a prerequisite for its clinical use, but clinical usefulness depends on some more factors, like the efficiency and robustness of the algorithms employed.

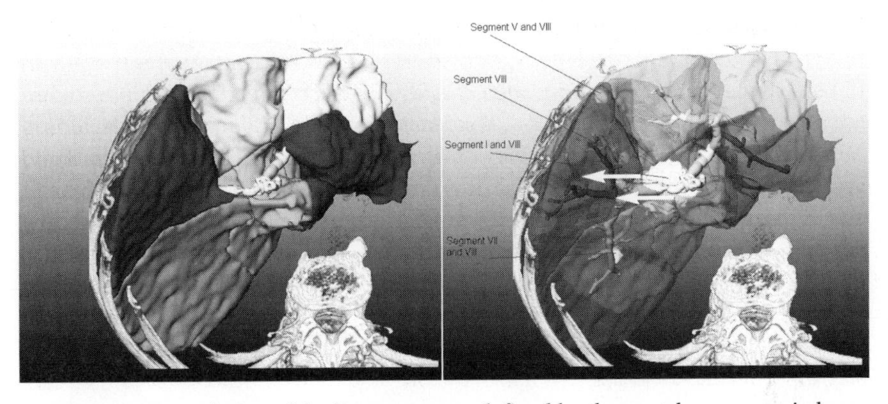

Fig. 5. *Left:* 3D rendering of the liver segments defined by the portal venous main branches in Figure 4, right.
Right: Rendering of the same dataset with some transparent segments revealing the location in relation to embeded tumors (see the lines) and vessels.

Anatomical Evaluation

The validation of the approximation of the segments is based on an anatomical study on vascular corrosion casts of the human liver. The portal vein of eight unembalmed cadavers were injected in situ with a casting resin. After hardening of the injected liquid, the liver was extracted from the body and corroded. One of the resulting casts is shown in Figure 6, left.

The scheme of the validation procedure is shown in Figure 6, right. Contrary to in vivo data (e.g. Figure 3 and 4) CT scans of the casts make it possible to extract branches with a diameter of about 1mm (Figure 6a). This yields sufficient branching generations and makes it possible to determine very accurately the location and geometry of the portal segments of the liver. The voids between the branches are closed with morphological dilation and erosion operations (Figure 6c). The resulting solid portal segments provide a very precise approximation of the true anatomical segments. To simulate the incomplete portal trees obtained from in vivo radiological data, we systematically pruned the trees obtained from the casts (Figure 6b). To validate the approximation methods for the liver segments we compared the approximations made for the pruned casts (Figure 6d) with the exact segment anatomy of the casts (Figure 6c). For all eight casts we found a correct volumetric overlap of 80 percent on the average (range between 75 percent and 85 percent), which is the accuracy we expect for clinical data [3].

Surgical Evaluation

In addition to the study based on casts our methods have been evaluated in the clinical environment for more than 300 patients until now. For the preoperative planning of liver resections in patients with liver tumors, the liver, tumors, arteries, portal and hepatic vein were extracted from biphasic spiral CT data and visualized in 3D. This requires between 60 to 90 minutes. The interactive rotation

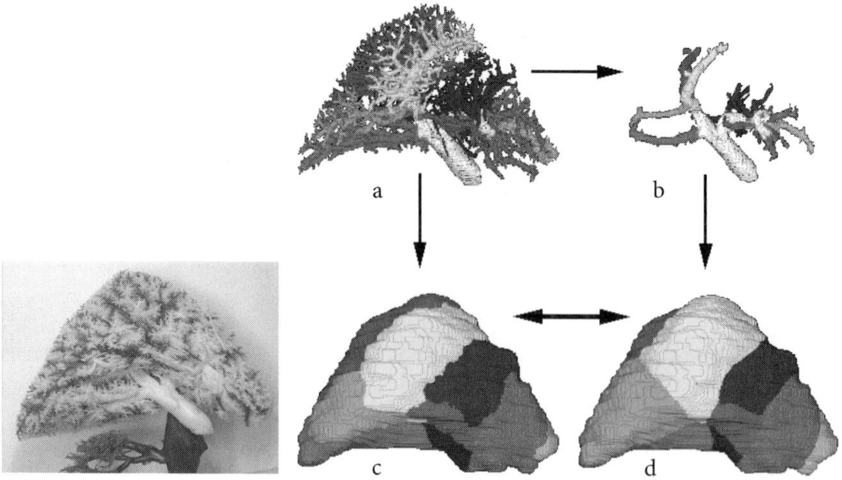

Fig. 6. *Left:* Corrosion cast of the portal- (yellow) and hepatic vein (black).
Right: Validation of the approximation methods: (a) Rendering of the portal vein obtained from a CT scan of a human liver cast. The main subtrees are assigned different colors revealing the liver segments. (b) Pruned version of the portal vein shown in (a). (c) Morphological operations are used to fill the holes in the portal cast (a) so as to obtain solid authentic anatomical segments. (d) Approximated liver segments based on the pruned portal vein in (b).

of the anatomic and pathologic structures enables a precise 3D understanding of the location of tumors within the liver and in relation to the intrahepatic vessels systems. It has been shown that these visualizations make it possible to interactivelely plan liver resections and thus to improve the preparation especially of complex liver resections. The intraoperative findings agree with the 3D visualizations [8].

Discussion

In current practice the preoperative identification of the individual liver anatomy is assessed by the inspection of 2D tomographic images. The intrahepatic vessels serve as guiding structures for a schematic understanding of the anatomy (like the Couinaud model, Figure 1). Studying the casts of eight human livers we found that the portal tree and therefore also the portal segments have a high degree of interindividual variability. This is in agreement with previous findings [4], [2], [9] and confirms the need for new radiological methods that take into account more accurately the individual characteristics. The Nearest Neighbor and the Laplacian method provide a new way to computationally predict the individual 3D geometry of the portal segments. Based on clinical data containing three or more orders of branches, the expected volumetric overlap is more than 80 percent.

Superior to "mental" and schematic reconstructions from 2D images is a 3D visualization of the individual vessels, segments and tumors. The interactive rotation of the structures and the possibility to assign colors and transparency values to each structure, makes it possible to clearly perceive the spatial relations [10], [3], [11]. This enables an easier interpretation of diagnostic findings for the surgeon and serves as a basis for discussions on the planning of complex resections within the surgical team [8]. Also quantitative measures, such as the volume of tumors and segments, are essential to estimate the remaining liver parenchyma and the length of branches for planning operative vessel reconstructions.

The techniques and methods described in this paper are dedicated to liver surgery. However, very similar techniques can be employed for a range of other application areas, like interventions to remove lung cancer. Again, the high variability of the individual structures requires a planning based on individual data. To assess the spatial relations between a hierarchical vessel system and some lesions, is the most crucial issue in preoperative planning.

Acknowledgments

We want to thank all members of the following institutions for their substantial support and intensive cooperation: the Department of Diagnostic Radiology, Philipps University Marburg, Germany (K.J. KLOSE), the Department of Morphology, University Medical Center, Geneva, Switzerland (J.H.D. FASEL), the Department of Diagnostic Radiology, Medical School Hannover, Germany (M. GALANSKI) and the Department of General and Transplantation Surgery, University Hospital Essen, Germany (K.J. OLDHAFER). We also want to acknowledge the support of our collaboraters at MeVis: H. Bourquain, M. Hindennach, A. Littmann and O. K. Verse.

References

[1] Couinaud L, *Le Foie – Etudes anatomiques et chirurgicales.* Masson, Paris 1957

[2] Fasel JHD, Selle D, Gailloud P, Evertsz CJG, Terrier F, Peitgen H-O et al. "Segmental anatomy of the liver: poor correlation with CT", *Radiology,* 1998; 206(1): 151–156

[3] Selle D, Preim B, Schenk A, Peitgen H-O. "Analysis of Vasculature for Liver Surgery Planning", IEEE Transactions on Medical Imaging, 2002; 21 (11), 1344–1357

[4] Platzer W, Maurer H. "Zur Variabilität der Lebersegmente", Chir. Praxis, 1966; 10: 499–505

[5] Priesching A. *Leberresektionen,* Urban & Schwarzenberg, München, 1986

[6] Nelson RC, Chezmar JL, Sugarbaker PH, Murray DR, Bernardino ME. "Preoperative localization of focal liver lesions to specific liver segments: utility of CT during arterial portography", *Radiology,* 1990; 176: 89–94

[7] Soyer P, Bluemke DA, Bliss DF, Woodhouse CE, Fishman EK. "Surgical anatomy of the liver: demonstration with spiral CT during arterial portography and multiplanar reconstruction", *American journal of Radiology,* 1994; 163: 99–103

[8] Oldhafer KJ, Högemann D, Stamm G, Raab R, Peitgen H-O, Galanski M. "Dreidimensionale Visualisierung der Leber zur Planung erweiterter Leberresektionen", *Chirurg,* 1999; 70(3): 233–238

[9] Leppek L, Selle D, Habermalz E, Klose KJ, Nies C, Evertsz CJG, Jürgens H, Relecker S, Peitgen H-O. "Computerized segmental analysis of liver parenchyma in vivo", To appear in Radiology

[10] Zahlten C, Jürgens H, Evertsz CJG, Leppek R, Peitgen H-O, Klose K-J. "Portal vein reconstruction based on topology", *European Journal of Radiolology,* 1995; 19: 96–100

[11] Selle D, Evertsz CJG, Peitgen H-O, Jürgens H, Klose KJ, Fasel J. "Computer aided preoperative planning of segment oriented liver surgery: radiological perspectives", In: Broelsch CE, Izbicki JR, Bloechle C, Gawad KA (ed). Monduzzi Editore, Bologna 1997, pp 253–257

Recommended Readings

R.C. Nelson, J.L. Chezmar, P.H. Sugarbaker, D.R. Murray, M.E. Bernardino (1990) Preoperative localization of focal liver lesions to specific liver segments: utility of CT during arterial portography, *Radiology,* 176, pp. 89–94

P. Soyer, D.A. Bluemke, D.F. Bliss, C.E. Woohouse, E.K. Fishman (1994) Surgical anatomy of the liver: demonstration with spiral CT during arterial portography and multiplanar reconstruction, *American journal of Radiology,* 163, pp. 99–103

Authors

Achille Basile	Dipartimento di Matematica e Statistica, Università "Federico II", Napoli, Italy
Dominik Böhm	MeVis, Center for Medical Diagnostic System and Visualization, University of Bremen, Germany
Jochen Brüning	Institut für Mathematik, Humboldt-Universität, Berlin, Germany
Enrico Casadio Tarabusi	Dipartimento di Matematica, Università degli Studi "La Sapienza", Roma, Italy
Capi Corrales Rodriganez	Departamento de Algebra, Universidad Complutense, Madrid, Spain
Michel Darche	Centre-Sciences CCSTI, Orleans, France
Camillo Dejak	Dipartimento di Chimica Fisica, Università Ca' Foscari, Venezia, Italy
Luciano Emmer	Film-maker
Michele Emmer	Dipartimento di Matematica, Università degli Studi "La Sapienza", Roma, Italy
Enrico Giusti	Dipartimento di Matematica, Università Statale, Firenze, Italy
Peter Greenaway	Film-maker
Giorgio Israel	Dipartimento di Matematica, Università degli Studi "La Sapienza", Roma, Italy
Harold W. Kuhn	Department of Mathematics, Princeton University, USA
Alessandro Languasco	Dipartimento di Matematica Pura e Applicata, Università, Padova, Italy
Marco Li Calzi	Dipartimento di Matematica Applicata, Università Cà Foscari, Venezia, Italy
Richard Mankiewicz	Centre for the Cultural and Historical Aspects of Mathematics (CHAsM), Middlesex University, U.K.
Gustavo Mosquera R.	Film-maker

Piergiorgio Odifreddi	Dipartimento di Matematica, Università degli Studi, Torino, Italy
Roberto Pastres	Dipartimento di Chimica Fisica, Università Ca' Foscari, Venezia, Italy
Heinz-Otto Peitgen	MeVis, Center for Medical Diagnostic System and Visualization, University of Bremen, Germany
Alberto Perelli	Dipartimento di Matematica, Università, Genova, Italy
Achille Perilli	Artist
Konrad Polthier	Fachbereich Mathematik, Technische Universität, Berlin, Germany
Bernhard Preim	Fakultät für Informatik / Institut für Simulation und Graphik, Otto-von-Guericke-Universität Magedeburg, Germany
Claudio Procesi	Dipartimento di Matematica, Università degli Studi "La Sapienza", Roma, Italy
Trân Quang Hai	Département de Musique, CNRS, Musée de l'Homme, Parigi, France
Lucilla Ravà	Dipartimento di Matematica, Università degli Studi "Tor Vergata", Roma, Italy
Carla Rossi	Dipartimento di Matematica, Università degli Studi "Tor Vergata", Roma, Italy
Lucio Russo	Dipartimento di Matematica, Università degli Studi "Tor Vergata", Roma, Italy
Andrea Schenk	MeVis, Center for Medical Diagnostic System and Visualization, University of Bremen, Germany
Dirk Selle	MeVis, Center for Medical Diagnostic System and Visualization, University of Bremen, Germany
Wolf Spindler	MeVis, Center for Medical Diagnostic System and Visualization, University of Bremen, Germany
Silvano Tagliagambe	Dipartimento di Studi Filosofici ed Epistemologici, Università degli Studi "La Sapienza", Roma, Italy
Laura Tedeschini Lalli	Dipartimento di Matematica, Facoltà di Architettura, Università degli Studi "Roma Tre", Roma, Italy